Assessing and Improving Prediction and Classification

Theory and Algorithms in C++

Timothy Masters

Apress®

Assessing and Improving Prediction and Classification: Theory and Algorithms in C++

Timothy Masters
Ithaca, New York, USA

ISBN-13 (pbk): 978-1-4842-3335-1   ISBN-13 (electronic): 978-1-4842-3336-8
https://doi.org/10.1007/978-1-4842-3336-8

Library of Congress Control Number: 2017962869

Cover image by Freepik (www.freepik.com)

Managing Director: Welmoed Spahr
Editorial Director: Todd Green
Acquisitions Editor: Steve Anglin
Development Editor: Matthew Moodie
Technical Reviewers: Massimo Nardone and Matt Wiley
Coordinating Editor: Mark Powers
Copy Editor: Kim Wimpsett

Distributed to the book trade worldwide by Springer Science+Business Media New York, 233 Spring Street, 6th Floor, New York, NY 10013. Phone 1-800-SPRINGER, fax (201) 348-4505, e-mail orders-ny@springer-sbm.com, or visit www.springeronline.com. Apress Media, LLC is a California LLC and the sole member (owner) is Springer Science + Business Media Finance Inc (SSBM Finance Inc). SSBM Finance Inc is a **Delaware** corporation.

For information on translations, please e-mail rights@apress.com, or visit www.apress.com/rights-permissions.

Apress titles may be purchased in bulk for academic, corporate, or promotional use. eBook versions and licenses are also available for most titles. For more information, reference our Print and eBook Bulk Sales web page at www.apress.com/bulk-sales.

Any source code or other supplementary material referenced by the author in this book is available to readers on GitHub via the book's product page, located at www.apress.com/9781484233351. For more detailed information, please visit www.apress.com/source-code.

Printed on acid-free paper

This book is dedicated to Master Hidy Ochiai with the utmost respect, admiration, and gratitude. His incomparable teaching of Washin-Ryu karate has raised my confidence, my physical ability, and my mental acuity far beyond anything I could have imagined.

For this I will ever be grateful.

Table of Contents

About the Author

Timothy Masters has a PhD in mathematical statistics with a specialization in numerical computing. He has worked predominantly as an independent consultant for government and industry. His early research involved automated feature detection in high-altitude photographs while he developed applications for flood and drought prediction, detection of hidden missile silos, and identification of threatening military vehicles. Later he worked with medical researchers in the development of computer algorithms for distinguishing between benign and malignant cells in needle biopsies. For the past 20 years he has focused primarily on methods for evaluating automated financial market trading systems. He has authored eight books on practical applications of predictive modeling.

- *Deep Belief Nets in C++ and CUDA C: Volume III: Convolutional Nets (CreateSpace, 2016)*

- *Deep Belief Nets in C++ and CUDA C: Volume II: Autoencoding in the Complex Domain* (CreateSpace, 2015)

- *Deep Belief Nets in C++ and CUDA C: Volume I: Restricted Boltzmann Machines and Supervised Feedforward Networks* (CreateSpace, 2015)

- *Assessing and Improving Prediction and Classification* (CreateSpace, 2013)

- *Neural, Novel, and Hybrid Algorithms for Time Series Prediction* (Wiley, 1995)

- *Advanced Algorithms for Neural Networks* (Wiley, 1995)

- *Signal and Image Processing with Neural Networks* (Wiley, 1994)

- *Practical Neural Network Recipes in C++* (Academic Press, 1993)

About the Technical Reviewers

Massimo Nardone has more than 23 years of experience in security, web/mobile development, cloud computing, and IT architecture. His true IT passions are security and Android.

He currently works as the chief information security officer (CISO) for Cargotec Oyj and is a member of the ISACA Finland Chapter board. Over his long career, he has held many positions including project manager, software engineer, research engineer, chief security architect, information security manager, PCI/SCADA auditor, and senior lead IT security/cloud/SCADA architect. In addition, he has been a visiting lecturer and supervisor for exercises at the Networking Laboratory of the Helsinki University of Technology (Aalto University).

Massimo has a master's of science degree in computing science from the University of Salerno in Italy, and he holds four international patents (related to PKI, SIP, SAML, and proxies). Besides working on this book, Massimo has reviewed more than 40 IT books for different publishing companies and is the coauthor of *Pro Android Games* (Apress, 2015).

Matt Wiley is a tenured associate professor of mathematics with awards in both mathematics education and honor student engagement. He has earned degrees in pure mathematics, computer science, and business administration through the University of California and Texas A&M systems. He serves as director for Victoria College's quality enhancement plan and managing partner at Elkhart Group Limited, a statistical consultancy. With programming experience in R, C++, Ruby, Fortran, and JavaScript, he has always found ways to meld his passion for writing with his joy of logical problem solving and data science. From the boardroom to the classroom, Matt enjoys finding dynamic ways to partner with interdisciplinary and diverse teams to make complex ideas and projects understandable and solvable.

Preface

This book discusses methods for accurately and appropriately assessing the performance of prediction and classification models. It also shows how the functioning of existing models can be improved by using sophisticated but easily implemented techniques. Researchers employing anything from primitive linear regression models to state-of-the-art neural networks will find ways to measure their model's performance in terms that relate directly to the practical application at hand, as opposed to more abstract terms that relate to the nature of the model. Then, through creative resampling of the training data, multiple training sessions, and perhaps creation of new supplementary models, the quality of the prediction or classification can be improved significantly.

Very little of this text is devoted to designing the actual prediction/classification models. There is no shortage of material available in this area. It is assumed that you already have one or more models in hand. In most cases, the algorithms and techniques of this book don't even care about the internal operation of the models. Models are treated as black boxes that accept observed inputs and produce outputs. It is your responsibility to provide models that perform at least reasonably well, because it is impossible to produce something from nothing.

The primary motivation for writing this book is the enormous gap between academic researchers developing sophisticated prediction and classification algorithms and the people in the field actually using models for practical applications. In many cases, the workers-in-the-field are not even aware of many powerful developments in model testing and enhancement. For example, these topics are addressed in detail:

- It's commonly known that cross validation is a simple method for obtaining practically unbiased estimates of future performance. But few people know that this technique is often just about the worst such algorithm in terms of its error variance. We can do better, occasionally a lot better.

- Mean squared error (MSE) is an extremely popular and intuitive measure of prediction accuracy. But it can be a terrible indicator of real-world performance, especially if thresholded predictions will be used for classification. This is common in automated trading of financial instruments, making MSE worthless for this application.

- Many people choose a single performance indicator to use throughout the life of the development process. However, true experts will often optimize one measure during a hill-climbing training operation, a second measure to choose from among several competing models, and a third to measure the real-world performance of the final model.

- The importance of consistent performance is often ignored, with average performance being the focal point instead. However, in most cases, a model that performs fairly well across the majority of training cases will ultimately outperform a model that performs fabulously most of the time but occasionally fails catastrophically. Properly designed training sets and optimization criteria can take consistency into account.

- It is well known that a training set should thoroughly encompass a variety of possible situations in the application. But few people understand the importance of properly stratified sampling when a dataset will be used in a critical application involving confidence calculations.

- Many classifiers actually predict numerical values and then apply threshold rules to arrive at a class decision. Properly choosing the ideal threshold is a nontrivial operation that must be handled intelligently.

- Many applications require not only a prediction but a measure of confidence in the decision as well. Some developers prefer a hypothesis-testing approach, while others favor Bayesian methods. The truth is that whenever possible, both methods should be used, as they provide very different types of information. And most people ignore a critical additional step: computing confidence in the confidence figure!

- Often, an essential part of development is estimating model parameters for examination. The utility of these estimates is greatly increased if you can also compute the bias and variance of the estimates, or even produce confidence intervals for them.

- Most people will train a model, verify its reasonable operation, and place it into service. However, there are iterative algorithms in which certain cases in the training set are given more or less weight as a sequence of trained models is generated. These algorithms can sometimes behave in an almost miraculous fashion, transforming a weak classifier into one with enormous power, simply by focusing attention on specific training cases.

- The final step in many model development applications is to pool all available data and train the model that will be deployed, leaving no data on which to test the final model. Monte Carlo permutation tests can offer valuable performance indications without requiring any data to be sacrificed. These same tests can be used early in the development to discover models that are susceptible to overfitting.

- A common procedure is to train several competing models on the same training set and then choose the best performer for deployment. However, it is usually advantageous to use all of the competing models and intelligently combine their predictions or class decisions to produce a consensus opinion.

- It is often the case that different models provide different degrees of performance depending on conditions that do not serve as model inputs. For example, a model that predicts the pharmacological effect of a drug cocktail may perform best when the expression of one gene dominates, and another model may be superior when another gene expression is dominant. Automated gating algorithms can discover and take advantage of such situations.

- It is not widely known that the entropy of a predictor variable can have a profound impact on the ability of many models to make effective use of the variable. Responsible researchers will compute the entropy of every predictor and take remedial action if any predictor has low entropy.

- Many applications involve wildly throwing spaghetti on the wall to see what sticks. A researcher may have literally hundreds or even thousands of candidates for predictor variables, while the final model will employ only a handful. Computations involving mutual information and uncertainty reduction between the candidates and the predicted variable provide methods for rapidly screening candidates, eliminating those that are least likely to be useful while simultaneously discovering those that are most promising.

- One exciting subject in information theory is causation, the literal effect of one event causing another, as opposed to their just being related. Transfer entropy (Schreiber's information transfer) quantifies causative events.

These and many more topics are addressed in this text. Intuitive justification is a focal point in every presentation. The goal is that you understand the problem and the solution on a solid gut level. Then, one or more references are given as needed if you want to see rigorous mathematical details. In most cases, only the equations that are critical to the technique are provided here. Long and tedious proofs are invariably left to the references. Finally, the technique is illustrated with heavily commented C++ source code. All code is supplied on my web site (TimothyMasters.info) as complete working programs that can be compiled with the Microsoft Visual Studio compiler. Those sections of code that are central to a technique are printed in the book for the purpose of explication.

The ultimate purpose of this text is threefold. The first goal is to open the eyes of serious developers to some of the hidden pitfalls that lurk in the model development process. The second is to provide broad exposure for some of the most powerful model enhancement algorithms that have emerged from academia in the past two decades, while not bogging you down in cryptic mathematical theory. Finally, this text should provide you with a toolbox of ready-to-use C++ code that can be easily incorporated into your existing programs. Enjoy.

CHAPTER 1

Assessment of Numeric Predictions

- Notation
- Overview of Performance Measures
- Selection Bias and the Need for Three Datasets
- Cross Validation and Walk-Forward Testing
- Common Performance Measures
- Stratification for Consistency
- Confidence Intervals
- Empirical Quantiles as Confidence Intervals

Most people divide prediction into two families: classification and numeric prediction. In classification, the goal is to assign an unknown case into one of several competing categories (benign versus malignant, tank versus truck versus rock, and so forth). In numeric prediction, the goal is to assign a specific numeric value to a case (expected profit of a trade, expected yield in a chemical batch, and so forth). Actually, such a clear distinction between classification and numeric prediction can be misleading because they blend into each other in many ways. We may use a numeric prediction to perform classification based on a threshold. For example, we may decree that a tissue sample should be called malignant if and only if a numeric prediction model's output exceeds, say, 0.76. Conversely, we may sometimes use a classification model to predict values of an ordinal variable, although this can be

© Timothy Masters 2018
T. Masters, *Assessing and Improving Prediction and Classification*,
https://doi.org/10.1007/978-1-4842-3336-8_1

dangerous if not done carefully. Ultimately, the choice of a numeric or classification model depends on the nature of the data and on a sometimes arbitrary decision by the experimenter. This chapter discusses methods for assessing models that make numeric predictions.

Notation

Numeric prediction models use one or more *independent variables* to predict one or more *dependent variables*. By common convention, the independent variables employed by a model are designated by some form of the letter x, while the dependent variables are designated by the letter y. Unless otherwise stated, scalar variables are represented in lowercase italic type with an optional subscript, and vectors are represented in lowercase bold type. Thus, we may refer to the vector $\mathbf{x}=\{x_1, x_2, ...\}$. When we want to associate $\mathbf{x}$ with a particular case in a collection, we may use a subscript as in $\mathbf{x}_j$. Matrices are represented by uppercase bold type, such as $\mathbf{X}$.

The parameters of a prediction model are Greek letters, with theta (θ) being commonly employed. If the model has more than one parameter, separate Greek letters may be used, or a single bold type vector ($\boldsymbol{\theta}$) may represent all of the parameters.

A model is often named after the variable it predicts. For example, suppose we have a one-parameter model that uses a vector of independent variables to predict a scalar variable. This model (or its prediction, according to context) would be referred to as $y(\mathbf{x}; \theta)$.

A collection of observed cases called the *training set* is used to optimize the prediction model in some way, and then another independent collection of cases called the *test set* or *validation set* is used to test the trained model. (Some schemes employ three independent collections, using both a test set and a validation set. One is used during model development, and the other is used for final verification. We'll discuss this later.) The error associated with case i in one of these collections is the difference between the predicted and the observed values of the dependent variable. This is expressed in Equation (1.1).

$$e_i = y(\mathbf{x}_i; \theta) - y_i \tag{1.1}$$

Overview of Performance Measures

There are at least three reasons why we need an effective way to measure the performance of a model. The most obvious reason is that we simply need to know if it is doing its job satisfactorily. If a model's measured performance on its training set is unsatisfactory, there is no point in testing it on an independent dataset; the model needs revision. If the model performs well on the training set but performs poorly on the independent set, we need to find out why and remedy the situation. Finally, if the performance of a model is good when it is first placed in service but begins to deteriorate as time passes, we need to discover this situation before damage is done.

The second reason for needing an effective performance criterion is that we must have an objective criterion by which the suitability of trial parameter values may be judged during training. The usual training procedure is to have an optimization algorithm test many values of the model's parameter (scalar or vector). For each trial parameter, the model's performance on the training set is assessed. The optimization algorithm will select the parameter that provides optimal performance in the training set. It should be apparent that the choice of performance criterion can have a profound impact on the value of the optimal parameter that ultimately results from the optimization.

The final reason for needing an effective performance criterion is subtle but extremely important in many situations. When we compute the parameter value that optimizes a performance criterion or when we use a performance criterion to choose the best of several competing models, it is a little known fact that the choice of criterion can have a profound impact on how well the model generalizes to cases outside the training set. Obviously, a model that performs excellently on the training set but fails on new cases is worthless. To encourage generalization ability, we need more than simply good average performance on the training cases. We need *consistency* across the cases.

In other words, suppose we have two alternative performance criteria. We optimize the model using the first criterion and find that the average performance in the training set is quite good, but there are a few training cases for which the trained model performs poorly. Alternatively, we optimize the model using the second criterion and find that the average performance is slightly inferior to that of the first criterion, but all of the training cases fare about equally well on the model. Despite the inferior average performance, the model trained with the second criterion will almost certainly perform better than the first when it is placed into service.

A contrived yet quite realistic example may clarify this concept. Suppose we have a performance criterion that linearly indicates the quality of a model. It may be production hours saved in an industrial process, dollars earned in an equity trading system, or any other measurement that we want to maximize. And suppose that the model contains one optimizable parameter.

Let us arbitrarily divide a data collection into three subsets. We agree that two of these three subsets constitute the training set for the model, while the third subset represents data observed when the trained model is placed into service. Table 1-1 depicts the average performance of the model on the three subsets and various combinations of the subsets across the range of the parameter.

Examine the first column, which is the performance of the first subset for values of the parameter across its reasonable range. The optimal parameter value is 0.3, at which this subset performs at a quality level of 5.0. The second subset has a very flat area of optimal response, attaining 1.0 for parameter values from 0.3 through 0.7. Finally, the third subset attains its optimal performance of 6.0 when the parameter equals 0.7. It is clear that this is a problematic training set, although such degrees of inconsistency are by no means unknown in real applications.

Table 1-1. *Consistency Is Important for Generalization*

Parameter	1	2	3	1+2	1+3	2+3
0.1	-2	-1	-3	-3	-5	-4
0.2	0	0	-2	0	-2	-2
0.3	5	1	-1	6(/4)	4(/6)	0
0.4	4	1	0	5(/3)	4(/4)	1(/1)
0.5	3	1	3	4(/2)	6(/0)	4(/2)
0.6	0	1	4	1(/1)	4(/4)	5(/3)
0.7	-1	1	6	0	5(/7)	7(/5)
0.8	-2	0	0	-2	-2	0
0.9	-3	-1	-2	-4	-5	-3

Now let us assume that the training set is made up of the first two subsets, with the third subset representing the data on which the trained model will eventually act. The fourth column of the table shows the sum of the performance criteria for the first two subsets. Ignore the additional numbers in parentheses for now. A training procedure that maximizes the total performance will choose a parameter value of 0.3, providing a grand criterion of 5+1=6. When this model is placed into service, it will obtain a performance of –1; that's not good at all!

Another alternative is that the training set is made up of the first and third subsets. This gives an optimal performance of 3+3=6 when the parameter is 0.5. In this case, the performance on the remaining data is 1; that's fair at best. The third alternative is that the training data is the second and third subsets. This reaches its peak of 1+6=7 when the parameter is 0.7. The in-use performance is once again –1. The problem is obvious.

The situation can be tremendously improved by using a training criterion that considers not only total performance but consistency as well. A crude example of such a training criterion is obtained by dividing the total performance by the difference in performance between the two subsets within the training set. Thus, if the training set consists of the first two subsets, the maximum training criterion will be (3+1)/(3–1)=4/2=2 when the parameter is 0.5. The omitted subset performs at a level of 3 for this model. I leave it to you to see that the other two possibilities also do well. Although this example was carefully contrived to illustrate the importance of consistency, it must be emphasized that this example is not terribly distant from reality in many applications.

In summary, an effective performance criterion must embody three characteristics: *Meaningful* performance measures ensure that the quantity by which we judge a model relates directly to the application. *Optimizable* performance measures ensure that the training algorithm has a tractable criterion that it can optimize without numerical or other difficulties. *Consistent* performance measures encourage generalization ability outside the training set. We may have difficulty satisfying all three of these qualities in a single performance measure. However, it always pays to try. And there is nothing wrong with using several measures if we must.

Consistency and Evolutionary Stability

We just saw why asking for consistent performance from a model is as important as asking for good average performance: A model with *consistently* decent performance within its training set is likely to generalize well, continuing to perform well outside the training set. It is important to understand that this property does not apply to just the model at hand this moment. It even carries on into the future as the model building and selection process continues. Let us explore this hidden bonus.

In most situations, the modeling process does not end with the discovery and implementation of a satisfactory model. This is just the first step. The model builder studies the model, pondering the good and bad aspects of its behavior. After considerable thought, a new, ideally improved model is proposed. This new model is trained and tested. If its performance exceeds that of the old model, it replaces the old model. This process may repeat indefinitely. At any time, the model being used is an evolutionary product, having been produced by intelligent selection.

An informed review of the normal experimental model-building procedure reveals how easily it is subverted by evolutionary selection. First, a model is proposed and trained, studied, and tweaked until its training-set performance is satisfactory. A totally independent dataset is then used to verify the correct operation of the model outside the training set. The model's performance on this independent dataset is a completely fair and unbiased indicator of its future performance (assuming, of course, that this dataset is truly representative of what the model will see in the future). The model is then placed into service, and its actual performance is reasonably expected to remain the same as its independent dataset performance. No problem yet.

We now digress for a moment and examine a model development procedure that suffers from a terrible flaw; let's explore how and why this flaw is intrinsic to the procedure. Suppose we have developed two competing models, one of which is to be selected for use. We train both of them using the training set and observe that they both perform well. We then test them both on the test set and choose whichever model did better on this dataset. We would like to conclude that the future performance of the chosen model is represented by its performance on the test set. This is an easy conclusion to reach, because it would certainly be true if we had just one model. After all, the test set is, by design, completely independent of the training set. However, this conclusion is seriously incorrect in this case. The reason is subtle, but it must be understood by all responsible experimenters. The quick and easy way to explain the problem is to note that because the test set was used to select the final model, the former test set is actually now part of the training set. No true, independent test set was involved. Once this is understood, it should be clear that if we want to possess a fair estimate of future performance *using only these two datasets* the correct way to choose from the two models is to select whichever did better on the *training set* and then test the winner. This ensures that the test set, whose performance is taken to be indicative of future performance, is truly independent of the training set. (In the next section we will see that there is a vastly better method, which involves having a third dataset. For now we will stick with the dangerous two-dataset method because it is commonly used and lets us illustrate the importance of consistency in model evolution.)

That was the quick explanation. But it is worth examining this problem a bit more deeply, because the situation is not as simple as it seems. Most people will still be confused over a seeming contradiction. When we train a model and then test it on independent data, this test-set performance is truly a fair and unbiased indicator of future performance. But in this two-model scenario, we tested both models on data that is independent of the training set, and we chose the better model. Yet suddenly its test-set performance, which would have been fair if this were the only model, now magically becomes unfair, optimistically biased. It hardly seems natural.

The explanation for this puzzling situation is that the final model is not *whichever of the two models was chosen.* In actuality, the final model is the *better of the two models.* The distinction is that the final model is not choice *A* or choice *B*. It is in a very real sense a third entity, the *better* model. If we had stated in advance that the comparison was for fun only and we would keep, say, model *A* regardless of the outcome of the comparison, then we would be perfectly justified in saying that the test-set performance of model *A* fairly indicates its future performance. But the test-set performance of whichever model happens to be chosen is not representative of that mysterious virtual model, *the winner of the competition.* This winner will often be the *luckier* of the two models, not the better model, and this luck will not extend into future performance. It's a mighty picky point, but the theoretical and practical implications are very significant. Take it seriously and contemplate the concept until it makes sense.

Enough digression. We return to the discussion of how evolutionary selection of models subverts the unbiased nature of test sets. Suppose we have a good model in operation. Then we decide that we know how to come up with a new, improved model. We do so and see that its training-set performance is indeed a little bit better than that of the old model. So we go ahead and run it on the test set. If its validation performance is even a little better than that of the old model, we will certainly adopt the new model.

However, if it happens that the new model does badly on the test set, we would quite correctly decide to retain the old model. This is the intelligent thing to do because it provides maximum expected future performance. However, think about the digression in the previous paragraph. We have just followed that nasty model-selection procedure. The model that ends up being used is that mysterious virtual entity: *whichever performed better on the test set*. As a result, the test set performance is no longer an unbiased indicator of future performance. It is unduly optimistic. How optimistic? In practice, it is nearly impossible to say. It depends on many factors, not the least of which is the nature of the performance criterion itself. A simple numerical example illustrates this aspect of the problem.

Suppose we arbitrarily divide the test set on which the model selection is based into four subsets. In addition, consider a fifth set of cases that comprises the as-yet-unknown immediate future. Assume that the performance of the old and new models is as shown in Table 1-2.

Table 1-2. *Inconsistent Performance Degrades Evolutionary Stability*

	Old Model	New Model
Subset 1	10.0	20.0
Subset 2	-10.0	15.0
Subset 3	-10.0	10.0
Subset 4	20.0	-30.0
Subset 5	15.0	5.0

Start with the situation that the first four subsets comprise the decision period and the fifth is the future. The mean performance of the old model in the decision period is (10-10-10+20)/4=2.5, and that for the new model is 3.75. Based on this observation, we decide to replace the old model with the new. The future gives us a return of 5, and we are happy. The fact that the old model would have provided a return of 15 may be forever unknown. But that's not the real problem. The problem is that the arrangement just discussed is not the only possibility. Another possibility is that the fifth subset in that table is part of the decision set, while the fourth subset represents the future. Recall that the entries in this table are not defined to be in chronological order. They are simply the results of an arbitrary partitioning of the data. Under this revised ordering, the new model still wins, this time by an even larger margin. We enthusiastically replace the old model. But now the future holds an unpleasant surprise, a loss of 30 points. I leave it as an exercise for you to repeat this test for the remaining possible orders. It will be seen that, on average, future performance is a lot worse than would be expected based on historical performance. This should be a sobering exercise.

Now we examine a situation that is similar to the one just discussed, but different in a vital way. If we had trained the models in such a way as to encourage performance that is not only good on average but that is also consistent, we might see a performance breakdown similar to that shown in Table 1-3.

Table 1-3. *Consistent Performance Aids Evolutionary Stability*

	Old Model	New Model
Subset 1	10.0	8.0
Subset 2	12.0	12.0
Subset 3	8.0	10.0
Subset 4	10.0	8.0
Subset 5	10.0	11.0

Letting the last subset be the future, we compute that the mean performance of the old model is 10, and the new model achieves 9.5. We keep the old model, and its future performance of 10 is exactly on par with its historical performance. If the reader computes the results for the remaining orderings, it will be seen that future performance is, on average, only trivially inferior to historical performance. It should be obvious that consistent performance helps ameliorate the dangers inherent in evolutionary refinement of models.

Selection Bias and the Need for Three Datasets

In the prior section we discussed the evolution of predictive models. Many developers create a sequence of steadily improving (we hope!) models and regularly replace the current model with a new, improved model. Other times we may simultaneously develop several competing models that employ different philosophies and choose the best from among them. These, of course, are good practices. But how do we decide whether a new model is better than the current model, or which of several competitors is the best? Our method of comparing performance can have a profound impact on our results.

There are two choices: we can compare the training-set performance of the competing models and choose the best, or we can compare the test-set performance and choose the best. In the prior section we explored the impact of consistency in the performance measure, and we noted that consistency was important to effective evolution. We also noted that by comparing training-set performance, we could treat the test-set performance of the best model as an unbiased estimate of expected future performance. This valuable property is lost if the comparison is based on test-set performance because then the supposed test set plays a role in training, even if only through choosing the best model.

With these thoughts in mind, we are tempted to use the first of the two methods: base the comparison on training-set performance. Unfortunately, comparing training-set performance has a serious problem. Suppose we have two competing models. Model A is relatively weak, though decently effective. Model B is extremely powerful, capable of learning every little quirk in the training data. This is bad, because Model B will learn patterns of noise that will not occur in the future. As a result, Model B will likely have future performance that is inferior to that of Model A. This phenomenon of an excessively powerful model learning noise is called *overfitting*, and it is the bane of model developers. Thus, we see that if we choose the best model based on training-set performance, we will favor models that overfit the data. On the other hand, if we avoid this problem by basing our choice on an independent dataset, then the performance of the best model is no longer an unbiased estimate of future performance, as discussed in the prior section. What can we do?

To address this question, we now discuss several concepts that are well known but often misunderstood. In the vast majority of applications, the data is contaminated with noise. When a model is trained, it is inevitable that some of this noise will be mistaken by the training algorithm for legitimate patterns. By definition, noise will not reliably repeat in the future. Thus, the model's performance in the training set will exceed its expected performance in the future. Other effects may contribute to this phenomenon as well, such as incomplete sampling of the population. However, the learning of noise is usually the dominant cause of this undue optimism in performance results. This excess is called *training bias*, and it can be severe if the data is very noisy or the model is very powerful.

Bias is further complicated when training is followed by selection of the best model from among several competitors. This may be a one-shot event early in the development cycle, in which several model forms or optimization criteria are placed in competition, or it may result from continual evolution of a series of models as time passes. Regardless, it introduces another source of bias. If the selection is based on the performance of the competing models in a dataset that is independent of the training set (as recommended earlier), then this bias is called *selection bias*.

A good way to look at this phenomenon is to understand that a degree of luck is always involved when models are compared on a given dataset. The independent data on which selection is based may accidentally favor one model over another. Not only will the truly best model be favored, but the luckiest model will also be favored. Since by definition luck is not repeatable, the expected future performance of the best model will be on average inferior to the performance that caused it to be chosen as best.

Note by the way that the distinction between training bias and selection bias may not always be clear, and some experts may dispute the distinctions given here. This is not the forum for a dispute over terms. These definitions are ones that I use, and they will be employed in this text.

When the training of multiple competing models is followed by selection of the best performer, an effective way to eliminate both training and selection bias is to employ three independent datasets: a *training set*, a *validation set*, and a *test set*. The competing models are all trained on the training set. Their performance is inflated by training bias, which can be extreme if the model is powerful enough to learn noise patterns as if they were legitimate information. Thus, we compute unbiased estimates of the capability of each trained model by evaluating it on the *validation set*. We choose the best validation-set performer, but realize that this performance figure is inflated by selection bias. So, the final step is to evaluate the selected model on the *test set*. This provides our final, unbiased estimate of the future performance of the selected model.

In a time-series application such as prediction of financial markets, it is best to let the training, validation, and test sets occur in chronological order. For example, suppose it is now near the end of 2012, and we want to find a good model to use in the upcoming year, 2013. We might train a collection of competing models on data through 2010 and then test each of them using the single year 2011. Choose the best performer in 2011 and test it using 2012 data. This will provide an unbiased estimate of how well the model will do in 2013.

This leads us to yet another fine distinction in regard to model performance. The procedure just described provides an unbiased estimate of the 2010–2011 model's true ability, *assuming that the statistical characteristics of the data remain constant.* This assumption is (roughly speaking) called *stationarity*. But many time series, especially financial markets, are far from stationary. The procedure of the prior paragraph finished with a model that used the most recent year, 2012, for validation only. This year played no part in the actual creation of the model. If the data is truly stationary, this is of little consequence. But if the data is constantly changing its statistical properties, we would be foolish to refrain from incorporating the most recent year of data, 2012, in the model creation process. Thus, we should conclude the development process by training the competing models on data through 2011. Choose the best performer in 2012, and then finally train that model on data through 2012, thereby making maximum use of all available data.

It's important to understand what just happened here. We no longer have an unbiased estimate of true ability for the model that we will actually use, the one trained through 2012. However, in most cases we can say that the performance obtained by training through 2010, selecting based on 2011, and finally testing on 2012, is a decent stand-in for the unbiased estimate we would really like. Why is this? It's because what our procedure has actually done is test our *modeling methodology*. It's as if we construct a factory for manufacturing a product that can be tested only by destroying the product. We do a trial run of the assembly line and destructively test the product produced. We then produce a second product and send it to market. We cannot actually test that second product, because our only testing method destroys the product in order to perform the test. But the test of the first product is assumed to be representative of the capability of the factory.

This leads to yet another fine point of model development and evaluation. Now that we understand that what we are often evaluating is not an actual model but rather an entire methodology, we are inspired to make this test as thorough as possible. This means that we want to perform as many train/select/test cycles as possible. Thus, we may choose to do something like this:

- Train all competing models on data through 1990

- Evaluate all models using 1991 validation data and choose the best

- Evaluate the best model using 1992 data as the test set

- Train all competing models on data through 1991

- Evaluate all models using 1992 validation data and choose the best

- Evaluate the best model using 1993 data as the test set

Repeat these steps, walking forward one year at a time, until the data is exhausted. Pool the test-set results into a grand performance figure.

This will provide a much better measure of the ability of our model creation procedure than if we did it for just one train/select/test cycle.

Note that if the cases in the dataset are independent (which often implies that they are not chronological), then we can just as well use a procedure akin to ordinary cross validation. (If you are not familiar with cross validation, you may want to take a quick side trip to the next section, where this topic is discussed in more detail.) For example, we could divide the data into four subsets and do the following:

- Train all competing models on subsets 1 and 2 together.

- Evaluate all models on subset 3 and choose the best.

- Evaluate the best model using subset 4 as the test set.

Repeat these steps using every possible combination of training, test, and test sets. This will provide a nearly (though not exactly) unbiased estimate of the quality of the model-creation methodology. Moreover, it will provide a count of how many times each competing model was the best performer, thus allowing us to train only that top vote-getter as our final model.

To quickly summarize the material in this section:

- The performance of a trained model in its training set is, on average, optimistic. This *training bias* is mostly due to the model treating noise as if it represents legitimate patterns. If the model is too powerful, this effect will be severe and is called *overfitting*. An overfit model is often worthless when put to use.

- If the best of several competing models is to be chosen, the selection criterion should never be based on performance in the training set because that overfit models will be favored. Rather, the choice should always be based on performance in an independent dataset often called a *validation set*.

13

- The performance of the best of several competing models is, on average, optimistic, even when the individual performances used for selection are unbiased. This *selection bias* is due to luck playing a role in determining the winner. An inferior model that was lucky in the validation set may unjustly win, or a superior model that was unlucky in the validation set may unjustly lose the competition.

- When training of multiple models is followed by selection of the best, selection bias makes it necessary to employ a third independent dataset, often called the *test set*, in order to provide an unbiased estimate of true ability.

- We are not generally able to compute an unbiased estimate of the performance of the actual model that will be put to use. Rather, we can assess the average performance of models created by our model-creating methodology and then trust that our assessment holds when we create the final model.

Cross Validation and Walk-Forward Testing

In the prior section we briefly presented two methods for evaluating the capability of our model-creation procedure. These algorithms did not test the capability of the model ultimately put into use. Rather, they tested the process that creates models, allowing the developer to infer that the model ultimately produced for real-world use will perform with about the same level of proficiency as those produced in the trial runs of the factory.

In one of these algorithms we trained a model or models on time-series data up to a certain point in time, optionally tested them on chronologically later data for the purpose of selecting the best of several competing models, and then verified performance of the final model on data still later in time. This procedure, when repeated several times, steadily advancing across the available data history, is called *walk-forward testing*.

In the other algorithm we held out one chunk of data, trained on the remaining data, and tested the chunk that was held out. To accommodate selection of the best of several competing models, we may have held out two chunks, one for selection and one for final testing. When this procedure is repeated many times, each time holding out a different chunk until every case has been held out exactly once, it is called cross validation.

Both of these algorithms are well known in the model-development community. We will not spend undue time here rehashing material that you will probably be familiar with already and that is commonly available elsewhere. Rather, we will focus on several issues that are not quite as well known.

Bias in Cross Validation

First, for the sake of at least a remedial level of completeness, we should put to rest the common myth that cross validation yields an unbiased estimate of performance. Granted, in most applications it is very close to unbiased, so close that we would usually not be remiss in calling it unbiased. This is especially so because its bias is nearly always in the conservative direction; on average, cross validation tends to under-estimate true performance rather than over-estimate it. If you are going to make a mistake, this is the one to make. In fact, usually the bias is so small that it's hardly worth discussing. We mention it here only because this myth is so pervasive, and it's always fun to shatter myths.

How can it be biased? On the surface cross validation seems fair. You train on one chunk of data and test on a completely different chunk. The performance on that test chunk is an unbiased measure of the true performance of the model that was trained on the rest of the data. Then you re-jigger the segregation and train/test again. That gives you another unbiased measure. Combining them all should give you a grand unbiased measure.

There is a simple explanation for the bias. Remember that the size of a training set impacts the accuracy of the trained model. If a model is trained on one million randomly sampled cases, it will probably be essentially perfect. If it is trained on two cases, it will be unstable and probably do a poor job when put to use. Cross validation shrinks the training set. It may be small shrinkage; if the training set contains 500 cases and we hold out just one case at a time for validation, the size of these training sets will be 99.8 percent of that presumably used to train the final model. But it is nevertheless smaller, resulting in a small diminishment of the tested model's power. And if we do tenfold cross validation, we lose 10 percent of the training cases for each fold. This can produce noticeable performance bias.

Overlap Considerations

There is a potentially serious problem that must be avoided when performing walk-forward testing or cross validation on a time series. Developers who make this error will get performance results that are anti-conservative; they overestimate actual performance, often by enormous amounts. This is deadly.

This issue must be addressed because when the independent variables used as predictors by the model are computed, the application nearly always looks back in history and defines the predictors as functions of recent data. In addition, when it computes the dependent variable that is predicted, it typically looks ahead and bases the variable on future values of the time series. At the boundary between a training period and a test period, these two regions can overlap, resulting in what is commonly termed *future leak*. In essence, future test-period behavior of the time series leaks into the training data, providing information to the training and test procedures that would not be available in real life.

For example, suppose we want to look at the most recent 30 days (called the *lookback length*) and compute some predictors that will allow us to predict behavior of the time series over the next 10 days (the *look-ahead length*). Also suppose we want to evaluate our ability to do this by walking the development procedure forward using a training period consisting of the most recent 100 days.

Consider a single train/test operation. Maybe we are sitting at Day 199, training the model on Days 100 through 199 and then beginning the test period at Day 200. The training set will contain 100 cases, one for each day in the training period. Think about the last case in the training set, Day 199. The predictors for this case will be computed from Day 170 through Day 199, the 30-day lookback length defined by the developer. The predicted value will be computed based on Day 200 through Day 209, the 10-day look-ahead length also defined by the developer.

Now think about the first case in the test set, Day 200. Its predictors will be computed from Days 171 through 200. The overlap with the prior case is huge, as these two cases share 29 of their 30 lookback days. Thus, the predictors for the first test case will likely be almost the same as the predictors for the final training case.

That alone is not a problem. However, it does become a problem when combined with the situation for the predicted variable. The first test case will have its predicted value based on Days 201 through 210. Thus, nine of its ten predicted-value look-ahead days are shared with the final case in the training set. As a result, the predicted value of this first test case will likely be similar to the predicted value of the final training case.

In other words, we have a case in the training set and a case in the test set that must be independent in order to produce a fair test but that in fact are highly correlated. When the model training algorithm learns from the training set, including the final case, it will also be learning about the first test case, a clear violation of independence! During training, the procedure had the ability to look ahead into the test period for information.

Of course, this correlation due to overlap extends beyond just these two adjacent cases. The next case in the test set will share 28 of its 30 lookback days with the final training case, will share 8 of its 10 look-ahead days, and so forth. This is serious.

This same effect occurs in cross validation. Moreover, it occurs for interior test periods, where the end of the test period abuts the beginning of the upper section of the training period. Thus, cross validation suffers a double-whammy, being hit with this boundary correlation at both sides of its test period, instead of just the beginning.

The solution to this problem lies in shrinking the training period away from its border (or borders) with the test period. How much must we shrink? The answer is to find the minimum of the lookback length and the look-ahead length and subtract 1. An example may make this clear.

In the application discussed earlier, the walk-forward training period ended on Day 199, and the test period began on Day 200. The 10-day look-ahead length was less than the 30-day lookback length, and subtracting 1 gives a shrinkage of 9 days. Thus, we must end the training period on Day 190 instead of Day 199. The predicted value for the last case in this training set, Day 190, will be computed from the 10 days 191 through 200, while the predicted value for the first case in the test set (Day 200) will be based on Days 201 through 210. There is no overlap.

Of course, the independent variable for the first test case (and more) will still overlap with the dependent variable for the last training case. But this is no problem. It does not matter if just the independent periods overlap, or just the dependent periods. The problem occurs only when both periods overlap, as this is what produces correlated cases.

With cross validation we must shrink the training period away from both ends of the test period. For example, suppose our interior test period runs from Day 200 through Day 299. As with walk forward, we must end the lower chunk of the training period at Day 190 instead of at Day 199. But we must also begin the upper chunk of the training period at Day 309 instead of Day 300. I leave it as an exercise for you to confirm that this will be sufficient but not excessive.

Assessing Nonstationarity Using Walk-Forward Testing

A crucial assumption in time-series prediction is stationarity; we assume that the statistical characteristics of the time series remain constant. Unfortunately, this is a rare luxury in real life, especially in financial markets, which constantly evolve. Most of the time, the best we can hope for is that the statistical properties of the series remain constant long enough to allow successful predictions for a while after the training period ends.

There is an easy way to use walk-forward testing to assess the practical stationarity of the series, at least in regard to our prediction model. Note, by the way, that certain aspects of a time series may be more stationary than other aspects. Thus, stationarity depends on which aspects of the series we are modeling. It is possible for one model to exhibit very stationary behavior in a time series and another model to be seriously nonstationary when applied to the same time series.

The procedure is straightforward. Perform a complete walk-forward test on the data using folds of one case. In other words, begin the training period as early as the training set size and lookback allow, and predict just the next observation. Record this prediction and then move the training set window forward one time slot, thus including the case just predicted and dropping the oldest case in the training set. Train again and predict the next case. Repeat this until the end of the dataset is reached, and compute the performance by pooling all of these predictions.

Next, do that whole procedure again, but this time predict the next *two* cases after the training period. Record that performance. Then repeat it again, using a still larger fold size. If computer time is not rationed, you could use a fold size of three cases. My habit is to double the fold size each time, using fold sizes of 1, 2, 4, 8, 16, etc. What you will see in most cases is that the performance remains fairly stable for the first few, smallest fold sizes and then reaches a point at which it rapidly drops off. This tells you how long a trained model can be safely used before nonstationarity in the time series requires that it be retrained.

Of course, statistical properties rarely change at a constant rate across time. In some epoch the series may remain stable for a long time, while in another epoch it may undergo several rapid shifts. So, the procedure just described must be treated as an estimate only. But it does provide an excellent general guide. For example, if the performance plummets after more than just a very few time slots, you know that the model is dangerously unstable with regard to nonstationarity in the series, and either you should go back to the drawing board to seek a more stable model or you should resign yourself to frequent retraining. Conversely, if performance remains flat even for a large fold size, you can probably trust the stability of the model.

Nested Cross Validation Revisited

In the "Selection Bias and the Need For Three datasets" section that began on page 9, we briefly touched on using cross validation nested inside cross validation or walk-forward testing to account for selection bias. This is such an important topic that we'll bring it up again as a solution to two other very common problems: finding an optimal predictor set and finding a model of optimal complexity.

In nearly all practical applications, the developer will have in hand more predictor candidates than will be used in the final model. It is not unusual to have several dozen promising candidates but want to employ only three or so as predictors. The most common approach is to use stepwise selection to choose an optimal set of predictors. The model is trained on each single candidate, and the best performer is chosen. Then the model is trained using two predictors: the one just chosen and each of the remaining candidates. The candidate that performs best when paired with the first selection is chosen. This is repeated as many times as predictors are desired.

There are at least two problems with this approach to predictor selection. First, it may be that some candidate happens to do a great job of predicting the noise component of the training set, but not such a great job of handling the true patterns in the data. This inferior predictor will be selected and then fail when put to use because noise, by definition, does not repeat.

The other problem is that the developer must know in advance when to stop adding predictors. This is because every time a new predictor is added, performance will improve again. But having to specify the number of predictors in advance is annoying. How is the developer to know in advance exactly how many predictors will be optimal?

The solution is to use cross validation (or perhaps walk-forward testing) for predictor selection. Instead of choosing at each stage the predictor that provides the best performance when trained on the training set, one would use cross validation or walk-forward testing within the training set to select the predictor. Of course, although each individual performance measure in the competition is an unbiased estimate of the true capability of the predictor or predictor set (which is why this is such a great way to select predictors!), the winning performance is no longer unbiased, as it has been compromised by selection bias. For this reason, the selection loop must be embedded in an outer cross validation or walk-forward loop if an unbiased performance measure is desired.

Another advantage of this predictor selection algorithm is that the method itself decides when to stop adding predictors. The time will come when an additional predictor will cause overfitting, and all competing performance measures decrease over the prior round. Then it's time to stop adding predictors.

A similar process can be used to find the optimal complexity (i.e., power) of a model. For example, suppose we are using a multiple-layer feedforward network. Deciding on the optimal number of hidden neurons can be difficult using simplistic methods. But one can find the cross validated or walk-forward performance of a very simple model, perhaps with even just one hidden neuron. Then advance to two neurons, then three, etc. Each time, use cross validation or walk-forward testing to find an unbiased estimate of the true capability of the model. When the model becomes too complex (too powerful), it will overfit the data and performance will drop. The model having the maximum unbiased measure has optimal complexity. As with predictor selection, this figure is no longer unbiased, so an outer loop must be employed to find an unbiased measure of performance.

Common Performance Measures

There are an infinite number of choices for performance measures. Each has its own advantages and disadvantages. This section discusses some of the more common measures.

Throughout this section we will assume that the performance measure is being computed from a set of n cases. One variable is being predicted. Many models are capable of simultaneously predicting multiple dependent variables. However, experience indicates that multiple predictions from one model are almost always inferior to using a separate model for each prediction. For this reason, all performance measures in this section will assume univariate prediction. We will let y_i denote the true value of the dependent variable for case i, and the corresponding predicted value will be $\hat{y}_i$.

Mean Squared Error

Any discussion of performance measures must start with the venerable old *mean squared error* (MSE). MSE has been used to evaluate models since the dawn of time (or at least so it seems). And it is still the most widely used error measure in many circles. Its definition is shown in Equation (1.2).

$$MSE = \frac{1}{n}\sum_{i=1}^{n}\left(\hat{y}_i - y_i\right)^2 \tag{1.2}$$

Why is MSE so popular? The reasons are mostly based on theoretical properties, although there are a few properties that have value in some situations. Here are some of the main advantages of MSE as a measure of the performance of a model:

- It is fast and easy to compute.

- It is continuous and differentiable in most applications. Thus, it will be well behaved for most optimization algorithms.

- It is very intuitive in that it is simply an average of errors. Moreover, the squaring causes large errors to have a larger impact than small errors, which is good in many situations.

- Under commonly reasonable conditions (the most important being that the distribution is normal or a member of a related family), parameter estimates computed by minimizing MSE also have the desirable statistical property of being *maximum likelihood* estimates. This loosely means that of all possible parameter values, the one computed is the most likely to be correct.

We see that MSE satisfies the theoretical statisticians who design models, it satisfies the numerical analysts who design the training algorithms, and it satisfies the intuition of the users. All of the bases are covered. So what's wrong with MSE? In many applications, plenty.

The first problem is that only the *difference* between the actual and the predicted values enters into the computation of MSE. The *direction* of the difference is ignored. But sometimes the direction is vitally important. Suppose we are developing a model to predict the price of an equity a month from now. We intend to use this prediction to see if we should buy or sell shares of the stock. If the model predicts a significant move in one direction, but the stock moves in the opposite direction, this is certainly an error, and it will cost us a lot of money. This error correctly registers with MSE. However, other errors contributing to MSE are possible, and these errors may be of little or no consequence. Suppose our model predicts that no price move will occur, but a large move actually occurs. Our resulting failure to buy or sell is a lost opportunity for profit, but it is certainly not a serious problem. We neither win nor lose when this error appears. Going even further, suppose our model predicts a modest move and we act accordingly, but the actual move is much greater than predicted. We make a lot more money than we hoped for, yet this incorrect prediction contributes to MSE! We can do better.

Another problem with MSE is the squaring operation, which emphasizes large errors at the expense of smaller ones. This can work against proper training when moderately unusual outliers regularly appear. Many applications rarely but surely produce observed values of the dependent variable that are somewhat outside the general range of values. A financial prediction program may suffer an occasional large loss due to unforeseen market shocks. An industrial quality control program may have to deal with near catastrophic failures in part of a manufacturing process. Whatever the cause, such outliers present a dilemma. It is tempting to discard them, but that would not really be right because they are not truly invalid data; they are natural events that must be taken into account. The process that produces these outliers is often of such a nature that squashing transformations like *log* or *square root* are not theoretically or practically appropriate. We really need to live with these unusual cases and suffer the consequences because the alternative of discarding them is probably worse. The problem is that by squaring the errors, MSE exacerbates the problem. The training algorithm will produce a model that does as much as it can to reduce the magnitude of these large errors, even though it means that the multitude of small errors from important cases must grow. Most of the time, we would be better off minimizing the errors of the "normal" cases, paying equal but not greater attention to the rare large errors.

Another problem with MSE appears only in certain situations. But in these situations, the problem can be severe. In many applications, the ultimate goal is classification, but we do this by thresholding a real-valued prediction. The stock price example cited earlier is a good example. Suppose we desire a model that focuses on stocks that will increase in value soon. We code the dependent variable to be 1.0 if the stock increases enough to hit a profit target, and 0.0 if it does not. A set of results may look like the following:

Predicted:	0.2	0.9	0.7	0.3	0.1	0.9	0.5	0.6	0.1	0.2	0.9	0.4	0.9	0.7
Actual:	0.0	1.0	0.0	1.0	0.0	1.0	0.0	1.0	1.0	0.0	1.0	1.0	1.0	0.0

On first glance, this model looks terrible. Its MSE would bear out this opinion. The model frequently makes small predictions when the correct value is 1.0, and it also makes fairly large predictions when the correct value is 0.0. However, closer inspection shows this to be a remarkable model. If we were to buy only when the prediction reaches a threshold of 0.9, we would find that the model makes four predictions this large, and it is correct all four times. MSE completely fails to detect this sort of truly spectacular performance. In nearly all cases in which our ultimate goal is classification through thresholding a real-valued prediction, MSE is a poor choice for a performance criterion.

Mean Absolute Error

Mean absolute error (MAE) is identical to mean squared error, except that the errors are not squared. Rather, their absolute values are taken. This performance measure is defined in Equation (1.3).

$$MAE = \frac{1}{n}\sum_{i=1}^{n}|\hat{y}_i - y_i|$$ (1.3)

The use of absolute value in place of squaring alleviates the problems of squaring already described. This measure is often used when training models for negative-feedback controllers. However, the other problems of MSE remain. MAE is generally recommended only in special applications for which its suitability is determined in advance. Also note that MAE is not differentiable. This can be a problem for some optimization algorithms.

R-Squared

If the disadvantages of MSE discussed earlier are not a problem for a particular application, there is a close relative of MSE that may be more intuitively meaningful. *R-squared* is monotonically (but inversely) related to MSE, so optimizing one is equivalent to optimizing the other. The advantage of R-squared over MSE lies purely in its interpretation. It is computed by expressing the MSE as a fraction of the total variance of the dependent variable, then subtracting this fraction from 1.0, as shown in Equation (1.4). The mean of the dependent variable is the usual definition, shown in Equation (1.5).

$$R^2 = 1 - \frac{\sum_{i=1}^{n}(\hat{y}_i - y_i)^2}{\sum_{i=1}^{n}(y_i - \bar{y})^2}$$ (1.4)

$$\bar{y} = \frac{1}{n}\sum_{i=1}^{n}y_i$$ (1.5)

If our model is completely naive, always using the mean as its prediction, the numerator and denominator of Equation (1.4) will be equal, and R-squared will be zero. If the model is perfect, always predicting the correct value of the dependent variable, the numerator will be zero, and R-squared will be one. This normalization to the range zero to one makes interpreting R-squared easy, although pathologically poor predictions can result in this quantity being negative. R-squared is a common performance measure, especially for models that are expected to perform almost perfectly. Just remember that R-squared shares all of the disadvantages of MSE that were discussed earlier. Also, if performance even slightly beyond total naivete is considered good, R-squared is so tiny that it is nearly always worthless as a performance measure.

RMS Error

RMS stands for *root-mean-square*. *RMS error* is the square root of mean squared error. Why would we want to use the square root of MSE as a performance measure? The reason is that the squaring operation causes the MSE to be a nonlinear function of the errors, while RMS error is linear. Doubling all of the errors causes the RMS error to double, while the MSE quadruples. RMS error is commensurate with the raw data. For example, suppose we had a miracle model that could predict the life span of people. If we were told that the model has an RMS error of five years, this figure is immediately interpretable. But if we were told that the model has a mean squared error of 25, extra effort would be required to make sense of the figure. The difference between MSE and RMS error may seem small, but one should never discount the importance of interpretability.

Nonparametric Correlation

Naturally, it is always nice to be able to accurately predict a dependent variable. But sometimes we are better off being satisfied with doing nothing more than getting the relative magnitudes correct. In other words, we can often be happy if relatively large predicted values correspond to relatively large actual values, and small predictions correspond to small realizations. This is usually sufficient in applications in which our only goal is to identify cases that have unusually large or small values of the dependent variable. This goal is a significant relaxation of what is normally expected of a model. Why should it be considered? There are at least two situations in which expressing performance in terms of relative ordering is not only useful but perhaps even mandatory for good operation.

First, there are some models for which accurate prediction is an unrealistic expectation by the very nature of the model. The classic example is the *General Regression Neural Network (GRNN)* when it is applied to a very difficult problem. (The GRNN is defined in [Specht, 1991] and discussed in [Masters, 1995a] and [Masters, 1995b].) When the training set contains numerous cases for which similar values of the independent variable correspond to widely different values of the dependent variable, the smoothing kernel will cause the predictions to cluster around the mean of the dependent variable for most values of the independent variable. Good relative ordering is still attainable in many situations, but the model's R-squared will be practically zero. In this case, training to optimize MSE or a close relative is nearly hopeless, while optimizing correlation can give surprisingly good results.

The other situation in which it is good to relax the model's fitting requirement is when the dependent variable can occasionally have unusual values, especially if these unusual values are due to heavy-tailed noise. For example, suppose the majority of values lie between –100 and 100, but every now and then a burst of noise may cause a value of 500 or even more to appear. Such values could be arbitrarily truncated prior to training, but this is a potentially dangerous operation, especially if the values may be truly legitimate. Asking a model to make such unusual predictions will almost always cause the training procedure to make deleterious compromises. The prediction error will be dominated by these few special cases, and the effort expended to cater to them will cause performance among the majority to suffer. In such applications, we can often be content requiring only that the model make unusually large predictions for cases that have unusually large values of the dependent variable. This way, the training algorithm can focus equally on all cases.

If ordinary linear correlation were used as a performance criterion, the first of the two situations just discussed would be accommodated. Models that are inherently unable to accurately predict the dependent variable, despite that variable having a reasonable distribution, could be effectively trained by optimizing linear correlation between the predicted and true values. However, the second situation would still be problematic because outliers distort linear correlation. This situation is best treated by considering only the relative order of the predicted and true values. A measure of the degree to which the relative orders correspond is called *nonparametric correlation*. The most common measure of nonparametric correlation is *Spearman rho*. It is easy to compute and has excellent theoretical and practical properties. Source code for computing Spearman rho is included on my web site.

Success Ratios

In many applications, a single poor prediction can cause considerable damage. Suppose we have two models. Model *A* has significantly better average performance than Model *B*. Model *A* obtains its good performance by virtue of having many excellent predictions, although it does suffer from a few terrible predictions. In contrast, Model *B* has almost no exceptionally good or bad predictions, but the vast majority of its predictions are fairly decent. In some applications we would prefer one model, and in other applications we would prefer the other model. For applications that prefer the consistent but slightly inferior Model *B*, any of several success ratios may be an appropriate performance criterion.

The key point of a success ratio is that it considers successful and unsuccessful predictions separately. Success for a case must be indicated by a positive performance criterion for that case, and failure must be indicated by a negative criterion. The degree of success or failure for the case is indicated by the magnitude of the criterion. For an automated market trading system, the obvious criterion is the profit (a positive number) or loss (a negative number) for each trade. For applications in which there is a less obvious indicator of prediction quality, a little imagination may be needed to devise a means of indicating the quality of each individual prediction. Once a suitable casewise indicator is defined, there are at least two effective ways to describe the performance of the model on a set of cases.

A standard performance indicator for automated trading systems is the *profit factor*, sometimes called the *success factor* in more general applications. This is computed by dividing the sum of the successes by the sum of the failures, as shown in Equation (1.6).

$$success\ factor = \frac{\sum_{x_i>0} x_i}{-\sum_{x_i<0} x_i} \tag{1.6}$$

The problem with the success factor is that the denominator can be zero when the model is superb. This is easily remedied by slightly rearranging the terms to produce a criterion that varies from zero for a terrible model to one for a perfect model. This success ratio is shown in Equation (1.7).

$$success\ ratio = \frac{\sum_{x_i>0} x_i}{\sum_{x_i>0} x_i - \sum_{x_i<0} x_i} \tag{1.7}$$

When either of these performance criteria is used, it is often important that some additional minimal performance constraint be imposed. For example, it may be possible for the training algorithm to find a model whose performance is consistent (almost always positive) but so mediocre that it is practically worthless. This situation can be avoided by including in the training criterion a severe penalty for the model failing to meet a reasonable minimum average performance. Once this minimum threshold is satisfied, the success ratio can be allowed to dominate the training criterion.

Another special situation is when the prediction is used to make a thresholded classification decision. This topic will be discussed in detail later, but for now understand that an application may choose to act if and only if the prediction exceeds a specified threshold. If action is taken for a case, a reward or penalty is earned for that case. In such an application, we should ensure that the threshold is such that a minimum number of cases exceed it and hence trigger action. This encourages honesty in the criterion by preventing a few lucky cases from dominating the calculation. If the threshold is so restrictive that only a few cases trigger action, random luck may cause all of them to have a positive return, leading to false optimism in the quality of the model. To have a reasonable sample, a sufficient number of cases should trigger action.

Alternatives to Common Performance Measures

A good deal of this chapter so far has been devoted to criticizing common performance measures. So where are the recommended alternatives? Many are in Chapter 2, "Assessment of Class Predictions." The reason is that applications for which performance measures like MSE are particularly inappropriate have a strong tendency to be applications in which a numeric prediction is used to make a classification decision. This hybrid model falls in a gray area between chapters, and it could technically appear in either place. However, the best performance measures for such models generally lie closer to classification measures than to numeric measures. See page 60 for some excellent alternatives to MSE and its relatives in many applications.

Stratification for Consistency

It has been seen that good average or total performance in the training set is not the only important optimization consideration. Consistent performance is also important. It encourages good performance outside the training set, and it provides stability as models are evolved by selective updating. An often easy way to encourage consistency is to stratify the training set, compute for each stratum a traditional optimization criterion like one of those previously discussed, and then let the final optimization criterion be a function of the values for the strata. What is a good way to stratify the training set? Here are some ideas:

- The optimal number of strata depends on the number of training cases because the individual subsets must contain enough points to provide a meaningful criterion measurement. Somewhere between 5 and 20 strata seem to be generally effective. In rare cases, each case might be considered a stratum, producing as many strata as there are training cases.

- If the data is a time series, the obvious and usually best approach is to keep it in chronological order and place dividing lines approximately equally along the time line. This helps the model to downplay nonstationary components while searching for commonality.

- Sometimes it is suspected that an extraneous variable may impact the model. However, for some reason, this variable will be impractical to measure when the model is used. Therefore, the variable is not used as an input to the model. If this troubling variable is measurable historically, it may be wise to do so and assign training cases to strata according to values of this extraneous variable. This way, the training procedure will be encouraged to find a model that is effective for all values of the variable.

- If no obvious stratification method exists, just assign cases randomly. This is better than no stratification at all.

The power of stratification can sometimes be astounding. It is surprisingly easy to train a model that seems quite good on a training set only to discover later that its performance is spotty. It may be that the good average training performance was based on spectacular performance in part of the training set and mediocre performance in the rest. When the conditions that engender mediocrity appear in real life and the model fails to perform up to expectations, it is only then that the researcher may study its historical performance more closely and discover the problem. It is always better to discover this sort of problem early in the design process.

Once the training set has been intelligently stratified and the individual performance criteria computed, how can these multiple measurements be combined into one grand optimization criterion? There are an infinite number of possibilities, and you are encouraged to use your imagination to find a method that fits itself to the problem at hand. Some methods are better than others.

Dividing the mean performance across strata by the standard deviation is an obvious candidate. This is the essence of the famous *Sharpe ratio* so dear to economists. However, the presence in the denominator of a quantity that can theoretically equal zero, and that often nearly does just that in practice, leads to dangerous instability. A model that has small average performance and tiny standard deviation can generate an inappropriately large optimization criterion. This is not recommended.

One general family of methods that can be useful is to sort the criteria for the strata and employ a function of the sorted values. A method that often works extremely well is to use the mean of the criteria after a small adjustment: discard the best stratum and replace it with the worst. This effectively doubles the contribution of the worst stratum, while totally ignoring the contribution of the best. In this way, models whose good average performance is primarily due to one exceptionally good stratum are eliminated. At the same time, models that perform exceptionally poorly in any stratum are discouraged. Some applications may need to discourage poor performance even more strongly, so feel free to over-weight the second-poorest stratum, or employ other variations of differential weighting.

Finally, stratification may reveal weakness in the model. If a time series behaves inconsistently, it is nonstationary and major revision is necessary. If different values of a stratification variable produce inconsistent results, then it may be necessary to employ a separate specialist model for each value. Such situations should never be ignored.

Confidence Intervals

When we use a trained model to make a prediction, it is always nice to have an idea of how close that prediction is likely to be to the true value. It is virtually never the case that we can state with absolute certainty that the prediction will be within some realistic range. All such statements are necessarily probabilistic. In other words, we can almost never say with certainty something like, "The prediction is 63, and the true value is guaranteed to be somewhere between 61 and 65." But what we can often say is, "There is at least a 95 percent probability that the true value is between 61 and 65." It should be intuitively obvious that there must be a trade-off between the probability and the range of the confidence interval. For a given model and prediction, we may be able to choose from among a variety of confidence intervals of varying certainty. For example, we may be able to assert that there is at least a 95 percent chance that the true value is between 61 and 65, a 99 percent chance that the true value is between 60 and 66, a 90 percent chance that the true value is between 62 and 64, and so forth. Wider intervals have higher confidence.

When we compute confidence intervals, it is clearly beneficial to have the confidence interval as narrow as possible for any specified probability. It most likely does no good to learn that there is at least a 90 percent chance that the true value is between 20 and 500 if this range is ridiculously wide in the context of the application. There is a trade-off between the assumptions we are willing to make about the statistical distribution of the errors and the width of the confidence intervals that can be computed. This section discusses *parametric* confidence intervals. These intervals make significant (and often unrealistic) assumptions about the error distribution, and in return they deliver the narrowest possible intervals for any desired confidence level. In a later section we will explore a good method for computing confidence intervals when we are not willing to make any assumptions about the distribution, although the cost will be somewhat wider intervals.

The Confidence Set

No matter what method is used to compute confidence intervals, the first step is the same. A *confidence set* of cases must be collected. This is a collection of cases not unlike the training set and the test set (and perhaps the validation set used in some training schemes). It must be independent of all of the cases that were used in any way to train the model, as the training process causes the model to perform unnaturally well on the data on which its training was based. As with the other data sets, the confidence set must be fairly and thoroughly sampled from the population of cases in which the model will ultimately operate. But note that it is especially important that the confidence set be representative of the cases that will appear later when the model is put to work. Suppose that careless sampling results in some operational situations being over-represented and others failing to appear in the set as often as they are actually expected to appear. If this problem happens during collection of the training set, the model will probably perform very well in the former operational situations but underperform in the latter areas.

This is unfortunate, but probably not disastrous. If this problem taints the test set, the computed error by which the model is judged will be slightly distorted. Again, this is not good, but it is probably not too terrible. However, if the confidence set is tainted by bad sampling, every confidence statement made for the rest of the life of the model will be incorrect. In most cases, this is very undesirable. It pays to devote considerable resources to acquiring a fair and representative confidence set.

There is a dangerous error that is commonly made in collecting a confidence set: many people don't do it. They reuse the test set to compute confidence information. On the surface, this seems reasonable and economical. The test set is (presumably!) independent of the training set, so the single most important criterion is met. And perhaps the experimenter was particularly careful in collecting the test set, making its use as the confidence set especially appealing. What is vital to realize is that in a very real sense, the test set is actually part of the training set. Granted, it did not explicitly take part in the model's optimization. Yet it did take part in a subtle way. The test set is generally used as the ultimate arbiter of the model's quality. After the model is trained, it is tested. If its performance on the test set is acceptable, the model is kept. If its performance is poor, the model is redesigned. This means that the selection of the model ultimately deployed depends on the test set. Understand that actual rejection and redesign *does not need to have even happened.* Just the supposition that poor performance on the test set would trigger this response is sufficient to taint the test set. Many people who do not have training in statistics have trouble comprehending that the mere *possibility* of an event that never actually occurs can taint a dataset. Nevertheless, it is true. The only way that we can safely assert that the test set is untainted and usable for the confidence set is if we stalwartly proclaim *in advance* that the model will be deployed, no matter how poorly it is found to perform on the test set. This is almost always a silly procedure, and it is certainly not encouraged here. The point is that the use of any dataset in the final model selection decision taints that dataset. The fact that a model will be accepted if and only if it performs well on the test set means that the model is biased to perform better on this set than it would perform in the general population. The end result is that confidence intervals computed from the test set will be unnaturally narrow, the worst possible type of error. The actual probability that the true value lies outside the computed confidence interval will be higher than the computed probability. This is unacceptable. The independent confidence set must be used only after the model has passed all tests. And to be strictly honest in deployment, it must be agreed in advance that poor performance on the confidence set will not result in reformulation of the model, although failure at the confidence stage probably justifies firing the engineer in charge of validation!

A crucial assumption in any confidence calculation is that the error measurements on which the computation is based must be *independent and identically distributed.* This means that the prediction error for each case must not be influenced by the error that happened on another case. It also means that the probability distribution of each prediction error, both in the confidence set and in the future, must be the same. Violations of this pair of assumptions will impact the computed confidence intervals, perhaps insignificantly, or perhaps dangerously. Be sure to consider the possibility that these assumptions may not be adequately met.

Serial Correlation

One common way in which the independence assumption may be violated is if the application is a time series and the error series exhibits serial correlation. In other words, it may be that an unusually large error at one point increases the probability of an unusually large (or perhaps small) error at the next point. The net effect of this violation is to decrease the effective number of degrees of freedom in the dataset, producing a value lower than that implied by the number of cases in the confidence set. This makes the confidence calculations less stable than they would otherwise be. However, in practice this may not be as serious a problem as expected. A good model will not produce significant serial correlation in its errors. In fact, if the experimenter notices serial correlation in the errors, this is a sign that the model needs more work so that it can take advantage of the predictive information that leads to serial correlation.

Consider that the goal of any prediction model is to extract as much predictive information as possible from the data. If the predictor variables and/or the predicted variable contain serial correlation that may leak into the prediction errors, the model is not doing its job well. It is ignoring this important source of information. A good model will produce prediction errors that are pure random numbers, white noise with no serial dependencies whatsoever. The degree to which prediction errors exhibit serial correlation is the degree to which the model is shirking its duty by failing to account for the correlation in the data when making predictions.

Multiplicative Data

There is a common situation in which although the errors do not satisfy the *identically distributed* assumption, they can be transformed to satisfy this assumption by a simple procedure. This situation arises when the natural variation in the predicted variable is linearly related to the value of the variable. The most common example of this phenomenon is stock prices. When a stock is trading around a price of 100, its day-to-day variation will probably be about twice as great as when it was trading around 50. In such cases, an intelligently designed model will be predicting the log of the price, rather than the price itself. There is a good chance that the errors will be identically distributed enough to allow reliable computation of confidence intervals. However, suppose that the model is unavoidably predicting the price itself. If the price covers a wide range, the errors will definitely not be identically distributed. The solution is to compute the log of the predicted price and the log of the true price. Define the error as the difference between these two logs. This error may nearly always be used to compute confidence intervals for the difference in logs. Exponentiate the computed bounds to find the bounds for the price itself. This simple technique can be applied to many situations.

Normally Distributed Errors

There is a powerful theorem called the *Central Limit Theorem* that (roughly) decrees that if a random variable that we measure has been produced by summing or averaging many other independent random variables, then our measured variable will tend to have a normal distribution, no matter what the distribution of the summed components happens to be. It is a frequent fact of life that things we measure have come about as the result of combining the effects of many component processes. For this reason, variables measured from physical processes often have an approximately normal distribution. The assumption of normality pervades statistics.

If we can safely assume that the model's prediction errors follow a normal distribution, we can easily compute a confidence interval that is optimal in that, for any specified confidence probability, the interval is as narrow as it could possibly be. This is a desirable property for which we should strive.

How can we verify that the errors follow a normal distribution? There is no method by which we can examine a sample (generally the confidence set's errors) and definitively conclude that the population is normal. Normality tests only allow us to conclude that the sample is probably *not* normal. Thus, one procedure is to apply several such tests. If none of the tests indicates non-normality, we may hesitantly conclude that the population is probably normal. If the normality tests focus on the types of non-normality most likely to cause problems in computing confidence intervals, we have additional insurance against serious errors.

Unfortunately, there is a serious practical problem with using a rigorous automated method such as performing multiple statistical tests designed to reject the assumption of normality. The problem is that such tests are usually too sensitive. If the confidence set is large (which it really should be), statistical tests of normality will quite possibly indicate non-normality when the degree of departure from normality is so small as to have little practical impact on the confidence intervals. For this reason, the commonly employed explicit tests of skewness and kurtosis, and even the relatively sophisticated Kolmogorov-Smirnov or Anderson-Darling tests, are not recommended for other than the most stringent applications.

The easiest and often most relevant test for normality is simple visual examination of a histogram of the errors. The histogram should have a classic bell curve shape. If one or more errors lie far from the masses, normality is definitely suspect. If more errors lie to the right (or left) of the center than lie on the other side, this lack of symmetry is a serious violation of normality. Either of these situations should cause us to abandon confidence intervals based on a normal distribution. More sophisticated statistical tests will often detect asymmetry and heavy tails that are not visible on a histogram. However, this is one of those situations in which many experts agree that what you can't see can't hurt you. Departures from normality that are not plainly visible on a histogram will probably not seriously degrade the quality of the confidence intervals. Nevertheless, if the application demands high confidence in the quality of the confidence intervals, rigorous testing may be advised.

Once we are willing to accept the assumption that the model's prediction errors follow a normal distribution, computing confidence intervals for the errors is straightforward. Let the errors for the n cases in the confidence set be x_i, $i = 1, ..., n$. The mean of the errors is computed with Equation (1.8), and the estimated standard deviation of the errors is computed with Equation (1.9).

$$\bar{x} = \frac{1}{n}\sum_{i=1}^{n} x_i \tag{1.8}$$

$$s_x = \sqrt{\frac{1}{(n-1)}\sum_{i=1}^{n}(x_i - \bar{x})^2} \tag{1.9}$$

If we knew the exact mean and standard deviation of the error population, an exact confidence interval could be computed as the mean plus and minus a multiple of the standard deviation, where the multiple depends on the confidence desired. In practice, we almost never know these quantities, and they must be estimated using the two equations just shown. The confidence set is a random sample subject to variation, so the confidence intervals are only an approximation. However, as long as n is reasonably large, say more than a few hundred or so, the approximation is close enough for most work. The method of the next section can be used if more rigor in regard to this issue is required, though that method generally provides confidence intervals that are wider than these.

By the way, do not automatically assume that the mean of the errors must be zero. It is certainly true that most models tend to produce errors that are nearly centered at zero, and in rare instances we may want to force a mean of zero. However, some models deliberately generate slightly off-center errors in an effort to trade off bias for variance reduction. The safest plan is to actually compute the mean and use it. When this is done, the confidence interval is expressed in Equation (1.10).

$$P\{\bar{x} - z_{\alpha/2}s_x \leq x \leq \bar{x} + z_{\alpha/2}s_x\} = 1 - \alpha \qquad (1.10)$$

Values of z_α for various values of α can be found in any basic statistics text. For a 90 percent confidence interval, $z_{.05}=1.64$; for a 95 percent confidence interval, $z_{.025}=1.96$; and for a 99 percent confidence interval, $z_{.005}=2.58$. Note that t-distribution values would be essentially identical to these for the large confidence set sizes used in practice.

Empirical Quantiles as Confidence Intervals

Sometimes we are unable to comfortably assume that the prediction errors follow a normal distribution. When this is so, there is an excellent alternative method for computing confidence intervals for predictions. This method does not assume that the errors follow any particular distribution. It will work regardless of the distribution. In the strictest mathematical sense, these confidence intervals are not totally uninfluenced by the error distribution. There is no known way to compute completely distribution-free confidence intervals. However, in the vast majority of practical applications with a large confidence set, the influence of the error distribution is small and may be safely ignored. As the size of the confidence set grows, the computed intervals converge to the correct values. But always remember that the *independent and identically distributed* assumptions discussed earlier still apply. These assumptions really matter.

The technique described in this section is based on the belief that the distribution of errors observed in the confidence set will also be observed when the model is actually put to use. In other words, suppose the confidence set contains 100 cases, and we count 12 cases of these in which the prediction error exceeds, say, 1.7. This observation is extrapolated to make the assertion that when the model is used later, there is a 12 percent probability that an error will exceed 1.7. In the next section we will explore the safety of this bold assumption. For now we accept this as a reasonable and proper technique.

The *quantile* of order p of the distribution of the random variable X is defined as the value t such that $P\{X<t\}\leq p$ and $P\{X\leq t\}\geq p$. This two-part definition is needed to handle discrete distributions, and even then t may not be unique (i.e., the interval between two discrete values). It is easiest to think of quantiles in terms of continuous distributions. For example, if we know that there is a 20 percent (0.2) probability that a random case will be less than 3.7, we say that the 0.2 quantile of this distribution is 3.7. The method of this section computes the quantiles of the empirically observed confidence set, assumes that this observed sample is representative of the true population, and uses these empirical quantiles to compute whatever confidence intervals are desired.

Here is a tiny example to intuitively illustrate the use of empirical quantiles as confidence intervals. More rigor will soon follow. Suppose we collect ten cases for the (ridiculously small) confidence set. These error measurements, sorted in ascending order, are as follows: {–8, –6, –5, –1, 0, 2, 3, 5, 7, 11}. When a confidence interval for upcoming errors is computed, this interval is traditionally symmetric in probability. Thus, if an 80 percent confidence interval is desired, we generally compute an error value such that no more than 10 percent of the cases equal or exceed this number, and we also compute an error value such that no more than 10 percent of the cases equal or are less than this number. The former number is the upper confidence limit, and the latter is the lower limit. In this example, 10 percent of the case errors equal or exceed 11, so this is the upper limit. Similarly, 10 percent of them are less than or equal to –8, so this is the lower limit. We can similarly find that a 60 percent confidence interval for future errors is provided by {–6, 7}.

To be specific, confidence intervals for future prediction errors are computed from the n confidence set errors as follows: First, sort the errors in ascending order. Suppose we desire a $1-2p$ confidence interval. Compute $m=np$, and if m is not an integer, truncate the fractional part. This provides conservative (wider than the exact) intervals. (For unbiased but slightly less conservative intervals, let $m=(n+1)p$.) The m'th smallest and m'th largest values in the sample are the lower and upper confidence limits, respectively.

Several variations on this method are possible. The probability does not need to be symmetrically split. In fact, some applications may not care about errors on one side or the other. In this case, the entire probability could lie on the side in question. Also, sometimes it is more appropriate to work backward. Certain error limits may be particularly meaningful to the application. In this situation, the desired error limits could be specified, and the empirically computed confidence would be presented to the user. In the context of the current example, it may be that the error should be greater than –4 in order for the application to succeed. The program could report to the user that this happened 70 percent of the time (and, by implication, could be expected to continue to do so).

Note that the errors in this context are somewhat different from errors in other common contexts. The error of a prediction as used here is the difference between the predicted and the true value. The sign is retained. Thus, a "small" error may actually be serious. If an observed error is "less than –6," this means that perhaps the error is –8, which may be a significant error. This understanding of the context will be especially important in the next section, where errors less than some small (generally negative) value are precisely the errors about which we are concerned.

Confidence Bounds for Quantiles

The empirical quantiles used to compute confidence intervals are themselves subject to sampling error. They are entirely at the mercy of the confidence set. A different confidence collection would generally yield different confidence intervals, due to nothing more than random sampling variation. When we compute confidence intervals from empirical quantiles, it is important that we have some idea of the stability and reliability of the lower and upper confidence bounds. In other words, we would like to have confidence bounds for the confidence bounds. This is the subject of this section.

We start by making a rather remarkable statement. It has already been pointed out that confidence intervals derived from empirical quantiles, while very robust in regard to the distribution of the prediction errors on which they are based, are nevertheless affected by that distribution. This, in fact, is true for all known methods of computing effective confidence intervals. Despite this fact, we can make probabilistic statements about these confidence intervals that *are* independent of the error distribution. In other words, once we compute lower and upper confidence bounds for future prediction errors, we can go on to compute *distribution-free* confidence bounds for the error confidence bounds. This is immensely useful and should be standard practice in all applications in which confidences are particularly important.

Let us begin with the lower confidence bound for the prediction error. The logic used for the lower bound is easily inverted to handle the upper bound. Recall that we have n cases in the confidence set, and we desire a lower confidence bound such that there is only the small probability p that future errors will be less than or equal to this lower bound. The errors are sorted, and the $m=np$'th smallest error is used as our best estimate of the lower confidence bound. Although the distribution of this quantity is not independent of the distribution of the errors, it is very robust in that it is practically always reasonable, regardless of the error distribution. But how reasonable?

The traditional approach to computing confidence intervals for a point estimate is to provide lower and upper bounds for the estimate. Thus, we might be inclined to compute lower and upper bounds for the lower bound on the error. But this is usually not the best approach in this situation. For starters, we almost never care if the error's lower bound is too low. This only means that the computed confidence interval is too wide; in reality the future errors will be better than we expect. Only in the rare instance that undue pessimism might cause abandonment of a project would this be a problem. What we really care about is the possibility that the error's lower bound is too high. But it turns out that finding upper confidence bounds for the error's lower confidence bound is tricky and potentially unstable. See [Masters, 1995b] for more details. There is a better way.

When we specify p and compute the error's lower confidence bound as the $m=np$'th smallest error, what we should really worry about is how large p might truly be. For example, suppose we let p be 0.1, implying that we want a lower error bound such that only 10 percent of the time will the obtained error be as bad as or worse than (less than or equal to) this lower bound. The logical question is: "How bad might be the *actual* probability of lying at or outside this lower bound?" In other words, it may be that in the future, the obtained error will (to our dismay) be less than or equal to this lower bound with a probability that is higher than the 0.1 we desire and expect. It is possible to make probability statements about this situation.

The most straightforward probability statement, though not often the most useful, is the following: What is the probability that the m'th smallest error, which we have chosen for our lower bound on the prediction error, is really the q quantile (or worse, to be technically correct) of the error distribution, where q is disturbingly larger than the p we desire. We hope that this probability is small, because this event is a serious problem in that it implies that our chosen lower confidence limit for the error is more likely to be violated than we believe. This probability question is answered by the *incomplete beta distribution*. In particular, in a collection of n cases, the probability that the m'th smallest will exceed the quantile of order q is given by $1 - I_q(m, n-m+1)$. This function can be found tabulated in many reference books, and a subroutine for computing it is supplied on my web site. We will return to this topic shortly and see a few examples as well. But first, a small digression.

The technique just described makes a lot of sense and is sometimes useful. We choose some comfortable p and use the $m=np$'th smallest error as our lower confidence bound for future errors. We may then choose some $q>p$ at which our comfort sags and use the incomplete beta distribution to find the probability of having obtained the confidence set that gave us this bound even though the true likelihood of violation is dangerously higher (q) than the one we desire. If we find that this probability is small, we are comforted.

But in many or most instances, we would prefer to do things in the opposite order. Instead of first specifying some pessimistic q and then asking for the probability of observing this deceptive confidence set, we first specify a satisfactorily low probability of deception and then compute the q that corresponds to this probability. For example, we may specify a desired lower confidence bound on the error of $p=0.1$. Supposing $n=200$, we use the 20'th smallest error as our lower confidence bound. We might arbitrarily pick $q=0.12$ and evaluate $1-I_{.12}(20, 181)=0.2$. This tells us that there is a probability of 0.2 that our confidence set provided a lower confidence bound that is really likely to be violated with probability of at least 0.12 instead of the 0.1 we expect. But we are more inclined to want to work in the reverse order. We might specify that we want very low probability (only 0.001, say) of having collected a confidence set whose actual probability of violation is seriously greater than what we expect. Suppose we again set $p=0.1$ with $n=200$, thus choosing the 20'th smallest error as our lower error bound. We need to find q such that $1-I_q(20, 181)=0.001$. It turns out that $q=0.18$. This means that there is only the tiny chance of 0.001 that our lower error bound, which we expect to be violated 10 percent of the time, will actually be violated 18 percent or more of the time. Looked at another way, there is the near certainty of 0.999 that our supposed 0.1 confidence bound is in fact no worse than really a 0.18 bound.

This entire discussion so far has focused on evaluating confidence in the lower confidence bound for the error. Exactly the same technique is used for the upper confidence bound, because all we are really talking about is tail probabilities. Simply find the $m=np$'th largest (instead of smallest) error, and let this be our best estimate of the upper confidence bound. Then use the incomplete beta distribution to compute the confidence in a specified pessimistic q, or specify a confidence and compute the associated pessimistic q. Naturally, these are really referring to $1-q$ quantiles, but we still treat them the same way we did before because the pessimism is still working toward the center of the distribution. In fact, most people work with symmetric confidence intervals. Therefore, only one incomplete beta distribution computation is necessary to handle both sides of the confidence interval. For example, we just gave a sample problem in which there is only the tiny chance of 0.001 that our lower error bound, which we expect to be violated 10 percent of the time, will actually be violated more than 18 percent of the time. If we also compute the upper confidence limit on the error as the 20'th *largest* error, we may make exactly the same statement about violation of this upper bound.

My web site contains several subroutines that may be used to evaluate these probabilities. They are not listed in the text because they are long and not directly related to the subject of this text. The main subroutine is ibeta (param1, param2, p). It returns the incomplete beta distribution associated with p, the subscript in standard mathematical notation. The little subroutine orderstat_tail (n, q, m) does the simple calculation described earlier. It returns the probability that the m'th smallest case in a sample of n cases will exceed the q quantile of the distribution. Finally, quantile_conf (n, m, conf) solves the reverse problem. Given a small confidence level (0.001 in the last example), it returns the pessimistic q associated with this confidence.

Tolerance Intervals

The following example appeared earlier: We desire an 80 percent symmetric confidence bound for future prediction errors. This implies $p=0.1$ for each tail. Supposing $n=200$, we use the 20'th smallest error as our lower confidence bound, and we use the 20'th largest error as our upper confidence bound. We might specify that we want very low probability (only 0.001, say) of having collected a confidence set whose true probability of violation in a tail is seriously greater than what we expect. We need to find q such that $1-I_q(20, 181)=0.001$. We saw that $q=0.18$. This means that there is only the tiny chance of 0.001 that our lower error bound, which we expect to be violated 10 percent of the time, will actually be violated 18 percent or more of the time. This also means that there is only a probability of 0.001 that our upper error bound, which we similarly expect to be violated 10 percent of the time, will actually be violated more than 18 percent of the time. We may be tempted to combine these facts into the *incorrect* statement that there is a probability of $2*0.001=0.002$ that the confidence interval covers $1-2*0.18=64$ percent of the future errors. We might come to this erroneous conclusion by reasoning that there is a probability of 0.001 that the lower tail probability is really as large as .18, and the same is true of the upper tail. So, we sum these probabilities to find the probability of either violation occurring. The problem with this reasoning is that the two events are not independent. If one occurs, the other is not likely to occur. In other words, if one error bound happens to be violated, the sample that caused this unfortunate result will almost certainly not violate the other limit as well. In fact, the other confidence bound will probably be overly conservative. A different method is needed if we want to make simultaneous probability statements.

A *tolerance interval* is a pair of numbers that, with probability β, enclose at least the fraction γ of the distribution. Most people want a symmetric tolerance interval, in the sense that the m'th smallest and the m'th largest prediction errors are employed. In this symmetric case, a tolerance interval can be computed with Equation (1.11).

$$\beta = 1 - I_\gamma\left(n - 2m + 1, 2m\right) \tag{1.11}$$

The choice of whether to employ a tolerance interval or the *pessimistic q* method of the previous section depends on the application. If only one tail of the error confidence interval is important, compute a pessimistic q. But if the entire interval is important, the tolerance interval approach is probably superior.

We conclude this section with some examples of how one might compute confidence bounds for future predictions and assess the quality of these bounds. Assume that we have collected a confidence set consisting of 500 cases that are a fair representation of cases that we may encounter in the future. For each of these cases, we use our trained model to make a prediction. We then subtract the true value from each prediction, giving us a collection of 500 signed error measurements. The project specs require that future predictions be accompanied by confidence intervals. There are several possible ways that the requirements may be specified.

One possibility is that predictions that are too high (positive error) don't matter. Thus, only a lower bound (probably negative) for the error is needed. Suppose the project is stringent in this regard, demanding only a 0.01 chance that the prediction error is less than the lower bound. It may be, for example, that the lower bound drives a design consideration, and every time that the lower bound is violated, the company incurs a painful loss. The traditional route is to let $m=0.01*500=5$. Thus, we sort the errors and use the fifth-lowest as the lower bound for future errors. When management tells us that our department will be fed to the alligators if the lower bound is violated 1.5 percent of the time or more (instead of the 1.0 percent that is in the spec), we immediately use the subroutine orderstat_tail() or ibeta() to compute $1-I_{.015}(5, 496)=0.130$. This is the probability that we will all die in the jaws of the beast.

The advantage to using $m=np$ to choose the order statistic for the lower bound is that it is nearly unbiased. This roughly means that on average it will provide the most accurate true lower bound. Moreover, the tiny degree to which it is biased is in the conservative direction. Nonetheless, in the example just cited, we may find that the 0.130 probability of the department being thrown to the alligators is simply too high. Thus, we may choose to use the fourth smallest error as our lower bound, instead of the fifth. This gives us a more comfortable death probability of $1-I_{.015}(4, 497)=0.058$. Of course, this bending in the conservative direction may cause headaches for the design department when they are forced to deal with a still smaller lower bound. But that's their problem. Our eye is on the alligators.

Here's another way that the requirements may be specified. Perhaps under-predictions are not a problem. Only positive errors are serious. This calls for an upper bound on the errors. Moreover, it may be that the specs are a bit looser. A 0.1 confidence interval is all that is required, and even that is not firm. The only important thing is that we need to know with high probability the true probability associated with the upper bound. So, as usual, we let $m=0.1*500=50$. Since an upper bound is required, we sort the 500 errors and use the 50'th largest as the upper bound. Management may want to know with 99 percent certainty what the worst case is for the true upper bound. So we use quantile_conf() to compute the q such that $1-I_q(50, 451)=0.01$. This value turns out to be 0.133. Thus, we know that there is a 99 percent chance that our supposed 0.1 upper bound is in fact no worse than 0.133.

Finally, it may be that any sort of prediction error is problematic. The sign of the error is unimportant; only the magnitude matters. As usual, we set $m=np$ and use the m'th smallest and m'th largest errors as the lower and upper bounds, respectively. So, if the project requires a 90 percent confidence interval, $m=0.05*500=25$. To estimate the probability that the 90 percent coverage we expect is actually obtained, we must use a tolerance interval, computing $1-I_{.9}(451, 50)=0.522$. This is the probability that future errors will lie inside our confidence bounds at least 90 percent of the time. The fact that this probability turned out to be slightly greater than 0.5 should not be surprising. Remember that the choice of $m=np$ is nearly unbiased and slightly conservative. So, we would expect the coverage probability to be slightly better than the 0.5 we would obtain from an exactly unbiased interval.

Of course, having only slightly better than a 50-50 chance that our interval meets our specification is of concern. How should we deal with this? If we let m be smaller, we would increase this probability, but at the price of a confidence interval that is almost always too wide. In most applications, this is too high a price to pay. The real question regards the probability that the obtained interval actually provides less coverage than we expect. In this example, we want 90 percent coverage; we want the confidence bounds to be violated only 10 percent of the time. We can compute the probability that the coverage is at least 85 percent. According to Equation (1.11), this is $1-I_{.85}(451, 50)=0.9996$. Most people would be satisfied that our computed 90 percent tolerance interval has a near certainty of providing true coverage of at least 85 percent.

CHAPTER 2

Assessment of Class Predictions

- The Confusion Matrix
- ROC (Receiver Operating Characteristic) Curves
- Optimizing ROC-Based Statistics
- Confidence in Classification Decisions
- Confidence Intervals for Future Performance

The previous chapter focused on models that make numeric predictions. This chapter deals with models whose goal is classification. It must be understood that the distinction is not always clear. In particular, almost no models can be considered to be pure classifiers. Most classification models make a numeric prediction (of a scalar or a vector) and then use this numeric prediction to define a classification decision. Thus, the real distinction is not in the nature of the model but in the nature of the ultimate goal. The implications of this fact will resound throughout the chapter.

Also because of this relationship, you should understand that much of the material in Chapter 1 applies to this chapter as well. Most classification models are just prediction models taken one step further to suit the application. Thus, concepts such as consistency of error magnitude and selection bias apply to most classification problems as well as they do to prediction problems. You should see this chapter as an extension of Chapter 1, not a totally different topic.

© Timothy Masters 2018
T. Masters, *Assessing and Improving Prediction and Classification*,
https://doi.org/10.1007/978-1-4842-3336-8_2

The Confusion Matrix

The most basic measure of classification quality is the *confusion matrix*. This matrix contains as many rows and columns as there are classes, plus one additional column if the model allows the possibility of a *reject* category. Some applications require that the model have the option of indicating that a trial case most likely does not belong to any of the trained classes. This is the reject category. The confusion matrix entry at row *i* and column *j* contains the number of cases that truly belong to class *i* and have been classified as class *j*. Thus, we see that the confusion matrix for a perfect model contains positive entries only on the diagonal, with the off-diagonal elements being zero. Entries off the diagonal represent misclassifications.

The basic confusion matrix consists of case counts. There are two modifications to the basic version that can be useful. If the entries for each row are divided by the total of the entries for the row, each resulting number is the fraction of the cases in this row's class that were classified into the column's class. Sometimes it may be helpful to divide each entry by the total number of cases so as to assess the confusion relative to the entire population. In both cases, it may be advisable to multiply the fractions by 100 to express them as percents.

Expected Gain/Loss

The contents of an entire confusion matrix can sometimes be distilled into a single number. This is possible when each type of misclassification has its own well-defined cost. If the data collection on which the confusion matrix is based is meaningful in terms of its size and the proportion of cases in each class, the total cost incurred by the model is easily computed by multiplying the case count for each type of misclassification by the cost of that misclassification, and summing.

However, things are not usually this simple. There are two common problems with this naive approach. First, the proportion of cases in each class for the dataset that produced the confusion matrix may not equal the proportion to be expected in actual use. Second, the total number of cases is almost never meaningful, having been arbitrarily chosen on the basis of convenience. A measure that is almost always more meaningful is the expected cost per case. If we wish, we can then multiply this figure by the number of cases in a batch to be processed later in order to estimate the cost of processing that batch.

The expected cost per case is usually best computed in two steps. First, compute the expected cost per case for each class. This is done one row (true class) at a time. Divide each entry in the row by the row's total. This gives the fraction of the true class that lies in each classified class. Multiply each of these fractions by the cost associated with that entry, and sum across the row. The sum is the expected cost for a member of this class. Find this for all rows. Second, multiply these costs by the expected proportion of cases in each class and sum. This sum is the expected cost per case, regardless of class.

Here is an example of this computation. Table 2-1 shows the results for a three-class problem. Each of the 3×3=9 entries in the confusion matrix consists of three subentries. The upper one is the actual confusion count. The center entry is the fraction of each row, computed as already described. The bottom entry is the cost associated with that entry. Observe that, as is usually the case, the cost of a correct classification is zero.

The total number of cases in the first row is ten. Thus, the proportions of the first class in each classified class are 0.5, 0.2, and 0.3, respectively. These quantities, as well as the corresponding quantities for the other two rows, are shown in the table. The expected cost for a case in 0.5 the first class is $0 \cdot 0.5 + 4 \cdot 0.2 + 8 \cdot 0.3 = 3.2$. Similar calculations show the costs for the other two classes as 0.7 and 2.8, respectively. Suppose we expect 0.7 that in the future, 0.3 of the cases will be in class 1, 0.3 in class 2, and 3.2 0.4 in class 3. The expected cost per case is computed as $0.3 \cdot 3.2 + 0.3 \cdot 2.8 \ 0.7 + 0.4 \cdot 2.8 = 2.29$.

Table 2-1. *A Confusion Matrix with Associated Costs*

5	2	3
0.5	0.2	0.3
0	4	8
2	16	2
0.1	0.8	0.1
3	0	4
4	6	10
0.2	0.3	0.5
5	6	0

There is a potential danger in using expected cost as a performance indicator. This is particularly true if this quantity is optimized for training. The danger is that the expected cost can be sensitive to the specified costs and prior probabilities of the classes. In an industrial situation, where both the expected probability of each class and the cost of misclassifications may be known precisely, the expected cost can be an excellent performance measure. But suppose we have a medical application in which costs must be assigned to errors such as misdiagnosing a malignant tumor as benign. The cost will presumably be much higher than that for the opposite error. But if two different researchers assign different relative costs, their trained models may turn out to be quite different. Be wary of using expected costs as a performance indicator when either the cost or the probability of each class is subject to variation.

ROC (Receiver Operating Characteristic) Curves

This section focuses on a special but common classification situation. Many applications involve only two classes that are mutually exclusive (no case can belong to both classes) and exhaustive (each case must belong to one of the classes). The most common version of this situation is when there actually is only one class, called the *target class*, and each case either belongs to this class or does not. For example, a radar blip is a tank or it is something else (about which we do not care). A credit card transaction is either fraudulent or it is not.

Because this version of the problem is so common, it will form the basis of further discussions. Rather than treating each case as belonging to one of two classes, each case will be considered to be a member of the target class or not a member of that class. This naming convention has obvious military connotations. However, we may just as well call a malignancy a target, or we may call a successful financial trade a target, or we may call a fraudulent credit card transaction a target. The central idea is that there exists a dichotomy between a particular class of interest and everything else.

Hits, False Alarms, and Related Measures

There are several important definitions in this problem. It is obvious that the confusion matrix is two by two. Let us label the quantities in each of the four cells as follows:

True positive (TP): The number of targets correctly classified as targets

False negative (FN): The number of targets incorrectly classified as nontargets

True negative (TN): The number of nontargets correctly classified as nontargets

False positive (FP): The number of nontargets incorrectly classified as targets

The following definitions are commonly employed:

- The *hit rate* is the fraction of the targets (members of the target class) correctly classified as being in the target class. This is $TP / (TP + FN)$. It is also sometimes called *sensitivity*.

- The *false alarm rate* is the fraction of nontargets erroneously classified as being in the target class. Statisticians may call this the *Type I error rate*. This is $FP / (TN + FP)$.

- The *miss rate* is the fraction of the targets erroneously classified as not being in the target class. Statisticians may call this the *Type II error rate*. This is $FN / (TP + FN)$.

- The *specificity* is the fraction of nontargets correctly classified as not being in the target class. This quantity is $TN / (TN + FP)$.

Observe that the hit rate and the miss rate sum to one. Similarly, the false alarm rate and the specificity sum to one. Most military targeting applications employ the hit rate and false alarm rate. Sensitivity and specificity are commonly used in medical applications. The choice depends entirely on which measures happen to be more meaningful to the application at hand.

There are a few other performance measures that are not as commonly employed but that can be immensely valuable in some circumstances. These include the following:

- The usual goal is to detect targets. But sometimes we do not need to detect a large fraction of the targets. Instead, we need to be as certain as possible when a decision of "target" is made. The classic example involves trading financial markets. It does not matter if many large market moves are missed (a low hit rate). But when a large market move is predicted and a position is taken, the model had better be sure of itself. In this case, the *precision* is important. This is $TP / (TP + FP)$. When this measure is employed, we should also know what fraction of the population is represented by the denominator. This is $(TP + FP) / (TP + FP + FN + TN)$, which is the fraction of the cases that the model decided were targets.

- Occasionally, the converse goal is important. If a decision is made that the case is not a target, this decision must be correct with high probability. This arises, for example, in medical diagnosis. If a case is classified as not being diseased, this diagnosis definitely needs to be correct. In this case, the *negative precision*, sometimes also called the *null precision*, is $TN / (TN + FN)$. We then also need to consider how often this benign diagnosis occurs: $(TN + FN) / (TP + FP + FN + TN)$.

Computing the ROC Curve

Nearly all target/nontarget classification models make a numeric prediction and then compare the prediction to a predefined threshold. If the prediction meets or exceeds the threshold, the case is classified as a target. Otherwise, it is classified as a nontarget. It should be apparent that all of the performance measures just discussed are affected by the threshold. If the threshold is set at or below the minimum possible prediction, all cases will be classified as targets. This means that the hit rate will be a perfect 100 percent, but the false alarm rate will be an abysmal 100 percent. If the threshold is set above the maximum possible prediction, all cases will be classified as nontargets. This means that the hit rate will be zero, and the false alarm rate will be a perfect zero as well. The sweet spot lies somewhere in the middle. By sweeping the threshold from one extreme to the other, the hit rate and false alarm rate also sweep across their range from zero to 100 percent, one improving as the other deteriorates. The curve defined by these two rates (with the threshold serving only as an invisible parameter) is called the *Receiver Operating Characteristic (ROC)* curve.

When the ROC curve is plotted as a graph, the convention is to let the horizontal axis be the false alarm rate and let the vertical axis be the hit rate. Suppose the classification model is worthless, generating numeric predictions that are random, having nothing to do with the true class of a case. It is apparent that the hit rate and false alarm rate will be, on average, equal for all thresholds. The ROC curve will be an approximately straight line connecting the lower-left corner of the graph to the upper-right corner.

Now suppose the model is perfect. In other words, suppose there exists a threshold such that all targets produce a numeric prediction that equals or exceeds the magic threshold, while all nontargets generate a prediction that is less than the threshold. As the plot-generating parametric threshold moves from its minimum to the threshold of perfection, the false alarm rate will move from 100 percent to zero, with the hit rate remaining at its perfect value of 100 percent. As the parametric threshold continues on to its maximum, the hit rate will deteriorate from 100 percent to zero, while the false alarm rate remains at zero. The ROC curve will be a line running from the upper-right corner of the graph to the upper-left corner, then continuing on to the lower-left corner. Models that are good but less than perfect will have a ROC curve that lies somewhere between the two extremes of hugging the upper-left corner and a diagonal line. The more the curve is pulled away from the diagonal toward the upper-left corner, the better the model is performing. In fact, we will see later that the area under the ROC curve is a useful performance measure, since it considers the effectiveness of the model at all possible classification thresholds.

It is not often that simple visual examination of a graphical plot of a ROC curve proves useful. Much more information can be had by studying a table of the relevant performance measures. Table 2-2 is a sample ROC curve for a moderately effective classification model. It shows the values of the performance measures defined earlier, computed across the full range of thresholds.

Like many classification models, this one has been trained to produce an output of –1.0 for nontargets and 1.0 for targets. The ROC curve threshold (the rightmost column) accordingly covers this range. As expected, the hit rate and false alarm rate drop from 100 percent to zero as the threshold increases. Because this model has some predictive capability, the false alarm rate falls faster than the hit rate. In fact, the false alarm rate drops to 38.5 percent while the hit rate remains at a perfect 100 percent. The false alarm rate reaches a perfect zero while the hit rate is still at 60.0 percent. Unless the relative importance of the two types of error is extremely unbalanced, the ideal threshold will lie somewhere between these two points. Note that this chart also shows the specificity, the complement of the false alarm rate.

Table 2-2. *ROC Curve for a Moderately Effective Model*

Hit (35)	FA/Spec (26)	Mean Err	Target Prec.	Null Prec.	Threshold
100.0	100.0/0.0	50.0	57.4 (100.0)	0.0 (0.0)	-1.000
100.0	88.5/11.5	44.2	60.3 (95.1)	100.0 (4.9)	-0.900
100.0	84.6/15.4	42.3	61.4 (93.4)	100.0 (6.6)	-0.800
100.0	76.9/23.1	38.5	63.6 (90.2)	100.0 (9.8)	-0.700
100.0	76.9/23.1	38.5	63.6 (90.2)	100.0 (9.8)	-0.600
100.0	65.4/34.6	32.7	67.3 (85.2)	100.0 (14.8)	-0.500
100.0	61.5/38.5	30.8	68.6 (83.6)	100.0 (16.4)	-0.400
100.0	46.2/53.8	23.1	74.5 (77.0)	100.0 (23.0)	-0.300
100.0	38.5/61.5	19.2	77.8 (73.8)	100.0 (26.2)	-0.200
91.4	26.9/73.1	17.7	82.1 (63.9)	86.4 (36.1)	-0.100
85.7	26.9/73.1	20.6	81.1 (60.7)	79.2 (39.3)	0.000
82.9	23.1/76.9	20.1	82.9 (57.4)	76.9 (42.6)	0.100
74.3	11.5/88.5	18.6	89.7 (47.5)	71.9 (52.5)	0.200
68.6	3.8/96.2	17.6	96.0 (41.0)	69.4 (59.0)	0.300
60.0	0.0/100.0	20.0	100.0 (34.4)	65.0 (65.6)	0.400
51.4	0.0/100.0	24.3	100.0 (29.5)	60.5 (70.5)	0.500
42.9	0.0/100.0	28.6	100.0 (24.6)	56.5 (75.4)	0.600
34.3	0.0/100.0	32.9	100.0 (19.7)	53.1 (80.3)	0.700
22.9	0.0/100.0	38.6	100.0 (13.1)	49.1 (86.9)	0.800
11.4	0.0/100.0	44.3	100.0 (6.6)	45.6 (93.4)	0.900
0.0	0.0/100.0	50.0	0.0 (0.0)	42.6 (100.0)	1.000

ROC area = 0.9275
Normalized ROC area with sensitivity >= 0.9 = 0.6648

In many cases, the ideal threshold can be determined by simple inspection of the first two columns. Interpretation is helped if the number of targets and nontargets is also printed in the heading of the table. In this case, the numbers are 35 and 26, respectively.

If the cost of a miss and a false alarm are about the same, the mean error is a good single-number indicator of performance. This column is computed by finding the average of the false alarm rate and the miss rate (100 minus the hit rate). Of course, if misses and false alarms have significantly different cost, the average error loses much of its relevance.

In some applications, the precision (page 49) is of great importance. Usually we are most interested in the precision in regard to detecting targets. Occasionally we are interested in the precision in regard to confirming that a case is not a target. Table 2-2 provides both types of information. When the goal is correctly detecting targets, the column labeled *target precision* is most useful. Conversely, if the goal is to precisely identify nontargets, the column labeled *null precision* (or *negative precision* in some circles) is most useful. In each of these columns, the percent of cases classified as target or nontarget, respectively, is given in parentheses. We see, for example, that at a threshold of 0.1, 42.6 percent of the cases are classified as nontargets with an obtained (null or negative) precision of 76.9 percent. Similarly, at a threshold of 0.4, perfect target precision of 100 percent is obtained, although only 34.4 percent of the cases are classified as targets. Finally, the area under the ROC curve for this model is computed as a respectable 0.9275. This quantity is discussed in the next section.

Before leaving computation of the ROC curve, let us examine some code snippets that demonstrate an efficient way to perform the computations. In most cases, the user will specify a range of thresholds over which the ROC curve is to be listed, and the number of intervals in which the range is to be divided for the table. A sensible approach is to divide the range into the specified number of intervals, printing a line for each, plus one extra line at the end to cover all thresholds above the user's upper limit. There is almost never any interest in information below the lower limit. So, for example, Table 2-2 encompasses a range of thresholds from –1.0 to 1.0 divided into 20 intervals. The lowest interval contains thresholds from –1.0 to –0.9. The next handles –0.9 to –0.8, etc., until the 20th interval is for 0.9 to 1.0. The last line (number 21) counts cases whose threshold exceeds 1.0. It is customary to label each line with the lower bound of its threshold, since that is where the counting for each bin begins. Here is a code fragment for computing the information required to prepare this table:

```
for (i=0; i<nthresh; i++)
   hit[i] = fa[i] = 0.0;
hit_below = hit_above = fa_below = fa_above = 0.0;

factor = (double) nthresh / (stop - start);

for (icase=0; icase<n; icase++) {
   j = (int) (factor * (ind[icase] - start));

   if (cls[icase] > 0.0) {                   // Definition of target
      if (ind[icase] < start)                // First bin includes all thresholds below start
         hit_below += 1.0;
      else if (j >= nthresh)                 // Last bin includes all thresholds above stop
         hit_above += 1.0;
      else if (j >= 0)                        // Should always be true, but play it safe
         hit[j] += 1.0;
      }
   else {                                    // Non-target
      if (ind[icase] < start)                // First bin includes all thresholds below start
         fa_below += 1.0;
      else if (j >= nthresh)                 // Last bin includes all thresholds above stop
         fa_above += 1.0;
      else if (j >= 0)                        // Should always be true, but play it safe
         fa[j] += 1.0;
      }
   }
/*
   The bin counts are computed. Now cumulate.
*/

   t = hit_above;
   for (j=nthresh-1; j>=0; j--) {
      t += hit[j];
```

```
    hit[j] = t;
    }

n_targets = (int) (hit_below + t);
for (j=0; j<nthresh; j++)
    hit[j] /= (n_targets + 1.e-30);

hit_above /= (n_targets + 1.e-30);

t = fa_above;
for (j=nthresh-1; j>=0; j--) {
    t += fa[j];
    fa[j] = t;
    }

n_nontargets = (int) (fa_below + t);
for (j=0; j<nthresh; j++)
    fa[j] /= (n_nontargets + 1.e-30);
fa_above /= (n_nontargets + 1.e-30);
```

In this code, hit and fa are real arrays that will hold the hit rates and false alarm rates for the ROC curve. The _above and _below versions will cumulate the corresponding information for thresholds above and below the specified range, respectively. The user requested nthresh intervals ranging from start to stop. For each case, cls defines the class membership of the case, with values greater than zero indicating a target, and values less than or equal to zero being a nontarget. Also for each case, ind is the value of the indicator variable, which will be compared with the threshold in order to make a classification decision. Values of ind greater than or equal to the threshold will be classified as being a target.

Operation of the code is straightforward. It passes through all cases. For each, it checks the true class membership. The bin corresponding to the indicator variable is computed, and that bin count is incremented. If the indicator lies outside the user's specified range, the appropriate outer bin is incremented.

After the bins are cumulated, the algorithm sums downward. Remember that each bin covers cases for that bin and all above it (cases whose indicator exceeds the threshold). To get the total number of targets we add those in the below-the-limit bin.

Finally, divide all bin counts, as well as the extra bin on top, by the number of targets in order to convert them into fractions. The *1.e-30* ensures that we do not divide by zero in degenerate situations.

We can print or store the table by using code similar to that shown next. Refer to page 48 for definitions of the various ROC parameters that can be computed from these basic ingredients.

```
for (j=0; j<=nthresh; j++) {
  if (j < nthresh) {
    hj = hit[j];
    faj = fa[j];
    }
  else {
    hj = hit_above;
    faj = fa_above;
    }

  tp = hj * n_targets;
  fn = (1.0 - hj) * n_targets;
  tn = (1.0 - faj) * n_nontargets;
  fp = faj * n_nontargets;
  }
```

Area Under the ROC Curve

One disadvantage of the ROC curve table as a performance "measure" is that it contains a tremendous amount of superfluous detail. Sometimes a single number is all that is needed. We have already noted that a worthless model has a ROC curve that is a diagonal line, and a perfect ROC curve hugs the upper-left corner of the graph. It is apparent that a good model will have more area under its ROC curve than a poor model. A worthless model will have an area of 0.5 (because the graph runs from 0.0 to 1.0 in both axes). A perfect model will have an area of 1.0.

It is not difficult to compute the area under a ROC curve. The randomly sampled cases in the test set are themselves subject to statistical variation, so there is no need to attempt to achieve high accuracy in the numerical integration algorithm. Simple summation is generally sufficient. We'll jump right in with a code fragment and explain its operation later.

```
for (icase=0; icase<n; icase++) {
   indicator[icase] = ind[icase];                      // model's output
   is_target[icase] = (cls[icase] > 0.0) ? 1 : 0;      // Target vs. non-target
   }

qsortdsi (0, n-1, indicator, is_target);   // Sort outputs ascending, moving class

tp_count = 0;        // Counts true positives
fp_count = 0;        // Counts false positives
j = n-1;             // Will count down as icase counts up
area = 0;            // Will cumulate (unnormalized) ROC area here
ROC90 = -1.0;        // Will cumulate ROC area for hit rate >= .9

for (icase=0; icase<n; icase++) {
   if (is_target[j--]) {     // If this is truly a target
      ++tp_count;            // Then it is a true positive at this threshold
      area += fp_count;      // Count this thin rectangle

      if (tp_count / (double) n_targets >= 0.9) {
         if (ROC90 < 0.0)    // Start with partial rectangle
            ROC90 = (tp_count / (double) n_targets - 0.9) /
               (1.0 / (double) n_targets) * (n_nontargets - fp_count);
         else
            ROC90 += n_nontargets - fp_count; // Area to right of curve
         }
      }
   else                  // But if it is not a target
      ++fp_count;        // Then it is a false positive
   }

ROC area = 1.0 - (double) area / ((double) n_targets * (double) n_nontargets));
ROC90area = ROC90 / ((1.0 - 0.9) * (double) n_targets * (double) n_nontargets));
```

The first step is to copy the indicator and class (target versus nontarget) information to work vectors so we can rearrange them without annoying the calling routine. The qsortdsi routine sorts the n cases in ascending order of the indicator variable, simultaneously rearranging the class vector. Note that other than this sorting operation, the actual values of the indicator variable are not needed. The order in which the classes appear (ideally with the nontargets near the beginning, and the targets near the end after sorting) is all that is needed to compute the area under the ROC curve. Also note that if there are ties in the indicator variable, the algorithm does not employ any special tie-breaking procedures. Most applications do not produce ties, so this should not be a problem. But if ties are possible, it might be wise to call the algorithm several hundred times, using random tie breaking, and average the results. I am not aware of an algorithm for explicitly handling ties in a rigorous fashion (although such an algorithm probably does exist).

This code loops through the sorted data, working down from the highest indicator to the lowest. At each implied threshold, the class of the case is checked. If it is a target, the *true positive* counter is incremented. Otherwise, the *false positive* counter is incremented.

It helps to visualize the ROC curve as lying in a grid consisting of as many rows as targets, and as many columns as nontargets. The loop starts at the lower-left corner of the grid and advances upward and to the right as the implied threshold decreases. Every time the *true positive* counter is incremented, it moves up one row. When this happens, a new rectangle of unit height is defined. The width of this rectangle (on the left side of the ROC curve) is the current false-positive count, so we add this to area. In other words, we are summing the number of grid squares to the left of the ROC curve. After traversing the entire curve, we divide this sum by the total number of grid points and subtract from one to get the area below the ROC curve.

Another interesting area measurement is computed as we traverse the curve. In many applications, a model is useful only if it can obtain a very high hit rate. For example, if we are detecting incoming missiles, a perfect or nearly perfect hit rate is mandatory. If we are deciding whether a tissue sample is malignant, we'd better have an extremely high probability of identifying malignant specimens. In such applications, most of the ROC curve is irrelevant. The only important part is that which lies above a specified hit rate.

The code just shown computes that area under the ROC curve that is above a hit rate of 0.9. It does this by cumulating the area to the right of the ROC curve as soon as the hit rate reaches 0.9. The first rectangle is tricky, though, because it will in general be partial, with the 0.9 line splitting it. Thus, we initialize ROC90 to a negative number and use this as a flag. For the first rectangle we interpolate by using a height less than one unit. When the summation is complete, we compute the area by dividing by the total number of unit squares and normalize for the fraction of the total possible area covered.

Cost and the ROC Curve

The expected cost (gain or loss) of a classification model was discussed on page 46. Let us now specialize to the binary classification scenario in which each case is either a target or a nontarget. Suppose the prior probability that a case is a target is q. Let p_1 be the probability of a type I error, a false alarm, and let c_1 be the cost of this error. Let p_2 be the probability of a type II error, a miss, and let c_2 be the cost of this error. In most applications, q will be known from theory or experience, and the costs will be known at least approximately if not exactly. Therefore, these three quantities can be considered fixed constants. The two error probabilities, type I and II, are determined by the threshold. The expected cost is given by Equation (2.1). Algebraic manipulation of this equation leads to Equation (2.2).

$$COST = (1-q)p_1c_1 + qp_2c_2 \qquad (2.1)$$

$$p_1 = \frac{COST}{(1-q)c_1} - \frac{qc_2}{(1-q)c_1}p_2 \qquad (2.2)$$

Equation (2.2) makes it clear that for any fixed cost, the relation between p_1 and p_2 is linear. If this line is plotted on the ROC curve, its slope is determined by the prior probability that a case is a target and by the ratio of the costs of the two types of errors. As the expected cost is varied, a family of parallel lines is defined, with lines down and to the right representing higher cost. If the line intersects the ROC curve at two points, these intersection points represent two thresholds that provide equal but excessive cost. Lines that lie entirely above and to the left of the ROC curve represent performance that is unachievable by the model. The line that is precisely tangent to the ROC curve represents the best obtainable cost.

The preceding discussion should make it clear that the area under the ROC curve is not necessarily a good indicator of how well a classification model actually performs. Certainly the ROC area is a very good overall quality indicator. But when actual cost of operation is considered, ROC area can be deceptive. Look at Figure 2-1. This shows ROC curves for two models. They have identical areas under the curve. The prior probability and relative error costs in this example favor a low false alarm rate, even at the expense of a quite high miss rate. It is apparent that one model is very superior to the other in terms of cost, despite having the same ROC area. It should also be obvious that if the cost

59

situation were reversed, the other model would be favored. Therefore, do not be afraid to judge the power of a model based on the area under its ROC curve. But also do not fail to consider the actual cost of the model.

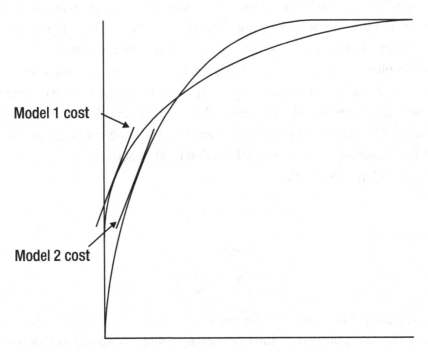

Figure 2-1. *The ROC area can be deceptive*

Optimizing ROC-Based Statistics

In many or most target/nontarget situations, the best possible results are obtained by optimizing a performance criterion that is directly based on some aspect of the ROC curve. The principal disadvantage of this approach is that such criteria almost never lend themselves to traditional optimization techniques. They are difficult or impossible to differentiate, eliminating any training methods that rely on gradients. They can exhibit troublesome discontinuities that may confound naive algorithms. In general, stochastic methods such as genetic algorithms or simulated annealing are the only practical training methods. Nevertheless, ROC- based statistics can be such effective optimization criteria that it is well worth any trouble involved.

A big hint that ROC-based statistics are useful comes from examining Figure 2-1. The two ROC curves shown in this figure have identical areas. It is likely that the models' performance as indicated by other traditional measures, such as MSE or R-squared, would also be similar or identical. Yet the difference in their actual costs is large. It obviously makes sense to design a training scheme that directly optimizes the cost. This section discusses several ROC-based optimization criteria that have widespread applicability.

Optimizing the Threshold: Now or Later?

ROC-based statistics depend not only on designing a good prediction model but also on finding an optimal threshold for defining the target/nontarget decision. How is this threshold obtained? There are three primary methods for doing so.

- The old-fashioned method is to totally ignore the threshold while training the model using some other optimization criterion. After the model is trained, compute the optimal threshold. If the result is satisfactory, stop. Otherwise, try a different model. This silly approach was needed when computer power and sophisticated optimization algorithms were scarce. It's barely excusable today.

- Treat the threshold as just one more optimizable parameter in the model. Append it onto the end of the parameter vector that is fed to the optimization algorithm. This method can work well. However, the addition of another parameter will significantly slow most optimization algorithms. This method should be avoided if possible unless computer time is freely available.

- Slightly increase the complexity of the criterion-evaluating function by including threshold optimization in this function. The main optimization algorithm supplies the criterion function with trial values of the model's parameters. After this trial parameter set is tested with all of the training cases, the threshold that optimizes the training criterion is computed, and the corresponding best performance measure is returned to the main optimizing algorithm. In other words, threshold optimization is performed with every function evaluation by the model optimizer. The main optimization algorithm optimizes only the model parameters, not the threshold. But for every trial parameter set, the function evaluating algorithm includes threshold optimization as part of its operation. This is usually the preferred method as it is relatively easy to implement, yet it effectively optimizes the threshold simultaneously with the model parameters.

The difference in training time using the second and third methods just described is sometimes great and sometimes insignificant. Most often it is great. Here is why. Suppose, for example, that the prediction model has three optimizable parameters. If we were to use the second method, the training algorithm would be effectively dealing with four optimizable parameters. Anyone who has worked in the field of numerical optimization knows that the speed difference between three and four parameters can be huge. Of course, the third method is not a free lunch. The price paid for allowing the main (complex and slow) optimizer to be burdened with only three parameters instead of four is that now the criterion function that it calls to evaluate performance must itself become a mini-optimizer. After setting the three trial parameters and evaluating this trial model for all training cases, the performance evaluator must traverse the ROC curve to find the optimal threshold. The key point is that in nearly all practical applications, applying the model to the training set is the slow part of a function evaluation. The time taken to traverse the ROC curve is usually insignificant. A very small price is paid to obtain a very large return.

Here is a sample code fragment that illustrates how the function-evaluating routine called by the main optimizer can compute the threshold that produces the best cost. This code also shows how to compute an effective optimization criterion. Simply optimizing the cost itself may not be the best approach. An absolute cost figure, while immensely informative in one sense, can be worthless for judging whether the model is really doing anything useful. If the two types of errors have very disparate costs, a naive model might just set the threshold at whichever extreme forces all misclassifications to be the cheaper type. Thus, a more informative performance criterion would compare the model's attained cost with that obtainable with a naive model. Finally, this code illustrates the fact that when the model is validated, it would not be fair to optimize the threshold. The previously trained threshold must be employed. This facility is demonstrated. In this code, the predicted values are in the *pred* array, and the true classifications are in the *trueval* array. In this array, true targets are flagged with 1.0, and nontargets are flagged with –1.0.

```
// A naive model would set the threshold at whichever extreme results
// in nothing but the cheaper error. Compute this naive cost so that
// we can scale the actual cost with it.

factor1 = factor2 = 0.0;        // Will be cost at each extreme threshold
for (i=0; i<n; i++) {           // Check each case
   if (trueval[i] < 0.0)        // If this is a non-target
      factor1 += fa_cost;       // It would be a false alarm at min thresh
```

```
    if (trueval[i] > 0.0)            // If this is a target
       factor2 += miss_cost;        // It would be a miss at max thresh
    } // For all cases

if (optimize_threshold) {           // If we are training (versus testing)
    sort (n, pred, trueval);        // Sort predicted values, keeping classes with them
    thr = pred[0];                  // Begin by setting trial threshold tiny
    n_missed = n_fa = 0;            // None are missed
    for (i=0; i<n; i++) {           // But pass through entire set (Redundant from above)
       if (trueval[i] < 0.0)        // And find all cases not a target
         ++n_fa;                    // These are false alarms
    } // This loop could be avoided by counting n_fa with factor1 above. Shown for clarity.

    best_cost = miss_cost * n_missed + fa_cost * n_fa;  // Also redundant: factor1!
                                    // Because n_missed is still zero

    for (i=1; i<n; i++) {           // Try all other possible thresholds
       if (trueval[i-1] < 0.0)      // If we just passed a non-target
         --n_fa;                    // We now have one less false alarm
       if (trueval[i-1] > 0.0)      // If we just passed a target
         ++n_missed;                // We now have one more miss
       if (pred[i-1] == pred[i])    // If we have a block of ties
         continue;                  // Only the first gives a valid cost
       test_cost = miss_cost * n_missed + fa_cost * n_fa;

       if (test_cost < best_cost) { // If it is the best so far
         best_cost = test_cost;     // Update the best cost
         thr = pred[i];             // And keep track of this threshold
       }
    } // For all possible thresholds beyond the lowest

    // There is one more possibility: Try a threshold beyond the max,
    // which results in nothing but misses.
    if (factor2 < best_cost) {
       best_cost = factor2;
       thr = 1.00000001 * pred[n-1];
    }
    } // If set_threshold
```

```
else {                          // Validating, so do not set threshold
  n_missed = n_fa = 0;
  for (i=0; i<n; i++) {         // Count all false alarms and misses
    if ((pred[i] >= thr) && (trueval[i] < 0.0))
      ++n_fa;
    if ((pred[i] < thr) && (trueval[i] > 0.0))  // Could make this logic slightly faster
      ++n_missed;               // But this is more clear
  }

  best_cost = miss_cost * n_missed + fa_cost * n_fa;
  } // Validating, so do not compute threshold

if (factor1 < factor2)          // Which naive model is cheaper?
  criterion = best_cost / factor1;   // Don't forget to prevent division by zero here!
else
  criterion = best_cost / factor2;
```

Note that the preceding code contains several redundant calculations. If speed were critical, they could be eliminated as noted, at the cost of a bit of clarity. Also note that the final division should protect against the pathological condition that the factor in the denominator is zero. This is avoided here for clarity.

Finally, this example considers only the cost of errors, and it compares this cost to that of a naive model. A gain-based method of threshold optimization is presented on page 273 as an alternative.

Maximizing Precision

This section has focused on the hit rate and the false alarm rate, largely because these two quantities are intimately related to the expected cost of using the model. But in many instances, the *precision* (page 49) is of paramount importance. Consider a stock-picking application that examines hundreds or thousands of equities, looking for a few good ones to purchase. The miss rate is of no great importance, because we do not particularly care if quite a few bargains are missed. Getting them all would be nice, but this is obviously an unrealistic expectation. The false alarm rate is of somewhat more importance, because we will certainly care to some degree what percentage of the losers are mistakenly chosen by the model for purchase. But even this figure can be misleading. What we really care about is how well the stocks that are picked by the model ultimately perform. In other words, of those that the model instructs us to purchase, what fraction actually does turn out to be good choices? This is defined as the precision of the model. Any good training algorithm should be prepared to maximize this quantity.

Sometimes it is not the target precision that is important, but rather the nontarget precision. Consider a medical model that classifies a tissue sample as benign or malignant. If one judges only the miss rate and the false alarm rate, the former is obviously more important than the latter. But what is of greatest importance is the answer to the question, "What fraction of those classified as benign truly are benign?" This is the nontarget precision (assuming that malignant is the target). Be prepared to optimize this quantity also.

Generalized Targets

So far, we have assumed that a target comes in exactly one form, and a nontarget comes in exactly one form. But in many applications, some target hits are better than others, and some misses or false alarms are worse than others. Understand that we are not talking about multiple-class problems. We are not explicitly trying to classify a case into one of three or more categories. We are still making a binary decision about whether a case is or is not the target being sought. It is just that there are different degrees of success and failure. The classification model may be able to use this information to increase its probability of hitting the most important targets and/or avoiding the worst false alarms.

One example of this situation is if the model examines passive infrared photographs of battle terrain, seeking to locate powered vehicles in a natural background. Missing an enemy Jeep would be bad, but missing an enemy tank with its gun pointed at you would be disastrous. Another example comes from automated market trading. Suppose the model defines a target as the morning of a day in which a market will rise substantially. Some false alarms will be characterized by the market closing about where it opened, producing no profit. Other false alarms will be characterized by the market plunging, producing a huge loss. The latter is clearly worse than the former. In situations like these, it may make sense to train the model to distinguish between errors of different severity. Examples of appropriate criteria are given in the next few sections. Note that these examples are geared toward maximizing a gain. They could easily be restructured to minimize a cost. This choice is purely due to common convention in this field.

Maximizing Total Gain

Assume that we have a model that makes a binary target/nontarget decision. When this model is applied to a case and the decision is made, the result will be a measured gain, which may be negative to indicate a loss. The model is applied to the entire training set, one case at a time, and the individual gains are tallied. Probably the most obvious performance figure to optimize is the total gain achieved on the training set. This is often a terrible choice.

The total gain ignores the variation in performance among the cases in the training set. In the previous chapter we saw that uniform performance helps generalization (page 3) and evolutionary stability (page 6). Optimizing total gain flies in the face of these extremely important considerations.

The situation becomes even worse for applications in which precision is more important than the hit rate or false alarm rate. Many applications have a gain of zero associated with declaring a case to be a nontarget, regardless of whether the decision is correct or incorrect. The chips go down only when a decision of *target* is made. A possibly large gain will be achieved if the decision proves to be correct, and a possibly large loss will be suffered if the decision is wrong. Since the decision threshold is probably being optimized along with the model parameters, different threshold values can generate significantly different numbers of target decisions. A parameter trial involving a large threshold may produce few target decisions, while another trial having a low threshold may produce many target decisions. This can have a profound impact on the total gain. Suppose one trial parameter set produces a large number of very large gains and a moderate number of very large losses. Then suppose another trial produces a decent number of moderate gains and no losses. In almost all situations, the second parameter set should be treated as preferable to the first, even though the first will have greater total gain. Models that generate consistently good results are nearly always better than models that generate wild performance swings ranging from fantastic to painful. Just ask any financial market trader. In fact, in the finance sector, consistent performance is particularly important because a prediction model that has extremely low probability of disaster can be leveraged to generate large returns, something that cannot be done if the prospect of a ruinous trade always looms nearby. Optimizing total gain is almost always problematic.

Maximizing Mean Gain

Assume that we are in the common situation of incurring a gain or loss on a case if and only if the case is classified as a target. This is nearly always true for automated market trading, because we open a trade if and only if the model detects the predefined target condition. If we do not open a trade, there is no gain or loss. Dividing the total gain by the number of target decisions provides the mean gain per decision. If the gain is one for a hit and zero for a miss, the mean gain is the precision (page 49) of the model.

Like the total gain, the mean gain ignores the consistency of the model's performance on individual cases. For this reason, mean gain shares most of the problems of total gain and is generally best avoided. However, mean gain can be good in some situations. For example, if a unique gain is always obtained for a hit and a smaller (or zero) unique gain is always obtained for a miss, the concept of consistency becomes irrelevant. Maximizing mean gain is equivalent to maximizing precision, which may be just what is desired. Even in this situation, care must be taken. The threshold will often be optimized along with the model's parameters. It will often happen that a very high threshold will provide just a few target decisions, all of which are correct. In fact, the threshold might be so high that just one target decision is made for the entire training set. If this happens to be a correct decision, which is likely, the precision will be a perfect 1.0, and the optimizer will be pleased. The users will not be nearly so pleased. This serious problem can be avoided by imposing a constraint that at least a specified minimum number of target decisions must be made. This prevents the trained threshold from being set unrealistically high by the optimizer.

Maximizing the Standardized Mean Gain

Two simple modifications can transform the mean gain from a poor optimization criterion to an excellent one. The worst problem with mean gain is that it ignores consistency. This problem is eliminated by computing the standard deviation of the gains and dividing the mean gain by this quantity. Inconsistent models will have a large standard deviation, which will lower this criterion accordingly. The other modification is to multiply by the square root of the number of target decisions. This rewards models that achieve their good performance by means of a large number of hits. Conversely, models that obtain good average performance based on very few good cases are penalized. This criterion, called a *t-score* by statisticians, is shown in Equation (2.3). In this equation, there are n target decisions, with the i'th target decision producing a gain of g_i. The mean gain is shown in Equation (2.4).

$$t = \frac{\sqrt{n}\,\bar{g}}{\sqrt{\dfrac{1}{(n-1)}\displaystyle\sum_{i=1}^{n}(g_i - \bar{g})^2}} \tag{2.3}$$

$$\bar{g} = \frac{1}{n}\sum_{i=1}^{n} g_i \tag{2.4}$$

This criterion is based on much more than intuition, although its mathematical properties are mostly beyond the scope of this text. Here are a few of the useful properties of this criterion:

- It does a good job of encouraging consistency. If a good mean gain is obtained by virtue of one or very few extremely large individual gains nestled amid mediocrity, this will inflate the denominator of Equation (2.3) as much as the numerator, nullifying the deception. But beware of models producing extremely consistent results. This can produce inflated values of the criterion, or even cause division by zero.

- Scaling according to the number of target decisions encourages repeatability. Suppose two competing good models have identical means and standard deviations, but one had two target decisions and the other had hundreds of target decisions. It should be obvious that the latter is to be preferred, because the former could have easily appeared as the result of pure luck. The scaling reflects this fact.

- The choice to scale by the *square root* of the number of target decisions has some compelling mathematical justifications. Under fairly general and realistic conditions, the criterion will have a standard deviation of approximately one when this scaling factor is used. This gives meaning to the criterion in absolute terms. Of course, a specifically optimized criterion loses all statistical meaning as far as probability statements are concerned. But it still retains its absolute meaning. And if the criterion is computed from an independent test set whose distribution is not terribly bizarre, meaningful statistical statements may be possible.

This criterion may often be legitimately applied to an independent test set to assess whether the mean gain of cases classified as targets is significantly greater than zero (or perhaps some other quantity). This information is often valuable. Consult any elementary statistics text for detailed information about how to use the *one-sample t-test* to decide if good results obtained on a test set are really good or if instead they might be nothing more than the product of luck.

Here is a summary of the important concepts of this test. A fundamental assumption that is theoretically necessary in order to make probability statements is that the gains must have a normal distribution. In practice, this may not be terribly restrictive. Even binary data, far from normal, generally gives acceptable results as long as there are a decent number of data points. The only cause for serious worry is if the distribution of the gains has one or two heavy tails. For example, if most gains range from –5 to 5, but a few gains of plus or minus 100 may occasionally appear, the t-test should not be used to make probability statements. We will discuss robust alternatives later in this text.

In many applications, a worthless model would be expected to obtain a mean gain (among cases classified as targets) of zero or less. If a positive mean gain is obtained, we would like to know the probability that a gain at least this good might have been the result of nothing more than luck. This probability may be found by consulting a table in nearly any statistics text. As a rough rule of thumb, obtaining a *t*-score in excess of 2.0 is quite good. If there are at least 60 cases classified as targets, the probability of this occurring by luck is about 0.025. A score over 3.0 is outstanding.

Finally, if there is a need to assess the model on a test set but the distribution of gains is too non-normal to trust the t-test, there is an alternative called the *bootstrap* test. See page 135 for details.

Confidence in Classification Decisions

It is often useful, or perhaps even vital, to have the ability to supplement a classification decision with a probability statement regarding confidence in that decision. This can be a surprisingly difficult task. The data collection process becomes significantly more stringent in that more and better data is needed. Several crucial assumptions must be satisfied. And even then, confidence statements may themselves be subject to troubling statistical variation due to random sampling error. This is a dangerous topic that must be approached with great respect. That said, let us begin.

Hypothesis Testing

The most straightforward approach to confidence calculation comes from standard statistical practice. Traditional nomenclature will be used here. Assume, as we have been doing, that the task is to classify a case as a target or as a nontarget. The model makes a numeric prediction, and the classification decision is based on the value of the prediction relative to a fixed threshold. If the model is effective, its predictions will usually be larger for target cases than for nontarget cases. Models are often trained to predict a value of 1.0 for targets and a value of –1.0 for nontargets. In most cases, the model's actual predictions will lie somewhere between these two extremes. Generalized targets, involving many possible levels, are also possible, though less common.

Traditional statisticians postulate an entity called the *null hypothesis*, which asserts that the condition we are seeking is not present. In the current context, we are most likely interested in being able to accompany a decision that a case is a target with a statement of confidence that this is so. Later, we will discuss the converse situation. So, right now our null hypothesis is that the case is a nontarget. The technique presented in this section will attempt to answer the question, "If this case truly is a nontarget (i.e., the null hypothesis is true), what is the probability that a prediction at least this large might have arisen from nothing more than random chance?" If this probability turns out to be quite small, we would be inclined to reject the null hypothesis in favor of the alternative that the case is a target. The reasoning in this situation goes like this:

1) The case is either a target or a nontarget.

2) If it is a nontarget, there is only a small probability that the model's prediction would be this high.

3) Therefore, it is probably a target.

There are two aspects of this confidence technique that must be fully understood. The first is rather subtle but vitally important. We do not start with an attained predicted value and then assert the probability that the case is a target or nontarget. The exact opposite occurs. We hypothesize the true class of the case and then compute the probability of this attained prediction under that hypothesis. In practice, most people will jump right to step 3 and assert the probability that the case is a target (the alternative hypothesis), based on the prediction. This is not always a deadly sin, but it can be. A thorough understanding of the route to the alternative hypothesis is needed.

Understanding the second aspect of this confidence technique is utterly crucial. It is a major source of error among careless practitioners, and the error is extremely serious. If one finds that there is a very small probability under the null hypothesis of obtaining a predicted value as large as that obtained, it is legitimate to conclude that there is a high likelihood that the null hypothesis is false, and hence that the alternative is true. However, the converse conclusion absolutely cannot be drawn.

I once more remind you that, as is common practice, we have trained the model with nontargets having a small (probably –1) value and targets having a large (probably +1) value. Suppose the prediction is at the low end of the scale, where most nontargets lie. We will naturally find that under the null hypothesis, a prediction of this magnitude is quite likely. (Computation of this probability will be discussed soon.) It is tempting to conclude that the null hypothesis is true. This is most emphatically *not* allowable.

In particular, suppose we have a means of computing probabilities associated with predictions made with nontarget cases. This is called the null hypothesis distribution. The model is presented with an unknown case, and its prediction is solidly in the nontarget territory (small). As much as we would like to conclude that there is a high confidence that this case is a nontarget, we *cannot* do so. For all we know, a large fraction of target cases may, to our annoyance, also produce predictions down here at the low end. The only way we can make statements about the confidence that a case is a *nontarget* is if we have information about the model's behavior for *targets*.

To summarize this vital and often overlooked concept, if the probability of an obtained prediction is small under the null hypothesis, we can legitimately have high confidence in the alternative. However, if the probability of an obtained prediction is large under the null hypothesis, we most certainly cannot conclude that the null hypothesis is true.

Now let's discuss the converse situation. When a case is presented to the model and the prediction is below the threshold, we classify the case as a nontarget. We may want to associate a probability with this decision. This is done with essentially the same technique as that just described, but in reverse. Now, the null hypothesis is that the case is truly a target. We must know the distribution of predictions for targets. We compute the probability of obtaining a prediction this small (or smaller) if the case really is a target. If the probability of a prediction so small is tiny, we can legitimately conclude that the case probably is a nontarget.

Once again, for this reverse situation, we see that the ability to make a confidence statement about one class hinges on our knowledge of the model's predictions for the *other* class. And once again, be warned that the opposite reasoning cannot be employed.

If we know the distribution of predictions for targets and we obtain a large prediction that lies solidly in the realm of target predictions, we cannot conclude that the case is a target. To do so would be a grave error. Our capabilities are limited to *rejecting* a hypothesis in favor of the alternative. We can never accept a hypothesis based on a distribution under that hypothesis.

The entire preceding discussion has hinged on our ability to know (or at least approximate) the distribution of the model's predictions under a hypothesis (the so-called null hypothesis, whether it be nontarget or target). In practice we will virtually never have direct theoretical knowledge of this distribution. It must be estimated by testing the model on a dataset. This dataset must satisfy several important requirements. In many applications, these requirements can often be difficult to meet. Failure to adequately satisfy these requirements can lead to erroneous confidence calculations. Be warned. Here are the requirements:

- The dataset must be totally independent of the data on which the model was trained. If the quality of a trained model was judged by its performance on a test set, with the idea that poor performance would lead to rejection of the model, this test set cannot be used to compute confidences. The reason is subtle but important: because the model was to be kept if and only if it performed well on the test set, there is a built-in prejudice toward good performance on this dataset. A completely new dataset must be collected.

- A large number of cases, typically at least several hundred, must be employed. This subject is discussed in more depth on page 76.

- All possible examples of the hypothesis population must be represented in the dataset, and they must appear in quantities proportional to their expected appearance rates in real life. In practice, this is the most difficult requirement to satisfy. If a case appears later, when the model is used, and this case was not represented in the confidence dataset, false rejection of the hypothesis may occur. For example, suppose the model will discriminate between benign and malignant skin lesions, and the null hypothesis is that the lesion is benign. Then the confidence set must include samples of all possible sorts of benign lesions that a doctor may excise, and they must be represented in at least approximately the same proportions that they would be encountered in real life. This is hard.

Once the confidence dataset is in hand, it is used to compute the function that converts a prediction to a confidence. This is most easily done by sorting the predictions and counting cases in the appropriate tail. The confidence in the alternative (which is *not* a probability, just a deliberately vague term) is estimated as one minus the fraction of cases in the tail.

For example, suppose we have a large and representative collection of nontarget cases. Our model has been trained to make small predictions (such as −1) for nontargets and large predictions (such as +1) for targets. To prepare for confidence evaluation, this trained model is invoked for each case in our nontarget collection. We preserve the predictions made for these nontarget cases, as they will form the basis of confidence calculations.

Later, an unknown case is tested, and it produces a prediction greater than the threshold. Therefore, we call it a target, and in addition to this classification decision, we would like a confidence measure. We now turn to the collection of predictions made for our nontarget confidence set, as described in the previous paragraph. Suppose that only a small fraction, say 0.03, of the nontarget collection has predictions this large or larger. Under the (stringent!) assumption that the confidence collection truly is representative of nontargets, we conclude that there is only a 3 percent chance of obtaining a prediction this large if the case is a nontarget. This lets us assert that we are 97 percent confident that the case is a target. Note, by the way, that we are most definitely *not* saying that there is a 97 percent *probability* that the case is a target. We are restricted to using the deliberately vague term *confidence*. This subject is discussed in more detail on page 82.

Alternatively, we might have a confidence collection containing a large number of representative target cases. At some point in the future, an unknown case is presented to the model, and the resulting prediction is less than the threshold. Therefore, we call this case a nontarget. Examining the confidence collection of targets, we may find that 0.4 of the cases in this collection have predictions this low or lower. We conclude that only 60 percent confidence in our nontarget decision is justified.

The best way to compute a hypothesis-test confidence is to break the process into two steps. First, the predictions in the confidence collection are sorted, and tail fractions are computed and stored. Second, as new unknown cases are encountered, a fast binary search is used to locate the correct position in the sorted array, and interpolate to return a probability. Naturally, this process is done separately for the target and nontarget collections. Targets and nontargets are not pooled into a single collection.

Here is a code fragment demonstrating the first step. We start with *ncases* predictions in the *x* array. This array is sorted, duplicates are removed, and the algorithm yields *n*, the number of unique values. Both tail probabilities (left and right) are computed here. In practice, only the right tail is used for computing target confidence from the nontarget collection, and only the left tail is used for computing nontarget confidences from the target collection.

```
sort (ncases, x);                              // Sort ascending
i = 0;                                         // Will index original predictions
n = 0;                                         // Will count unique values

while (i < ncases) {                           // Check all predictions in confidence collection
  x[n] = x[i];                                 // Copy first occurrence of this prediction
  rprob[n] = (double) (ncases - i) / ncases;   // Fraction greater or equal to x[n]
  while ((++i < ncases) && (x[i] == x[n]));    // Bypass repeats
  lprob[n++] = (double) i / ncases;            // Fraction less or equal to x[n]
  }
```

Once the confidence mapping function has been found by the algorithm just shown, all that remains is to perform a simple table lookup and interpolation to assign a confidence to an observed prediction. Here is code to do this for the left tail. This would be used when we have obtained a low prediction and want to compute confidence in our classification of the case as a nontarget. The confidence sample contained targets.

```
if (observed > x[n-1])                         // If the prediction is huge (rare!)
  return 0.0;                                  // We surely have no confidence

if (observed <= x[0])                          // If the prediction is tiny
  return 1.0 - lprob[0];                       // Inspires greatest reasonable confidence

lo = 0;                                        // Will always keep x[lo] < observed
hi = n - 1;                                    // And x[hi] >= observed

for (;;) {                                     // Binary search loop
  mid = (lo + hi) / 2;                         // Center of remaining search interval
  if (mid == low)                              // Are hi and lo adjacent?
    break;                                     // If so, we are there
  if (x[mid] < observed)                       // Replace appropriate end point with mid
    lo = mid;
```

```
else
    hi = mid;
}
```

```
tail = lprob[hi-1] + (observed - x[hi-1]) / (x[hi] - x[hi-1]) * (lprob[hi] - lprob[hi-1]);
return 1.0 - tail;
```

The code for processing the right tail (confidence in a target decision based on the nontarget samples) is practically identical to that just shown. The binary search and interpolation in rprob are exactly the same. The only difference is in the initial checks for boundary conditions. This is done as follows:

```
if (observed >= x[n-1])              // If the prediction is large
    return 1.0 - rprob[n-1];         // Inspires greatest reasonable confidence
if (observed < x[0])                 // If the prediction is tiny (rare!)
    return 0.0;                      // No confidence at all
```

Note that computing the tail probability via interpolation is overkill in many or most applications. It was shown here simply for completeness to cover those unfortunate (and potentially dangerous!) situations in which the data is highly granular. I generally employ the more conservative approach of using whichever of the endpoints of the interval is closer to the center of the distribution. This increases the tail area and hence decreases the confidence in the decision, though typically by only a tiny amount. The next section will assume that this more conservative method is used, although interpolation will not usually change the conclusions much at all.

Confidence in the Confidence

The quality of the confidence computations is inhibited by two sources of error. The most worrisome source of error is incomplete sampling. It is vitally important that all possible exemplars appear in the confidence set. Suppose, for example, that the goal is to detect a tank in an infrared image. A criminally careless designer might procure numerous examples of side views only. If such a collection is used for the training and target confidence sets, there will be problems. When a tank appears pointing directly at the sensor, it may well be classified as a nontarget if no representative examples appeared in the training set. When the low prediction is compared to the predictions in the target's confidence set, it may well lie far out in the left tail. This adds insult to injury, as there will be high confidence that this is not a tank. Such extreme negligence is not common, but lesser forms of inferior collection techniques are common, sometimes by accident and sometimes by laziness. Beware.

The second source of error in confidence calculations is unavoidable, though it can usually be controlled. This is random sampling error. Even if the designer is careful to give equal opportunity to all possible exemplars, bad luck may still intervene. Unknown to the designer, an inordinately large group of cases having unusually large or small predictions may find its way into the confidence set. This phenomenon can be difficult or even impossible to detect. Such an event will distort subsequent confidence calculations. The only way to protect against this situation is to collect a large enough confidence collection that any such anomalies will be diluted. How much error in the confidence might we expect from this uncontrollable random sampling error? It can be quantified with either of two simple mathematical tests.

The first method for estimating the degree of potential error in a confidence calculation is similar to a technique that we've seen before. On page 37 we learned how to compute confidence bounds for quantiles. Review that section if it is not familiar. In that section, we were interested in bounding confidence intervals for prediction errors. The same technique can be used to bound confidence in classifications obtained by thresholding predictions.

Suppose for the moment that we have collected a confidence set containing n cases from the target class. We choose a small probability p and compute $m=np$. We then take the m'th smallest prediction in the confidence set and use it as our classification threshold. Subsequent cases whose prediction is less than this value will be classified as nontargets, with the understanding that if the case truly is a target, there is only probability p that a prediction this small, which naturally results in misclassification, will be obtained. Our fear is that bad luck in collecting the confidence set results in the actual misclassification probability being some value q that exceeds p. We need to quantify the extent of this potential problem.

In particular, we need to answer the following question: What is the probability that the m'th smallest prediction in the target confidence set, which we have chosen for our classification threshold, is really the q quantile of the distribution, where q is disturbingly larger than the p we desire? We hope that this probability is small, because the event of q exceeding p is a serious problem in that it implies that the likelihood of misclassification is worse than we believe. This probability question is answered by the *incomplete beta distribution*. In particular, in a collection of n cases, the probability that the m'th smallest will exceed the quantile of order q is given by $1 - I_q(m, n-m+1)$.

As discussed in the "Confidence Bounds for Quantiles" section beginning on page 37, the bounds are symmetric, meaning that exactly the same method can be used for examining pessimistic quantile bounds for high thresholds such as would be used in a nontarget confidence set. Also in that section we saw three subroutines on my web site that can be useful for these calculations. These are ibeta (param1, param2, p) for the incomplete beta distribution, orderstat_tail (n, q, m) for the tail probability, and quantile_conf (n, m, conf) for solving the inverse problem. See that section for more details.

The technique just described is useful for quantifying potential problems with a classification threshold. However, much of this chapter has been concerned with the more general situation of computing a confidence in a decision, an actual number that, while not a probability of class membership, does nevertheless express confidence in a decision. As was the case for a classification threshold, this confidence calculation is based on predictions obtained in a sample called the confidence set. How sure can we be that our computed confidence accurately reflects the probability that an observed case arose from a hypothesized class? We now explore this subject.

The population (target or nontarget) from which the confidence collection is drawn has a *cumulative distribution function* (CDF) that we will call $F(x)$. By definition, when a randomly selected case from this population is presented to the model, the probability of getting a prediction less than or equal to any specified value x is $F(x)$. We do not know this function, but we approximate it by collecting a representative sample and using its CDF as shown in a previous section.

Let $S_n(x)$ be the proportion of the collection of n cases that are less than or equal to x. Suppose we let x traverse its entire range, and we compute the maximum difference between the true CDF and that defined by the collection. This worst error is shown in Equation (2.5).

$$D_n = \max_x \left| S_n(x) - F(x) \right| \qquad (2.5)$$

It is remarkable that the distribution of D_n does not depend on the nature of either the true or the sampled CDF. This facilitates a popular statistical test called the *Kolmogorov-Smirnov test*. For some typically small probability α, there exists a constant d_α such that there is a probability of only α that the worst error in the sample CDF exceeds d_α. This is expressed in Equation (2.6).

$$P\left\{ D_n = \max_x \left| S_n(x) - F(x) \right| > d_\alpha \right\} = \alpha \qquad (2.6)$$

The key point is that Equation (2.6) holds no matter what the true CDF is. We can rearrange this equation to cast it in a form that is immediately useful to the task at hand. Equation (2.7) shows how, given an observed sample CDF, we can find a probabilistic bound on the true CDF.

$$P\{S_n(x)-d_\alpha \leq F(x) \leq S_n(x)+d_\alpha\}=1-\alpha \tag{2.7}$$

Note that this equation provides confidence bounds for the *entire* distribution. Once we have acquired the confidence collection, we can specify a satisfactorily small α and then find the corresponding error limit.

Exact calculation of d_α from α is very complex, but a good approximation exists. This is shown in Equation (2.8).

$$d_\alpha \approx \sqrt{\frac{-\ln\left(\frac{\alpha}{2}\right)}{2n}} \tag{2.8}$$

In the tails, which is where we operate, this approximation is very good if n exceeds about 35, and it is almost exact if n exceeds 100. Since practical confidence sets will always be at least this large, and preferably much larger, there should be no problem using this equation.

Here is an example of this technique. Suppose we collect 100 cases for the confidence set. We decide that we will be satisfied if there is a 95 percent probability that the true confidence is within the computed bounds. This implies that α is 0.05. Equation (2.8) gives an error limit of 0.136. So we can be 95 percent sure that no matter what a computed confidence is, the actual confidence is within 0.136 of the computed value. Now suppose that an unknown case arrives and we use the hypothesis-testing technique to compute a confidence of, say, 80 percent. We can be 95 percent certain that the correct confidence figure, which we would know if we knew the true CDF, is no less than 66.4 percent, and no greater than 93.6 percent. This relatively wide range should inspire us to acquire a confidence collection that is as large as possible. Quadrupling the size of the collection will halve the confidence range.

One possible disadvantage of the Kolmogorov-Smirnov method is that absolute deviations between the empirical CDF and the true CDF usually reach their maximum near the center of the distribution. This test employs the same width for confidence bands across the entire range of the predicted value. Thus, the confidence interval for confidences is dominated by behavior near central values of the predictions, while many users are mostly interested in extreme predictions. This can even lead to anomalous behavior. For example, suppose the deviation computed by Equation (2.8) is 0.13, and our tail p-value is 0.04, meaning that our confidence in rejecting the null hypothesis is 0.96, plus or minus 0.13. It's fine to talk about a lower bound of 0.96–0.13=0.83, or 83 percent. But an upper bound on the confidence of 0.96+0.13=1.09 makes no sense, since a confidence cannot exceed 1.0. Despite this problem, Kolmogorov-Smirnov bounds are useful and should be computed in any critical application.

In many situations, we are not particularly interested in an upper bound on the confidence. As was the case for a pessimistic bound for the quantile associated with a decision threshold, we may be interested in only a lower bound on the confidence, but unlike the case for a single decision threshold, we want a universal lower bound, one that is valid for the entire range of possible predictions. In this case, one would double the tail p-value and use it in Equation (2.8). For example, let's again consider the situation of 100 cases in the confidence set and a desired confidence interval of 95 percent. A few paragraphs ago we found that $\alpha=0.05$ used in Equation (2.8) gave an error limit of plus or minus 0.136. But if we consider only errors in one direction, we set $\alpha=2*0.05=0.1$ and use this in Equation (2.8). This gives a distance of 0.122. So we can be 95 percent sure that no matter what a computed confidence is, the actual confidence is no more than 0.122 below the computed value. Observe that the bound is slightly tighter than the two-sided bound, which is always nice.

We end this section with a discussion of a small program supplied on my web site. The source code can be found in the file CONFCONF.CPP, and the statistical routines that it calls are in STATS.CPP. The program demonstrates the two confidence bound algorithms just discussed. It is called as follows:

```
ConfConf  ncases  pval  conf  nreps

    ncases - Number of cases in the sample
    pval - Probability value (<0.5) for quantile test
    conf - Desired confidence value (<0.5) for both tests
    nreps - Number of replications
```

To obtain the numbers used in the examples a few paragraphs ago, as well as a few other numbers, we could call the program as follows:

```
ConfConf  100  0.1  0.05  100000
```

This call specifies that the confidence set contains 100 cases. The rejection threshold is defined as that which provides a 0.1 probability of erroneous classification. We also specify a probability of 0.05 for both the pessimistic bound on the threshold and the Kolmogorov-Smirnov universal bound. Correct operation is verified by performing 100,000 simulations. The following output is produced:

```
If the dataset represents the null hypothesis, the threshold for rejecting
the null at p=0.1000 is given by the 91'th order statistic.
This is a conservative estimate of the 0.9000 quantile There is only a
0.0500 chance that it will really be the 0.8482 quantile or worse.

If the dataset represents the alternative hypothesis, the threshold for
rejecting the alt at p=0.1000 is given by the 10'th order statistic.
This is a conservative estimate of the 0.1000 quantile
There is only a 0.0500 chance that it will really be the 0.1518 quantile or
worse.

KS thresholds: two-tailed KS = 0.1358 one-tailed KS = 0.1224
Point failure (expected=0.0500) Lower=0.0502 Upper=0.0498
KS failure: two-tailed = 0.0464 NULL = 0.0476 ALT = 0.0459
```

We see from this output that the pessimistic bound for rejecting the null hypothesis is 0.8482 rather than the assumed 0.9, with this happening at the specified probability of 0.05. In other words, if we use the 91st order statistic as our rejection threshold, which would on average give us a false rejection rate of about 0.1, there is a 5 percent chance that the actual false rejection rate could be as high as 1.0–0.8482=0.1518 or worse. The symmetric result is obtained for rejecting the alternative hypothesis (which we often called the target set).

For the Kolmogorov-Smirnov test, this output shows that there is a 95 percent chance (1.0 minus our specified 0.05) that the correct confidence figure obtained from a test case will be the computed confidence plus or minus 0.1358. Alternatively, there is a 95 percent chance that the actual confidence will be at least the computed confidence minus 0.1224.

Finally, 100,000 replications of a simulated sample showed that the failure rates for the pessimistic bounds are almost exactly the expected value of 0.05. Failure rates for the three Kolmogorov-Smirnov possibilities are all slightly less than the 0.05 that we specified. This small inaccuracy is because of the combination of the facts that the $m=np$ order statistic is slightly biased in the conservative direction, and Equation (2.8) is only an approximation.

Bayesian Methods

When confidence collections are available for both the target and nontarget classes, it may seem silly to compute and examine *two* confidence figures. This is especially true when we observe that the two computed confidences do not generally sum to one. What does it mean to say that there is a 10 percent confidence that this case is a target and a 15 percent confidence that it is a nontarget? (Actually, it means a lot, and this topic is pursued on page 88.) It would seem to make more sense to compute a single confidence number that handles the other alternative by implication. In other words, we want to be able to say that there is an 80 percent chance that the case is a target, with this figure implying that there is a 20 percent chance that it is a nontarget. This ability is often useful. However, it will be seen that some disturbingly arbitrary and possibly dangerous assumptions are involved. Also, some important information is sacrificed. With these caveats, we now explore how to use Bayesian methods to compute a single confidence figure for a target/nontarget classification.

Even though this section focuses on binary classification, we will start by stating the multiple-class version of Bayes' theorem. This version will be referenced later. Suppose we have K mutually exclusive and exhaustive classes, and they are labeled $\{H_k, k=1, ..., K\}$. When a case is randomly sampled from a specified class and the model's prediction is made, this prediction is a random variable. Let the likelihood function of this prediction for class k be written $L(x|H_k)$. If the distribution happens to be discrete, the likelihood is simply the probability of x occurring. In the more usual situation that the distribution is continuous, the likelihood is the density function. Finally, assume that the prior probability of class H_k is p_k, remembering that $\Sigma p_k=1$. Given a case sampled from an unknown class, the probability that the case is from class H_k is given by Equation (2.9).

$$P\left(H_k|x\right) = \frac{p_k L\left(x|H_k\right)}{\sum_i \left(p_i L\left(x|H_i\right)\right)}$$
(2.9)

We need two pieces of information for each class in order to be able to use Equation (2.9) to compute the confidence (which here is the probability that the specified class is the source of the unknown case). We need the prior probabilities, and we need the likelihood functions. They are both problematic.

It is obvious from Equation (2.9) that the prior probabilities affect the computed confidence as much as the likelihoods. This is troubling because, in many applications, the prior probabilities are subjective at best and capricious at worst. What do we do when we have no objective evidence for assigning priors? The answer to this extremely important question was given by Bayes himself in what is known as *Bayes' Postulate* (sometimes also called the *Principle of Equidistribution of Ignorance*). He states that the prior probabilities should be assumed equal unless there is clear evidence to the contrary. Ignore this advice from the ultimate expert at your own peril.

Estimating the likelihood function is much more problematic than dealing with the priors. It may rarely be the case that the model's input variables can take on only a very few discrete values, and hence the predictions will also be discrete. In this situation, the likelihoods can be computed from the confidence sets by counting the number of occurrences of each discrete value and dividing this by the total number of cases. But it will almost never happen in real life that direct probabilities can be used this way. In most cases, the inputs are either continuous or they have so many possible discrete values that they might as well be continuous. It is necessary to estimate the density function of each class. This is not a trivial undertaking.

Probably the best method for estimating a density function from a collection of observations (model predictions here) is known as *Parzen's method*. Details can be found in many statistics texts, as well as in [Masters, 1993]. Only a summary of the technique is given here. Let the collection of n predictions from a class be $\{x_i, i=1, ..., n\}$. The density function of the population from which this collection was drawn can be estimated by Equation (2.10).

$$f(x) = \frac{c}{n\sigma} \sum_i e^{-\left(\frac{x-x_i}{\sigma}\right)^2} \qquad (2.10)$$

In this equation, c is a constant that is needed only to ensure that the density estimator integrates to unity. It does not concern us here because it appears in both the numerator and the denominator of Equation (2.9), and so it will cancel. We will ignore it. The problem is σ (sigma), the smoothing constant. Suppose we choose to let σ have a very tiny value. The sum in Equation (2.10) will be dominated by whichever x_i happens to be closest to x. All other summands will be negligibly small. The density estimate will consist of nothing more than a series of peaks, each corresponding to a case in the collection. This is clearly not satisfactory. Suppose instead that we choose to let σ have an extremely large value, many times greater than the variation within the collection. The density estimator will now have about the same relatively large value for any x near the collection, dropping off toward zero only when x is very far from any case in the collection. The contributions of the individual cases in the collection are largely ignored. This, too, is obviously not satisfactory. It is necessary that σ be some moderate value, large enough that no individual case dominates, but small enough that individual cases do make significant contributions to the density across the domain.

How is a good value of σ chosen? The goal is that for every possible x, some cases in the collection that are nearest to x make a significant contribution, a few more distant cases make a modest contribution, and the remainder of the collection is essentially ignored. This admittedly vague goal can usually be achieved by letting σ be somewhere between 0.05 times the standard deviation of the collection, up to perhaps 0.4 times the standard deviation. If the collection is very large (n is many hundreds of cases) and reasonably normal, the smaller limit is appropriate because it will squeeze maximum information from the individual cases. If the collection is not this large or if the distribution contains outliers, a large value of σ will provide better smoothing and more conservative behavior in the tails. In general, values of σ greater than about 0.4 times the standard deviation will blur the density excessively, while values much less than 0.05 times the standard deviation will produce unnatural peaks in the density estimator. Smaller values are appropriate only if the collection contains thousands of cases.

If the collection happens to contain a few relatively isolated predictions (not good!), it may be useful to manually compute the contribution of individual cases in advance. More importantly, it is informative to see how far away cases must be in order to reduce their contribution to a specific degree. Let d be the difference between the point x at which the density is being estimated and a case x_i in the collection. The contribution of this case to the sum in Equation (2.10) is given by Equation (2.11). Solving this for d gives Equation (2.12).

$$w = e^{-\left(\frac{d}{\sigma}\right)^2} \tag{2.11}$$

$$d = \sigma\sqrt{-\ln(w)} \tag{2.12}$$

A case exactly equal to the point x at which the density is being estimated will provide the maximum possible contribution to the sum, 1.0. Suppose we let $w=0.5$, implying that we are interested in seeing how far away from x a sample case must be in order for its contribution to drop to one half of that of a case exactly at x. This, of course, depends on σ. Assume that we have chosen σ to be 5.0 (a reasonable choice if the standard deviation happens to be from 50 to 100 or so). Equation (2.12) tells us that d=4.16. So for any given x, collection cases that happen to be at x–4.16 and x+4.16 will have half the contribution of a case that happens to be exactly at x. To take it a bit further, let w=0.1. We find that d=7.59. Thus, cases more distant from x than 7.59 will make an almost negligibly small contribution to the density estimate. This sort of information can be useful because it allows us to confirm that there are no "holes" in the collection. If, for example, the collection contained no cases with predictions from, say, 10 through 30, density estimates around x=20 would be extremely poor (near zero!). Sigma should be at least doubled, changing from 5 to 10. Values even larger would not be unreasonable.

There is a more rigorous and objective method for computing an optimum σ (or pair of sigmas if separate values are desired for the target and nontarget collections). This method may be overkill, and it may be dangerous to completely remove the experimenter from the selection process. Nevertheless, automation is handy, and the method appears to work well.

This method is based on defining a performance criterion that rates the effectiveness of any given σ (or pair). Once this criterion is available, any reasonable numeric optimization algorithm may be used to find the best sigma weight(s). An excellent way to rate this performance is to use cross validation. Here is the algorithm:

1) Remove the first case from the nontarget collection. Temporarily consider it an unknown.

2) Use Equation (2.10) to estimate the density at this "unknown" case in the nontarget collection and also in the target collection.

3) Use Equation (2.9) to compute the confidence that this case is a nontarget. It truly is a nontarget, so we ideally want the confidence to be exactly 1.0. It will generally be smaller than this. The difference is the error attributable to this case. Square this error and cumulate it into a sum.

4) Replace this case and now remove the second case in the nontarget collection. Use the two previous steps to compute its error and cumulate the total squared error.

5) Repeat this process for all cases in both the target and nontarget collections. Remember that for cases in the target collection, we want the target confidence to be 1.0, so compute the error accordingly.

The mean squared error across all cases appears to be an excellent indicator of the quality of the σ weight(s) used in Equation (2.10). In most practical applications, this function has a delightful bowl shape across a reasonable range of σ weights, so optimization is typically straightforward. Multiple broad local minima do sometimes appear, so a rough initial global search is needed to find the vicinity of the global minimum. However, local behavior of the error function is usually excellent.

In summary, here is how Bayesian confidences can be computed if we have large representative collections of cases from both the target and nontarget populations: The first step is to determine a reasonable sigma parameter for each population. The same value does not have to be used for both, although in practice they will usually be similar if not identical. If possible, make a graphical plot of the computed density functions. These plots may be interesting and informative. When an unknown case arrives, use the model to compute the prediction for this case. Apply Equation (2.10) to both the target and nontarget confidence collections to find the density estimate for each. Finally, plug these two values into Equation (2.9) to get the confidence figures. Note that this equation guarantees that the target and nontarget confidences will sum to one.

Do not be surprised if sometimes the predicted value relative to a preordained threshold gives a classification that has *lower* confidence than the other class. This can happen when the prediction is not clear cut because of bumpy density functions, especially if the relative error costs and/or prior probabilities are disparate. We'll see an example of this in Figure 2-4 on page 90.

Multiple Classes

So far, the subject of confidences has been limited to the simple case of target versus nontarget classification. This is by far the most common application. Nevertheless, it is sometimes necessary to be able to predict membership from among three or more classes. Confidence measurements become correspondingly more difficult in this case. The two methods already discussed can be generalized to more than two classes. However, issues become complex rapidly.

For the hypothesis testing method, we need a confidence collection for each class for which we want to be able to reject membership. An immediate problem is that rejection of one class no longer automatically implies confidence in the other class. For example, suppose we want to classify a tissue sample as benign, malignant, or atypical. If we have a good collection of atypical cases, we may be able to reject membership in the atypical class. But that alone tells us nothing about benign versus malignant. We need to judge this probability in the context of the other probabilities of rejection.

The problem gets even worse when one considers the fact that effective multiple-class classification almost always requires separate predictions for each class. The usual method is to train each prediction to be high for exactly one class, and low for all others. This means that we have the option of considering rejection probabilities for more than one distribution. The general consensus is that when more than two classes are involved, hypothesis testing is not a good choice for confidence computation.

There is an excellent way to use Bayes' method to compute confidences in multiple-class predictions. Train a single model with multiple outputs or a set of models, each making a prediction for a single class. The dependent variable is a large value for cases in that class, and a low value for cases in all other classes. Thus, the confidence sets will consist not of scalar predictions, but of vector predictions, with each vector having a length equal to the number of classes. Then, use these confidence sets as a training set to train a *probabilistic neural network* (PNN) to do the classification. A PNN intrinsically has confidences as its outputs, so it is ideal for this application. When an unknown case arrives, apply the prediction models in order to procure a vector of class predictions. Then submit this prediction vector as an input to the trained PNN. The outputs of the PNN are the computed class confidences. This technique has proven itself to be effective in many applications, and it is probably the most straightforward way of handling the multiple-class confidence problem. Probabilistic neural networks are beyond the scope of this text. They are introduced in [Specht, 1990a] and covered in depth in [Masters, 1995a]. A full discussion of confidence computation using PNNs can be found in [Masters, 1993].

Hypothesis Testing vs. Bayes' Method

Two very different techniques for computing classification confidences have been presented: hypothesis testing and Bayes' method. These confidences have dramatically different theoretical foundations, and they provide very different types of information. A comparison is in order.

First, it must be understood that we sometimes do not have the luxury of choosing between the two methods. The Bayesian method is much more stringent in its requirements. For hypothesis testing, a representative collection of cases needs to be available only for the hypothesis that we will attempt to reject in favor of the alternative. If one of the classes is not able to be adequately represented, we can still proceed with what we have. For Bayes' method, both the target and nontarget classes must be fully represented. Moreover, the quality of the representation is more critical for Baye's method than for hypothesis testing. Suppose a collection has an empty area in its distribution, a range of predicted values that, due to sampling error, contains few or no cases. We will be faced with an uncomfortable choice. Either σ will need to be chosen large enough to bridge the gap, causing excessive blurring and distortion everywhere else, or the computed density will be incorrectly tiny in this region, leading to incorrect confidence decisions later. Hypothesis testing uses no local information, only case counts from interior areas out to the end of the sorted collection. This makes it more tolerant of sampling problems.

Assuming that all necessary criteria have been met, let us now compare the meaning of confidence computed with hypothesis testing versus that computed using Bayes method. Figure 2-2 shows the estimated densities for a moderately good model. The left curve is the density function for the nontarget class, and the right curve is that for the target class. Where the two curves intersect at about 0.1, the Bayesian confidence in both classes is the same, 0.5 (assuming equal prior probabilities). Moreover, the area under the left curve to the right of this point is roughly equal to the area under the right curve to the left of this point. Thus, the hypothesis-testing confidences in both classes are also approximately equal, moderately large but nothing to get excited about. If a prediction less than about –0.75 appears, the Bayesian confidence in the nontarget class will clearly be essentially 1.0 because the target density is insignificant there. The area under the target curve to the left of this point is also just about zero, implying that the hypothesis-testing confidence is also nearly 1.0. It should be clear that no surprises lurk in this example.

Now look at Figure 2-3. This illustrates the density curves for a model with almost impossibly good performance. Predictions less than about –0.1 or greater than about 0.1 result in both confidence methods giving identical perfect confidence estimates. Still no surprises. But what if a prediction comes in just above zero, where the two density curves cross? The Bayesian confidences will be 0.5 for both classes. However, the hypothesis-testing confidences will be practically 1.0 for *both* classes! This is because *both* the target and nontarget hypotheses are rejected. The probability of a prediction of this value is nearly zero in both the target and nontarget populations. This conflict has great importance, as we will soon see.

And it gets worse. Problems like this do not occur only at sparse intersection points. Suppose a case produces a prediction of –0.1. The target density here is nearly zero, so Bayes' method will produce a confidence of virtually 1.0 that this case is a nontarget. But the area under the nontarget density to the right of this point is probably less than 0.01,

87

meaning that hypothesis testing based on the nontarget collection will give a confidence in excess of 0.99 that this case is a target! The two methods are (definitively!) saying exactly opposite things. What's going on?

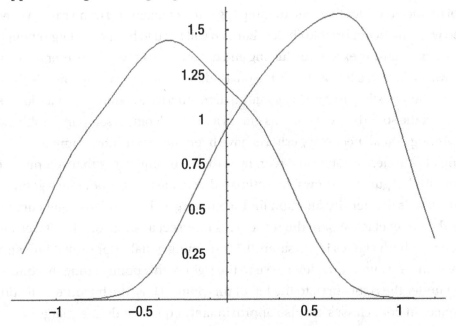

Figure 2-2. *Parzen density estimates for a fairly good model*

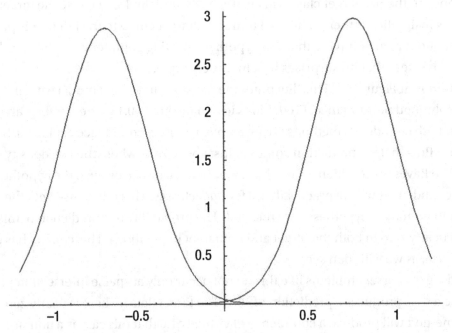

Figure 2-3. *Parzen density estimates for a fantasy model*

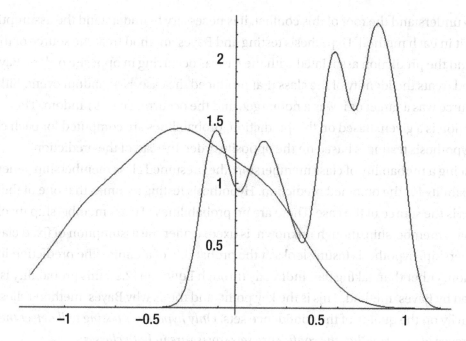

Figure 2-4. Parzen density estimates with poor targets

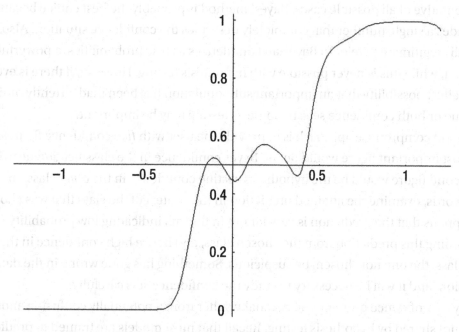

Figure 2-5. Bayes confidence when target definition is poor

To understand the root of this conflict, it is necessary to understand the assumptions implicit in each method. Hypothesis testing and Bayes' method treat the source of the case and the prediction associated with the case as occurring in opposite orders. Bayes' method treats the identity of the class that produced the case as a random event. Either the source was a target or it was a nontarget, and the occurrence was random. The prediction is a given. Based on this prediction, probabilities are computed for each class.

Hypothesis testing is based on the opposite order. Instead of the prediction producing a probability of class membership, the presumed class membership generates a probability for the obtained prediction. Hypothesis testing assumes that one of the two classes is the source of the case. There are no probabilities of class membership involved. The class membership, though unknown, is fixed. Under the assumption of fixed class membership, hypothesis testing looks at the probability of attaining the prediction in question. Other than taking part indirectly through Equation (2.9), this probability is ignored by Bayes' method. This is the key point, and this is why Bayes' method relies so heavily on the quality of the confidence sets. *Only hypothesis testing will detect the occurrence of an event that the confidence sets say is rare in both classes.*

If the experimenter is absolutely positive that both confidence sets are thoroughly representative of all possible cases, Bayes' method is probably the best choice because it provides a single number that completely describes the confidence situation. Also, it is usually legitimate to refer to Bayesian confidences as true probabilities, a powerful assertion, while this is never possible with hypothesis testing. However, if there is even the smallest possibility that an important subpopulation has been inadvertently omitted from one or both confidence sets, hypothesis testing may be important.

A good compromise approach is to provide the user with *two* confidence figures. The most important figure would be the Bayes confidence in the classification decision. The second figure would be the hypothesis-testing confidence in the *other* class. In other words, examine the attained prediction in the context of the class that was chosen. If it happens that the prediction is very far out in the tail, indicating low probability of obtaining this prediction from the chosen class, and hence high confidence in the *other* class, the one not chosen, be suspicious. Something has gone wrong in the data collection, and it will be necessary to study the confidence sets carefully.

Bayes confidence estimates occasionally suffer from a potentially confusing anomaly that is not shared by hypothesis testing. Recall that most models are trained to predict a large value (typically +1) for targets, and a small value (typically –1) for nontargets. This implies that we expect a monotonic mapping from the prediction to confidences: larger predictions map to larger confidences in the target class and smaller confidences in the nontarget class. This is precisely what happens in hypothesis testing. As the prediction increases, fewer cases exceed it in the nontarget distribution, leading to higher confidence in the target class. Simultaneously, more cases are smaller than it in the target distribution, leading to lower confidence in the nontarget class.

Unfortunately, this is not guaranteed to happen with Bayes confidence. Look at Figure 2-4 on page 89. This illustrates a target population density that is multimodal. Densities like this can arise when the target population contains several subclasses or when the sigma parameter for the Parzen estimator is too small. Observe how the target and nontarget densities cross repeatedly. Figure 2-5 shows the associated Bayes confidence as computed with Equation (2.9) on page 81. The horizontal axis in this plot is the model's prediction, and the vertical axis is the confidence in the target class, with 0.5 (equal confidence in the two classes) used as the horizontal axis. We would hope that this curve would steadily rise, indicating increasing confidence in the target class, as the prediction increases. But it doesn't. We are faced with the peculiar situation that at some places the confidence can flip-flop, with an increasing prediction causing the nontarget class to be favored over the target class!

Is this a serious problem? Specifically, should we base our classification decision on a threshold for the prediction, or should we base it on the Bayes confidence? They may not always give the same results. There is no easy answer. If the multiple modes in one or both densities are real and meaningful, then we should probably use the Bayes confidences to decide class membership, as they reflect the fact that certain regions are expected to be sparse. But before making this rather serious decision, you had better be positive that the multiple modes are truly representative of the nature of the underlying population. If instead they arise from multiple classes within what you are treating as a single class, you should investigate closely and consider a different modeling scheme such as vector prediction trained on each class. If multiple classes are ruled out (or must be ignored for technical reasons) then a larger smoothing parameter for the Parzen window should be used.

Final Thoughts on Hypothesis Testing

Hypothesis testing as a means of computing confidence in a classification decision seems straightforward. However, its indirect method of assertion makes it very susceptible to accidental abuse. For this reason, a final lecture on some of the subtleties of the logic is in order. We begin by repeating the logic by which one asserts confidence in a target classification when the model's prediction is high and we have information about the distribution of predictions in the nontarget population. This same logic could be inverted to assert confidence in a nontarget classification when we get a low prediction and we have information about the target distribution.

1) Our environment is mutually exclusive and exhaustive, so the case in question either is a target or is a nontarget. There are no other possibilities.

2) If we were to assume that the case is a nontarget, we would be faced with the fact that the probability of obtaining a prediction as large as or larger than that obtained is very small. In other words, if the case really were a nontarget, we would have just observed a highly unlikely event.

3) We thus conclude that the case is probably a target.

In many situations we can compute a quite accurate value for the probability of having observed such a large prediction from a nontarget case. The algorithm presented on page 75 provides decent estimates of this probability as long as the confidence set satisfies the required assumptions. We compute confidence in a target classification by subtracting the nontarget probability from one. So, for example, suppose we find that there is a probability of only 0.01 that a nontarget would produce a prediction this high. Subtracting this probability from one gives a confidence of 0.99 that the case is a target.

A common and potentially serious error is made if we call this 0.99 a probability. It is tempting, though wrong, to conclude that 99 percent of the time that we observe a prediction whose nontarget probability is 0.01 or less we will be correct in concluding that the case is a target. Looked at another way, consider, say, 100 instances in which we observe a case whose nontarget probability is 0.01 or less. These may be, for example, a batch of tissue samples that arrive at a pathology laboratory. Or they may be appearances of blips on a fighter aircraft radar screen. We are tempted to conclude that of these 100 occurrences of 0.01 or less nontarget probability, about 99 of them will in truth be targets. *This is terribly wrong!* The faulty logic goes like this: We just observed a case that is either a target or a nontarget. It had a high prediction, one that is very unlikely to have arisen from a nontarget. Hence, we classify it as a target. Then...

1) Either we are right (to call the case a target) or we are wrong. There are no other possibilities.

2) If we are wrong (the case really is a nontarget), we will observe a prediction this large only 1 percent of the time (0.01 probability).

3) Therefore there is a 99 percent probability (0.99) that we are right.

The flaw in this logic is that it has subtly reversed the order in which events occur. The correct order is that first we assume the fact that the case is a nontarget, and then we observe a prediction. We can legitimately compute a probability associated with the prediction. The phony logic just described says that we observed a prediction and then we try to compute a probability that the case is a target or nontarget. This probability can only be computed with Bayes' method, in which we have knowledge of the prediction distribution for both targets and nontargets and in which we are willing to assume prior probabilities for each class. Neither of these pieces of information is available to a hypothesis test.

If we want to subtract a probability of 0.01 from one and get a complementary probability of 0.99, this is legitimate. But we need to be careful to correctly define the meaning of this probability. It (0.99 in this example) is the probability that *when the case truly is a nontarget, we will observe a prediction less than that obtained for this case.*

To carry this further, suppose our task is to design a model that examines evidence regarding a credit card transaction that a customer is attempting to make at a distant store. If our model decides that the transaction is likely fraudulent, the store is instructed to refuse the purchase. Erroneous refusals will surely produce angry customers, so we decide to be extremely cautious. Suppose we have no information about the distribution of model predictions for the *fraudulent* class. But we do have lots of information about predictions for *nonfraudulent* transactions. In the interest of erring on the liberal side, we choose to reject a transaction only if the model's prediction is so high that its probability under the assumption of the nonfraudulent class is less than or equal to 0.001. This implies a confidence of 0.999 or more that the transaction is fraudulent. By choosing a threshold this high, we can legitimately make two statements about probabilities:

1) Only 0.1 percent of truly nonfraudulent transactions will be erroneously flagged as fraudulent.

2) 99.9 percent of all nonfraudulent transactions will be allowed to proceed.

There are two probability statements that we *cannot* make with a hypothesis test, no matter how much we would like to do so:

1) When we obtain a prediction so high that it reaches the 0.001 probability threshold, and hence we decide to forbid the transaction, we will be wrong to do so only 0.1 percent of the time.

2) When the model's prediction does not reach the 0.001 threshold and hence we allow the transaction to proceed, we will be wrong only 0.1 percent of the time.

The preceding two assertions reverse the order in which the probability is computed: a decision is made, and then a probability associated with the decision is asserted. This is incorrect. Most people would never make the mistake of believing the second incorrect statement. But an amazing number of people believe the first. The appeal of this illegal logic is strong; a clear-cut decision to reject the null hypothesis has just been made, and now you want to know the probability that the decision is correct. So, the loosely defined "confidence" in the decision is treated as if it's a probability. It happens all the time, and it's just plain wrong.

If you are comfortable with the preceding discussion and feel you solidly understand hypothesis testing, you may safely skip the remainder of this section. But if you remain confused, we will dig deeper still. The main thrust so far has been to explore the use of hypothesis tests to compute confidence figures for classification decisions and to vigorously proclaim that these confidences must not be interpreted as probabilities. But hypothesis testing does allow some probabilities to be asserted for classification decisions. The remainder of this section discusses the legality and illegality of such assertions.

We begin with a presentation of the official version of hypothesis testing as a means of indirectly making a claim. In practice, this may involve testing a weight-loss drug to decide if it is effective, or testing the profitability of a proposed trading system to see if it can make money. In the context of this chapter, hypothesis testing involves using a model's numeric prediction to make a claim about membership or nonmembership in a class. We recognize that random variation precludes all such claims from being definitive. Probabilities are involved. Our goal is to make as many probability claims as possible and to revert to weaker claims when necessary. Here is the official route of a hypothesis test:

1) We define a *null hypothesis*, which is often a "status quo" situation, though it need not be. We also define an *alternative hypothesis*, which is often a special situation on which we focus. In pharmaceutical research, the null hypothesis may be that a new drug is worthless, and the alternative hypothesis is that the drug works. In market trading the null hypothesis may be that there is no compelling reason to open a position now, while the alternative is that now is a good time to do so. Or the null hypothesis may be that the set of trades in a historical sample of a trading rule's activity represents the results of a worthless rule, while the alternative hypothesis is that these trades arose from a good rule. Often, the null hypothesis is simple and completely specified, implying that the distribution of test statistics (model predictions in the current context) is relatively easy to obtain. In contrast, the alternative hypothesis is often incompletely specified, making the alternative distribution difficult or impossible to obtain.

2) We agree that the null and alternative hypotheses are mutually exclusive and exhaustive, meaning that they cannot be true simultaneously and that together they cover all possible situations.

3) We choose in advance a small significance level usually called *alpha*. This will be, by definition, the probability that, IF the null hypothesis is true (emphasis on IF), we will erroneously reject it. Since we know the distribution of our statistic (the model's prediction here) under the null hypothesis, choice of alpha implies a threshold for our statistic. So we agree to reject the null hypothesis in favor of the alternative if the probability that we would observe a test statistic as extreme as what we actually observed is less than or equal to alpha. Equivalently, we reject if our test statistic is at or beyond the threshold. Note that one minus alpha is the probability that IF the null hypothesis is true, we will correctly fail to reject it. We saw how to estimate this probability on page 75. Also note that many people do not choose alpha in advance. Instead, they act as if the obtained probability had been exactly chosen in advance. The implications of this minor form of cheating are beyond the scope of this text.

So, for example, suppose we are willing to live with erroneously rejecting five percent of all null-hypothesis cases. In other words, we are willing to erroneously classify into the alternative-hypothesis class 5 percent of all cases that truly belong to the null-hypothesis class. If our confidence set consists of 100 cases (too small!), we would choose the fifth-largest prediction in this set of null-hypothesis cases. This would be our threshold for future classification decisions. In the language of this chapter, we classify a case as a member of the alternative class if and only if we have at least a 95 percent confidence in the decision. Subject to the sources of random variability already discussed, this would result in us correctly classifying about 95 percent of all future cases *that truly belong to the null-hypothesis class*. Only 5 percent of the members of this class would be incorrectly classified.

However, it must be strongly emphasized that we cannot make the converse assertion. Suppose a case elicits a prediction that meets the 95 percent confidence threshold. After assigning this case to the alternative-hypothesis class, we *cannot* assert that there is a 95 percent probability that we were correct to do so. That 100−5=95 percent probability of making a correct decision is the probability that we will correctly classify members of the null-hypothesis class. We know *nothing* about the behavior of the model under the alternative hypothesis, and hence we cannot make any general assertions about our ability to classify arbitrary cases. Our probability assertions are limited to members of the null-hypothesis class. If we do have knowledge of the model's behavior under the alternative hypothesis, we should probably be using Bayes' method as our primary decision maker. In the absence of this knowledge, the best we can do is say that we have a 95 percent *confidence* in our decision, a deliberately vague claim.

At the risk of being pedantic, let's make one last attempt to clarify this frequently confusing issue by means of a final example of hypothesis testing. Suppose we work in the quality-control department of a manufacturing facility. A particular gearbox is crucial to operation of the factory. Periodic maintenance is needed to keep this gearbox from failing catastrophically. This maintenance requires that the entire assembly line be shut down, so we want to put it off for as long as possible. Unexpected failure is troubling but not a disaster. A more serious problem is unneeded shutdowns for premature maintenance. Until you were hired, management simply let it run until the gearbox failed and then did the repairs. But you were hired to try to improve the situation. After some experimenting, you discover that the gearbox begins to emit abnormal noises when failure is immanent. So, you rig a sensor to the gearbox and feed its signal to a model whose output is trained to produce a large value when any of several unusual sounds appear. Normal sounds cause the model to make a small prediction.

Two things are immediately apparent. First, you will have no problem amassing a huge collection of model outputs for normal conditions, while it will be difficult to collect more than a few examples of the rare and varied abnormal noises. You barely have enough exemplars for your training and test sets. You certainly don't have enough for any kind of confidence set for the abnormal state.

Second, you see that failure to detect an impending breakdown is not a terribly serious problem. After all, this error simply brings the factory back to how it has operated all along: waiting for failure and then fixing the gearbox. But the opposite error, predicting failure too soon and hence causing premature shutdown of the factory, will be frowned upon by management. You wisely decide to be conservative in your decision.

With this in mind, you collect 1,000 samples of normal sounds, present each sample to your trained model, sort the model's predictions, and choose the tenth-largest prediction as your decision threshold. This means that you will call for shutdown and maintenance when the model provides an output so large that only 1 percent of normal samples exceed it. In the language of this chapter, you will flag immanent failure when your confidence in this decision is at least 99 percent. If you perform the test every morning, then one out of every 100 mornings in which the gearbox is normal you will incorrectly flag impending failure.

So far this example has been straightforward. Here's the monkey wrench. Suppose you are in charge of two such gearboxes. The new foreman in charge of one of them is a real cheapskate. When he replaces parts, he uses junk that typically fails in as few as ten days or so. The foreman in charge of the other gearbox goes to the other extreme. He just discovered a new titanium gear that is guaranteed for 50 years, and he put it in at the last repair. You, of course, haven't been made aware of any of this. The normal sound of both gearboxes is still the same. In fact, the failure sounds will also be about the same. Only the life of the gears is different. Very, very different.

Consider the probability of a correct decision when your model produces a prediction sufficiently high to reach the 99 percent confidence threshold and you therefore choose to sound the alarm. In the case of the cheapskate's gearbox, you will virtually always be correct to do so. If your model incorrectly flags impending failure only once out of every 100 or so good-condition days, but the gearbox fails every ten days, those (frequent!) times in which you flag failure will almost always be correct decisions. In fact, the probability of this being a correct decision will most likely exceed the confidence level of 99 percent! But the gearbox with the titanium gears will, for all practical purposes, never fail. So when your model unavoidably flags impending failure, it will always be wrong. You will have an essentially zero probability of having made the correct decision, despite the fact that you used the same correctly trained model on both gearboxes.

This example ideally makes clear the fact that a confidence figure for rejecting a null hypothesis cannot be interpreted as being the probability of having made a correct decision. It can be wildly inaccurate because of its dependence on the distribution of the prediction under the alternative hypothesis as well as prior probabilities. It's the widely disparate prior probabilities that caused problems in this example. Be warned.

Confidence Intervals for Future Performance

Once we have obtained a classification model that performs well, we will almost surely be interested in estimating how well it will do in the future. To estimate this future performance, we need a large collection of cases that satisfy the usual conditions for a confidence set. These conditions were set forth on page 73. Of course, all classes must be represented in the proportion in which they are expected in the future.

On page 35 we saw how empirical quantiles could be used as confidence intervals for future errors in numerical prediction. The situation is often more complex for classification, because performance for individual cases is often more granular than performance for numerical predictions. It may occasionally be that class decisions have continuous or nearly continuous results, such as returns for a market trading system. But more often, classification performance is binary: the predicted class is correct or incorrect. There may be numerical cost (or reward) values associated with decisions, but it is often the case that these costs are fixed, with one cost for an error and another for a correct decision. It makes no sense to talk about confidence intervals for individual case errors in this situation.

The granularity problem can be somewhat solved by grouping cases into sets whose mean performance (or other measure) is computed. For example, suppose we have a manufacturing process that produces two different grades of a product, with the grade of each sample being more or less random, depending on humidity, the mood of the control operator, and so forth. So there is a sensor monitoring the end of the line, and this sensor drives a model that grades each emerging sample. A correct grade incurs no cost. Misclassifying a truly Grade A product as Grade B incurs one cost, while the opposite error incurs another cost. There are three possible values associated with each test of a sample of the product: zero for a correct classification, and two different error values. It obviously makes no sense to talk about a confidence interval for a single decision by the grading model.

However, we could look at sets of ten samples and compute the mean cost incurred by the model per set of ten. If necessary, we could even bundle them in groups of 100 or more. This provides many more possible cost figures. Now we could use the methods presented earlier to compute confidence bounds for the mean cost per set of samples.

In particular, when we have such a collection of group performances, the technique of using empirical quantiles as confidence intervals, discussed on page 35, may be used to find confidence bounds for future costs or gains. Pessimistic bounds on the confidence bounds may also be computed. Even tolerance intervals may be computed, although they are less commonly useful. The main use of empirical quantiles in this context is to compute an approximate lower confidence bound for future gains or an upper bound for future loss.

One important use for empirical confidence intervals for future performance is in regard to ongoing verification of the model's correct operation. Suppose, for example, that we have determined that there is only a 10 percent probability that future gains will be less than –1000 (presumably a serious loss). Now suppose a string of actual losses worse than –1000 occurs. We should immediately suspect that the model is deteriorating. This is especially important when one is operating on a nonstationary time series.

CHAPTER 3

Resampling for Assessing Parameter Estimates

- Parameter Estimates
- Bias and Variance of Estimators
- Bootstrap Estimation of Bias and Variance
- Confidence Intervals
- Hypothesis Tests for Parameter Values
- Jackknife Estimates of Bias and Variance
- Bootstrapping Dependent Data

We often collect a random sample of cases from a population, let this sample interact with a model in some way, and then examine with interest a number (or several numbers) that result from the interaction. For example, we may use the random sample to train a model and then examine one or more of the model's learned parameters. More often, we apply a previously trained model to the cases and compute a measure of the model's performance so that we may judge the model's worth and perhaps even extrapolate its future performance. On page 35 we saw an excellent method for computing confidence bounds for individual prediction errors. On page 99 we saw that this same method could be used to compute confidence bounds for clusters of future gains or costs obtained from a classification scheme. The subject of this chapter is somewhat different, though nevertheless related. Here, we are not concerned with performance on individual or small groups of future cases. Rather, we compute a single measure that describes some aspect of the model, and then we judge the quality of this measurement.

© Timothy Masters 2018
T. Masters, *Assessing and Improving Prediction and Classification*,
https://doi.org/10.1007/978-1-4842-3336-8_3

Bias and Variance of Statistical Estimators

Let us suppose that we are interested in some population from which we are able to draw random samples. When we speak of random variables in the abstract sense, the capital letter X is traditionally employed. The random variable X in our population of interest has a probability distribution that we will call F, and the associated density function will be called f. X is usually a scalar in this chapter, although it may sometimes be a vector. The distinction will be clear from the context. When we speak of observed values of X, the lowercase x is traditionally used. If several cases are sampled, it may be necessary to distinguish the cases by means of a subscript. Thus, we may sample n cases and refer to them as $(x_1, x_2, ..., x_n)$.

It is useful to be able to talk about the average value of a random variable. Because the average is, in a rough sense, the usual value that we would expect to observe, statisticians employ the special term *expected value* or *expectation* of X. If the distribution of X is discrete, the expected value of X is simply the sum of all possible values weighted by the probability of each value. This is expressed in Equation (3.1). If the distribution is continuous, the integral shown in Equation (3.2) must be used.

$$E_F[X] = \sum_x xf(x) \tag{3.1}$$

$$E_F[X] = \int_{-\infty}^{\infty} xf(x)\,dx \tag{3.2}$$

In this chapter, we are interested in some parameter of F. Let us call this parameter θ, or sometimes $\theta(F)$ when we want to especially emphasize the fact that θ concerns F. In practice, $\theta(F)$ is often the mean or median of F, or perhaps its interquartile range (75'th percentile minus 25'th percentile). It may even be some complicated beast like the difference between two percentiles divided by the difference between two other percentiles. Our goal is to use a dataset sampled from the population to compute an estimate of θ as well as an estimate of the quality of this estimate.

Here are some examples of parameters in which we might be interested. Note that the second and third of these examples involve collecting a sample and then training a model and observing some aspect of the resulting model. The others start with a trained model and collect a sample to test the fixed model's behavior. Both are legitimate, although the latter is more common.

- The mean of the squared error (difference between true and predicted values) of a trained model

- The value of a particular regression coefficient in a model trained on a sample from the population

- The mean absolute value of a neural network's input weights for a particular predictor

- The probability that the value predicted by a trained model will exceed the corresponding true value by at least a specified amount

- The 90'th percentile of the absolute error of a trained model

- The correlation between an observed value of a variable that may or may not be a predictor and the concomitant error of the prediction of a trained model

Plug-in Estimators and Empirical Distributions

It is useful to think of our parameter of interest, θ, as a function of the population distribution F. Let us call this function $t(F)$. This notational convention is expressed in Equation (3.3).

$$\theta = t(F) \tag{3.3}$$

If we happened to know F, we would not have a problem to solve. What makes this chapter necessary is that in real applications we never know F. Everything that we know about F is embodied in a sample of n cases that we have randomly collected from F. This sample is denoted $(x_1, x_2, ..., x_n)$. In most situations, each x_i is a scalar. However, there is no reason each case cannot be a vector. For example, if θ is a correlation, F is bivariate.

Once we have in hand a collection of cases from F, this dataset defines an *empirical distribution function* (EDF). The EDF is a discrete distribution having probability $1/n$ for each x_i, $i=1, ..., n$. It is reasonable to assume that the EDF, which we will call $\hat{F}$, is to some useful degree representative of the underlying distribution F. Naturally, the degree to which $\hat{F}$ can be counted on to approximate F depends on n, with larger samples leading to better approximations.

Since we are willing to assume that the EDF approximates F, it seems that a good way to estimate θ, which is defined as $t(F)$, would be to apply the function t to the EDF, $\hat{F}$. This *plug-in estimator*, so named because we plug the EDF into the definition of θ, will be called $\hat{\theta}$. This is expressed in Equation (3.4).

$$\hat{\theta} = t(\hat{F}) \tag{3.4}$$

Most plug-in estimators do an excellent (and often optimal) job of approximating the true value of θ. This is especially true when we know nothing about F except what we can glean from the collected sample. Such will often be the case. The remainder of this chapter will make extensive use of plug-in estimators and the EDF. Nonetheless, it should be emphasized that the results of this chapter generally do not *depend* on $\hat{\theta}$ being a plug-in estimator. It is just that plug-in estimators are almost always excellent (or even optimal) as well as relatively easy to compute. This makes them popular.

Bias of an Estimator

When we collect a sample and use the empirical distribution function to compute $\hat{\theta}$ as an estimator of θ, we know that only with the wildest good luck will the estimate be exactly correct. Sometimes it will be too small, and sometimes it will be too large, all due to random sampling error. But what we would really like is if, on average, $\hat{\theta}$ were centered around the true value, θ. We would probably be dismayed if our computed $\hat{\theta}$ were consistently too large or too small. When our estimator consistently tends to be in error in one direction, it is said to be *biased*. Note, by the way, that the concept of bias has nothing to do with whether $\hat{\theta}$ is a plug-in estimator as described in the previous section, or it is some other completely arbitrary function of the empirical distribution function. Either type of estimator can be biased or unbiased.

To emphasize the point that a plug-in estimator is not mandatory for this discussion, let us define our estimator as being computed by some function s rather than t. This is expressed in Equation (3.5). Compare this to Equation (3.3), which is the official definition of our parameter. Remember that in a great many situations, s and t are exactly the same function because we wisely choose to use the plug-in estimator. But this is not universal. For example, we might want to sometimes use the *median* of the sample to estimate the mean of the population. In this example, t would be the mean and s the median.

$$\hat{\theta} = s(\hat{F}) \tag{3.5}$$

We can now define the bias of our estimator $\hat{\theta}$. The bias is the difference between $\hat{\theta}$'s expected value across all possible samples from the population and the true value. This is expressed in Equation (3.6).

$$\text{bias} = E_F\left[\hat{\theta}\right] - \theta = E_F\left[s(\hat{F})\right] - t(F) \tag{3.6}$$

In a few special cases, the bias of an estimator can be derived mathematically. But most of the time it cannot. Later in this chapter we will show a bootstrap method for approximating the bias of a statistical estimator. What does this do for us? The obvious answer is that if we know the bias, we can subtract it from our computed $\hat{\theta}$ to presumably create an estimate of θ that is closer to the correct value. The problem with this action is that very often the bootstrap estimate of the bias has dangerously high random variation. The act of subtracting this value from $\hat{\theta}$ has the effect of potentially introducing excessive error itself! Nonetheless, at least looking at a bias estimate is always useful, even if for no more reason than to have one more piece of information in our possession.

Variance of an Estimator

Bias is not generally the most serious problem we have when computing an estimate of a parameter. It can sometimes be very serious. But most often the worst problem is that bad luck in selecting the sample from the population provides an estimate of θ that is disturbingly far from the correct value. Contrast this problem with the problem of bias, in which an estimator tends to be *consistently* greater than or less than the correct value. The *variance* component of estimation error is random variation of $\hat{\theta}$ above or below its mean. This is expressed in Equation (3.7).

$$\text{variance} = E_F\left[\left(\hat{\theta} - E_F\left[\hat{\theta}\right]\right)^2\right] = E_F\left[\left(s(\hat{F}) - E_F\left[s(\hat{F})\right]\right)^2\right] \tag{3.7}$$

Let us examine this equation. It says that the variance of $\hat{\theta}$ is the expected value (under the population distribution F) of the squared difference between the computed value and the expected value of $\hat{\theta}$. For clarity, this equation also reminds us that $\hat{\theta}$ is defined by the function s acting on the empirical distribution. It is vital to note that the variance is around the mean of $\hat{\theta}$, not around θ. It will sometimes be the case that the estimator is unbiased, meaning that the mean of $\hat{\theta}$ is, in fact, equal to θ. But in the most general case, the two are not equal.

In summary, we see that the error in $\hat{\theta}$ when used to estimate θ can come from two completely different and independent sources. One source is the bias, in which $\hat{\theta}$ has a consistent tendency to be either greater than or less than $\hat{\theta}$. The other source is because the randomness inherent in the selection of the dataset causes $\hat{\theta}$ to vary around its mean value, whatever that mean value happens to be. In most practical situations, the latter significantly exceeds the former, making it the more troublesome source of error. This is why we are usually more interested in being able to approximate the variance of an estimator rather than its bias. However, methods for approximating both will be shown.

Bootstrap Estimation of Bias and Variance

The bootstrap algorithm allows us to compute estimates of the bias and the variance of any parameter estimate for any distribution. Understand that the foundation of the bootstrap is largely heuristic. There are no guarantees of good performance. It is possible to find some parameters of some distributions that respond very poorly to bootstrapping. But years of practical experience indicate that in the majority of practical applications, the bootstrap does a good job of estimating variance, and at least a fair job of estimating bias. When it's all you've got, you take it. Most of the time, it's all you've got. And what's really interesting is that in those rare situations in which explicit computation of bias and variance is possible, the bootstrap almost always comes respectably close to the mathematically correct estimates. This is a powerful argument in favor of routinely employing the bootstrap.

The idea behind the bootstrap is that the empirical distribution function implied by the random sample can serve as a proxy for the unknown true distribution. Inherent in this idea is the assumption that the cases comprising the sample are independent and identically distributed. Some applications violate this assumption, and use of the bootstrap in this case should be avoided. For example, suppose the cases in the sample are from a time series. They may exhibit serial correlation in that the value of one sample impacts the likely values of subsequent samples. Also, it is crucial that the sequence in which the cases were collected be irrelevant. The empirical distribution function must embody all knowledge contained in the collection of cases. If time order is of any importance, beware of using the ordinary bootstrap. It may not be valid.

Sometimes we find ourselves in a gray area regarding these assumptions. For example, suppose our data collection consists of individual profits obtained from a series of financial trades that occurred over a long time period. Some people would fear that serial correlation is possible, with changing market conditions being reflected in nonrandom strings of wins and losses. The ordinary bootstrap would not be valid in this situation. Other people might be willing to assume that each trade is an independent entity, allowing use of the bootstrap. Make the best judgment call that you can, perhaps with the aid of statistical tests like the runs test. On page 151 we will explore some variations on the ordinary bootstrap that can help when the data may contain serial dependencies.

Once we are comfortable assuming that the empirical distribution function fairly represents the population from which the sample was drawn, we are ready to continue. We compute $\hat{\theta}$ from the sample as our estimate of θ. In doing so, we accept the unavoidable fact that the nature of F and the nature of the definition of $\hat{\theta}=s(\hat{F})$ result in $\hat{\theta}$ being in error at least in regard to having variance around its mean, and possibly also in regard to being biased. So, how do we assess the type and degree of impact F and s have on $\hat{\theta}$? The answer is straightforward: We treat $\hat{F}$ as a population distribution, collect a sample from it, and compute a parameter estimate from this sample. Then we do it again and again. It is not unreasonable to believe that the effect F and s have on $\hat{\theta}$ will be reflected in the parameter computations from these repeated samplings from $\hat{F}$. If we repeat the samplings from $\hat{F}$ a great many times, we can study what happens and extrapolate the effect to our full-sample parameter estimate.

Some new notation is needed in order to rigorously present the bootstrap algorithms. Recall that we collected n cases, and these cases are denoted $(x_1, x_2, ..., x_n)$. This collection defines the EDF $\hat{F}$ as having equal probability at each sample point. Thus, a random sample, called a *bootstrap sample*, may be drawn from $\hat{F}$ by randomly selecting n cases from $(x_1, x_2, ..., x_n)$ with replacement. In all likelihood, some of the n original cases will appear more than once in the bootstrap sample, while others will not appear at all. Let us denote this bootstrap sample $(x_1{}^\star, x_2{}^\star, ..., x_n{}^\star)$. Just as the original sample defined an EDF $\hat{F}$, this bootstrap sample defines an EDF that we shall call $\hat{F}^\star$. We can apply the definition s of our parameter estimator to this EDF to compute the quantity shown in Equation (3.8).

$$\hat{\theta}^\star = s(\hat{F}^\star) \tag{3.8}$$

We will be performing this operation of randomly selecting a bootstrap sample and computing the parameter estimator from this sample many times. Let B be the number of such bootstrap replications, where B is typically anywhere from 500 to 5000 or so, depending on time constraints and accuracy requirements. The conventional nomenclature is to follow the $\star$ with the number of the replication. Thus, the EDF of bootstrap replication i is $\hat{F}^{\star i}$, and the result of applying s to this EDF is $\hat{\theta}^{\star i}$. The mean of $\hat{\theta}^{\star i}$ across all B replications is denoted $\hat{\theta}^{\star \bullet}$. This is expressed in Equation (3.9).

$$\hat{\theta}^{\star \bullet} = \frac{1}{B} \sum_{b=1}^{B} \hat{\theta}^{\star b} \tag{3.9}$$

We are now in a position of great power. In the fundamental problem we are weak because we do not know the source population F, and hence we do not know $s(F)$ that would be used to estimate $t(F)$. Moreover, we have only one sample of n cases from F at our disposal. But in this bootstrap scenario we know the source population $\hat{\theta}$ and we are free to take as many sets of n cases from this population as we like. We can study the behavior of s under $\hat{F}$ to our heart's content and use this knowledge to indicate how $\hat{\theta} = s(\hat{F})$ probably behaves under F.

The easiest property to evaluate with the bootstrap is the variance of the function s under $\hat{F}$ (and hence presumably its variance under F, which is what we really want to know). All we need to do is use the ordinary statistical estimate of the variance of $\hat{\theta}^{\star}$. This is expressed in Equation (3.10).

$$\mathrm{var}_{\hat{F}}\left(\hat{\theta}^{\star}\right) \approx \frac{1}{B-1} \sum_{b=1}^{B} \left(\hat{\theta}^{\star b} - \hat{\theta}^{\star \bullet}\right)^2 \tag{3.10}$$

This equation says that the variance of $\hat{\theta}^{\star}$ under sampling from the EDF $\hat{F}$ is estimated by the sum of squared deviations from the mean, divided by one less than the number of trials. Because $\hat{F}$ is a finite discrete distribution, we could actually compute the *exact* value of the variance if we wished and if we had enough computer power. This would be done by enumerating every possible bootstrap sample from $\hat{F}$. However, in the vast majority of practical situations, n is much too large for complete sampling to be possible. And even if it were possible, there would probably be no reason for doing so. This is because as long as B is reasonably large, the error in the estimate given by partial sampling and Equation (3.10) is almost certainly smaller than the error due to $\hat{F}$ being a less-than-perfect representation of F. Thus, in practice we choose B to be several hundred to several thousand and use Equation (3.10) as our bootstrap estimate of the variance of $\hat{\theta} = s(\hat{F})$ under F.

Estimating the bias of $\hat{\theta}$ is only slightly more difficult than estimating its variance. Look back at Equation (3.6) on page 105. This equation says that the actual bias, which we do not know but want to estimate, is the expected value under F of our parameter estimate, $\hat{\theta}=s(\hat{F})$, minus the true value of the parameter, $\theta=t(F)$. To use the bootstrap to estimate this quantity, we replace the unknown F with the known $\hat{F}$, and we estimate the expected value using bootstrap replications. The exact but generally impractical bias estimate is given by Equation (3.11), and the practical bootstrap implementation is given by Equation (3.12).

$$\text{bias}_{\hat{F}}\left(\hat{\theta}^{\star}\right)=E_{\hat{F}}\left[s\left(\hat{F}^{\star}\right)\right]-t\left(\hat{F}\right) \tag{3.11}$$

$$\text{bias}_{\hat{F}}\left(\hat{\theta}^{\star}\right)\approx\hat{\theta}^{\star\cdot}-t\left(\hat{F}\right) \tag{3.12}$$

As was the case for the variance, we could (in theory at least, given enough computer power) achieve exact equality in Equation (3.12) by enumerating all possible bootstrap samples rather than just B randomly selected samples when computing $\hat{\theta}^{\star\cdot}$. But there is no point in doing so, because the error due to incomplete bootstrap sampling is nearly always much less than the error due to random selection of the original sample on which $\hat{F}$ is based.

Code for Bias and Variance Estimation

This section presents some C++ code for computing bootstrap estimates of the bias and variance of a parameter estimator in the fully general (s and t different) case. In the next section, we will show how the plug-in case (s and t the same) can be handled in a better way. A listing of the bootstrap subroutine appears now, and an explanation of its operation follows. This code, along with a small demonstration program, can be found in the module BOOT_P_1.CPP on my web site.

```
void boot_bias_var (
   int n,                          // Number of cases in sample
   double *data,                   // The sample
   double (*user_s) (int, double *),    // Compute param being bootstrapped, s
   double (*user_t) (int, double *),    // Compute true param, t
   int nboot,                      // Number of bootstrap replications
```

```
   double *rawstat,              // Raw statistic of sample, theta-hat
   double *bias,                 // Output of bias estimate
   double *var,                  // Output of variance estimate
   double *work,                 // Work area n long
   double *work2                 // Work area nboot long
   )

{
   int i, rep, k;
   double stat, mean, variance, diff;

   mean = 0.0;

   for (rep=0; rep<nboot; rep++) {  // Do all bootstrap reps (b from 1 to B)

      for (i=0; i<n; i++) {          // Generate the bootstrap sample
         k = (int) (unifrand() * n); // Select a case from the sample
         if (k >= n)                 // Should never happen, but be prepared
            k = n - 1;
         work[i] = data[k];          // Put bootstrap sample in work
         }

      stat = user_s (n, work);       // Evaluate estimator for this boot rep
      work2[rep] = stat;             // Enables more accurate variance
      mean += stat;                  // Cumulate theta-hat star dot
      }                              // For all B bootstrap replications

   mean /= nboot;                    // Equation (3.9)
   variance = 0.0;
   for (rep=0; rep<nboot; rep++) {   // Cumulate variance
      diff = work2[rep] - mean;
      variance += diff * diff;
      }

   *rawstat = user_s (n, data);      // This is the final but biased estimate
   *bias = mean - user_t (n, data);  // Equation (3.12)
   *var = variance / (nboot - 1);    // Equation (3.10)
}
```

The outermost loop of this program performs all B (nboot here) bootstrap replications. For each replication, the first step is to randomly select a bootstrap sample. Note the check to make sure that the uniform generator has not produced a value that will place us past the end of the array. This should never happen with a good routine (unifrand() is always less than one), but a little insurance is always nice. Once the sample is in work, the parameter estimate for this bootstrap sample is computed by calling the routine that evaluates the s function. This result is saved in work2 for later use. Also, the sum is cumulated so that the mean $(\hat{\theta}^{\star\cdot})$ can be computed later. Naturally, the variance can be computed without using work2 as scratch storage. However, the formula required to do so can exhibit floating-point problems in some situations and is not recommend for quality applications.

After all bootstrap replications are complete, Equation (3.9) is used to compute $\hat{\theta}^{\star\cdot}$ and Equation (3.10) is used to compute the variance estimate. Also, $s(\hat{F})$ is computed and returned in *rawstat for the user's convenience.

The module BOOT_P_1.CPP on my web site contains a small program to exercise the subroutine just listed. This program will not be listed here because it would waste valuable space and it is fairly straightforward. However, a few words of explanation are in order because the program provides a useful template around which you can base your own tests.

The program performs two separate tests. One test uses the sample mean to estimate the population mean. Note that this is a plug-in situation, better handled with the method shown in the next section. Also, it is rather pointless to use a bootstrap to study the sample mean because its behavior is well known on a theoretical basis. However, this simple test confirms correct behavior. The population for this test is a standard normal distribution. When the program is run with reasonable parameters (such as 100 samples, 500 replications, and 10,000 tries), we will confirm that the mean of the parameter estimates is nearly zero, the variance of the parameter estimates is nearly the reciprocal of the number of samples, the mean bias estimate is nearly zero, and the estimated variance is almost equal to the observed parameter variance. Except for the last observation, these are completely expected results. The fact that the bootstrap estimate of the variance is practically equal to the actual variance is one of those wonderful properties of the bootstrap in that even when known theory makes the bootstrap unnecessary, its performance is usually on par with optimal methods.

The second test is more interesting. It uses the sample median to estimate the population mean. The population in this test is constructed by exponentiating a normal random variable. Thus, it has a long right tail and a short left tail. The function user_s() returns the median, and user_t returns the mean. With 100 samples, 500 replications, and 10,000 tries, the following results should appear:

- The mean of the parameter estimates is just about 1.0. This is expected because the median of F is exactly 1.0. (Remember that the median of the underlying normal distribution is zero and $\exp(0)=-1$.)

- The mean of the bias estimates is about –0.63. We would expect a fairly substantial negative bias because we are using the sample median to estimate the mean of a distribution that is strongly skewed in the positive direction.

- The standard deviation of the bias estimates is about 0.18. This is one of those fairly uncommon situations in which the bias estimate is large compared to its standard deviation, meaning that it would make sense to unbias the parameter estimate. Unfortunately, we don't often have the luxury of knowing this in advance like we do in a contrived test like this one.

- The mean variance estimate is about 0.018, not too far from the actual computed variance of 0.016. While not perfect, this is very good considering that no assumptions were necessary and the underlying distribution was extremely skewed.

The demonstration program itself is of relatively little direct interest. However, it clearly demonstrates some important ideas. Moreover, it serves as a platform from which you can leap. Experimentation is encouraged.

Plug-in Estimators Can Provide Better Bootstraps

So far, we have focused on the general parameter estimation problem, in which $\hat{\theta}=s(\hat{F})$ is used to estimate $\theta=t(F)$, where s and t are allowed to be different functions. The most recent example let s be the median and t be the mean. In many or most practical problems, this generality is not only unnecessary, but it is deleterious. Plug-in estimators (s is the same as t) usually have minimal bias as well as other desirable statistical properties. This section presents another reason for using a plug-in estimator whenever possible: the bootstrap estimate of the bias can be improved over the method of the previous section.

First, let us define what is meant by *improved*. Recall that when we perform a bootstrap to evaluate a property (such as bias) of a parameter estimator, there are two sources of error that impact our results. The first is always unavoidable. It is the fact that $\hat{F}$ comes from a single randomly selected collection of n cases from F. A different collection would almost certainly provide different results. The second source of error is that computational constraints virtually always force us to use just B bootstrap replications, as opposed to methodically selecting every possible bootstrap sample. The method presented in this section reduces the impact of the latter source of error. In other words, for any B, the method presented here will, with high probability, provide an estimate of the bias that is closer to the asymptotic value than the previous algorithm using the same B.

One implication of this fact is that if for some reason we cannot or will not use this improved algorithm, we can still use the old algorithm to obtain a bias estimate that is just as good. We will just have to use a lot more bootstrap replications with the old algorithm. Obviously, if the statistic is slow to compute, we have strong motivation to use this improved algorithm.

The level of this text permits only an intuitive explanation of how the algorithm works. If you're interested in a theoretical derivation, you should refer to [Efron and Tibshirani, 1993]. The idea is this: look back at Equation (3.12) on page 109. This equation defines the bias estimate as $\hat{\theta}^{\star \cdot}$ minus the parameter evaluated at the original sample, $\hat{F}$. Everything would be fine if $\hat{\theta}^{\star \cdot}$ had been computed using every possible bootstrap sample from $\hat{F}$. But only randomly selected B samples were used. Some cases are sampled many times, and others few times. Despite this incomplete sampling, the parameter against which $\hat{\theta}^{\star \cdot}$ is compared is in F fact computed from the entire original sample, $\hat{F}$, with each case taking part exactly once.

It would be fairer if this latter term were based on a perturbation of $\hat{F}$ more representative of the set of bootstrap samples chosen by the B replications. In other words, if some case is highly over-represented in the complete set of random bootstrap samples, this improved algorithm equally over-represents the same case in the "reference" sample. Unfortunately, the theoretical derivation of this algorithm holds water only when $t=s$. But we can live with this, since plug-in estimators are popular.

This new algorithm requires an additional piece of notation. Recall that the i'th bootstrap sample from $\hat{F}$ defined a new EDF that we called $\hat{F}^{\star i}$. We now define $\hat{F}^{\star \cdot}$ to be the mean (in the sense of mixture) of all B bootstrap EDFs. This quantity, which is itself an EDF, is defined as shown in Equation (3.13). In this equation, it should be understood that the equation represents a finite discrete distribution evaluated at each of the original sample points. This should be obvious, but it is mentioned to mollify purists who might balk at this simple and intuitive but not strictly perfect way of writing the definition.

$$\hat{F}^{\star\bullet} = \frac{1}{B} \sum_{b=1}^{B} \hat{F}^{\star i}$$

(3.13)

The new, improved definition of bias is now apparent. It is the equivalent of Equation (3.12), with $\hat{F}^{\star\bullet}$ replacing $\hat{F}$, and the plug-in estimator used for $\hat{\theta}^{\star\bullet}$ This is shown in Equation (3.14), where it is assumed that $\hat{\theta}^{\star\bullet}$ is computed using Equation (3.15).

$$\text{bias}_{\hat{F}}\left(\hat{\theta}^{\star}\right) \approx \hat{\theta}^{\star\bullet} - t\left(\hat{F}^{\star\bullet}\right)$$

(3.14)

$$\hat{\theta}^{\star\bullet} = \frac{1}{B} \sum_{b=1}^{b} t\left(\hat{F}^{\star b}\right)$$

(3.15)

The code for this improved bias estimation algorithm is very similar to that for the original algorithm. The main difference is that the user's subroutine for evaluating t must have the capability of operating directly on an EDF having non-integral counts. The original code only needed to be able to operate on the sample that implied the EDF. But now we not only do not have an actual sample representing $\hat{F}^{\star\bullet}$, but this EDF may not even represent integral counts of cases because it is an average. This should not present any significant difficulties in the majority of applications. Here is a subroutine for using the bootstrap to estimate the bias and variance of a parameter estimator. A discussion follows. This subroutine, along with a test program, may be found on my web site under the name BOOT_P_2.CPP.

```
void boot_bias_var (
    int n,                  // Number of cases in sample
    double *data,           // The sample
    double (*user_s) (int, double *, double *), // Compute param
    int nboot,              // Number of bootstrap replications
    double *rawstat,        // Raw statistic of sample, theta-hat
    double *bias,           // Output of bias estimate
    double *var,            // Output of variance estimate
```

```
   double *work,              // Work area n long
   double *work2,             // Work area nboot long
   double *freq               // Work area n long
   )
{
   int i, rep, k;
   double stat, mean, variance, diff;

   mean = 0.0;

   for (i=0; i<n; i++)
     freq[i] = 0.0;

   for (rep=0; rep<nboot; rep++) {  // Do all bootstrap reps (b from 1 to B)

     for (i=0; i<n; i++) {                 // Generate the bootstrap sample
       k = (int) (unifrand() * n);         // Select a case from the sample
       if (k >= n)                         // Should never happen, but be prepared
         k = n - 1;
       work[i] = data[k];                  // Put bootstrap sample in work
       ++freq[k];                          // Tally for mean frequency
       }

     stat = user_s (n, work, NULL); // Evaluate estimator for this rep
     work2[rep] = stat;             // Enables more accurate variance
     mean += stat;                  // Cumulate theta-hat star dot
     }

   mean /= nboot;
   variance = 0.0;
   for (rep=0; rep<nboot; rep++) {  // Cumulate variance
     diff = work2[rep] - mean;
     variance += diff * diff;
     }

   for (i=0; i<n; i++)              // Convert tally of usage
     freq[i] /= nboot * n;          // To mean frequency of use: Equation (3.13)
```

115

```
memcpy (work, data, n * sizeof(double));      // user_s may reorder, so preserve
*rawstat = user_s (n, data, NULL);            // Final but biased estimate
*bias = mean - user_s (n, work, freq);        // Equation (3.14)
*var = variance / (nboot - 1);                // Equation (3.10)
}
```

This subroutine has much in common with the previous routine. Some differences are in the parameter list. Only one parameter function is supplied because this is limited to plug-in use, where $s=t$. Also, this parameter function has one more calling parameter, the frequency vector. It is set to NULL if the function is to use only the data sample. Example user_s() routines for the sample mean and median appear in the module BOOT_P_2.CPP on my web site.

The bootstrap replication loop is exactly the same as before except that a count is kept of how many times each original sample point is used in a bootstrap sample. Later, this array is divided by nboot * n to convert it to an EDF. Finally, note that the original data array is saved in case the user function user_s() reorders it. It must remain in sync with the frequency array. This was not a concern in the prior example because there was no frequency array that corresponded to the original data. If your user function does not reorder the data, there is no need to preserve the data.

A Model Parameter Example

The prior examples considered a univariate distribution. The most common scenario for bootstrapping single variables is evaluating some aspect of the error of a trained model. We typically train a model and then collect an independent validation set and test the model. To judge the quality of the model, we compute some parameter of the distribution of the prediction errors. In days of old, this was the mean squared error. In these enlightened times, it is more often a custom error measure, perhaps based on quantiles or some measure of real-world performance. Then the bootstrap is used to qualify the performance measure.

In this section we use the bootstrap for an entirely different purpose. Rather than starting with a trained model, the training of the model itself is part of the bootstrap. The parameter we study is directly related to the model itself, as opposed to being a measure of error. For example, we may train a multiple-layer feedforward network and compute the sum of absolute values of all weights leading to a particular predictor as a measure of the importance of the predictor. The example here is much simpler. We attempt to predict a variable y as a linear function of a single predictor x. It is well known that the

116

factor β relating y to x as in Equation (3.16) is optimally estimated by Equation (3.17). We will bootstrap this parameter estimator so that we can compare an estimated value with its standard deviation. If we collected a sample and computed an estimate of, say, $\beta=3.4$, we might be impressed with the predictability of y from x. But if the bootstrap tells us that the standard deviation (square root of variance) of this estimate of β is 4.2, our enthusiasm would correctly wane.

$$y = \beta x + \gamma \tag{3.16}$$

$$\beta = \frac{\sum_i (x_i - \bar{x})(y_i - \bar{y})}{\sum_i (x_i - \bar{x})^2} \tag{3.17}$$

The code for this example is named BOOT_P_3.CPP, and it can be found on my web site. The parameter computation subroutine, which uses Equation (3.17), is listed afterward.

```
double param_beta (int n, double *x, double *y)
{
  int i;
  double xmean, ymean, xdif, ydif, xvar, covar;

  xmean = ymean = 0.0;
  for (i=0; i<n; i++) {
    xmean += x[i];
    ymean += y[i];
    }

  xmean /= n;
  ymean /= n;

  xvar = covar = 0.0;
  for (i=0; i<n; i++) {
    xdif = x[i] - xmean;
    ydif = y[i] - ymean;
    xvar += xdif * xdif;
    covar += xdif * ydif;
    }
```

```
    if (xvar != 0.0)
        return covar / xvar;
    return 0.0;
}
```

This example employs the general version of the bootstrap, even though the fact that the estimator is a plug-in means that the improved version could be used. This is fine because we are far more interested in the variance than the bias, and the variance is computed identically with both methods. Also, this version allows the parameter estimation routine to be simple by avoiding the need to deal with an explicit empirical distribution function. The code for this bivariate bootstrap combined with model training is now given.

```
void boot_bias_var (
    int n,                      // Number of cases in sample
    double *x,                  // Independent variable in sample
    double *y,                  // Dependent variable in sample
    double (*user_t) (int, double *, double *), // Compute parameter
    int nboot,                  // Number of bootstrap replications
    double *rawstat,            // Raw statistic of sample, theta-hat
    double *bias,               // Output of bias estimate
    double *var,                // Output of variance estimate
    double *xwork,              // Work area n long
    double *ywork,              // Work area n long
    double *work2               // Work area nboot long
    )
{
    int i, rep, k;
    double stat, mean, variance, diff;

    mean = 0.0;
    for (rep=0; rep<nboot; rep++) {  // Do all bootstrap reps (b from 1 to B)

        for (i=0; i<n; i++) {             // Generate the bootstrap sample
            k = (int) (unifrand() * n);   // Select a case from the sample
            if (k >= n)                   // Should never happen, but be prepared
                k = n - 1;
```

```
    xwork[i] = x[k];                // Put bootstrap sample in work
    ywork[i] = y[k];
    }

  stat = user_t (n, xwork, ywork);   // Evaluate estimator for this rep
  work2[rep] = stat;                 // Enables more accurate variance
  mean += stat;                      // Cumulate theta-hat star dot
  }

mean /= nboot;
variance = 0.0;
for (rep=0; rep<nboot; rep++) {     // Cumulate variance
  diff = work2[rep] - mean;
  variance += diff * diff;
  }

stat = user_t (n, x, y);            // This is the final but biased estimate
*rawstat = stat;
*bias = mean - stat;
*var = variance / (nboot - 1);
}
```

Little explanation should be needed for this example because it is practically identical to the example on page 109. Each bootstrap sample now consists of the *pair* of variables for the randomly selected case, and the parameter estimation routine is presented with this bivariate sample. More general implementations involving multiple predictors would probably merge all multivariate cases into a single array. The final bootstrap results are computed exactly as already described.

If you execute the demonstration program contained in BOOT_P_3.CPP, the following items will be noted:

- For a reasonably sized sample, say $n=100$, the bootstrap performs excellently. The bias estimate is essentially zero, and the variance estimate is right on target.

- For a small sample such as $n=10$, performance is still very good. The mean bias is essentially zero, and the variance estimate is close. The bootstrap estimate slightly over-estimates the true variance.

- For the ridiculously small sample size of $n=3$, performance falls apart. The bootstrap variance estimate is tremendously larger than the true value, and the bias standard deviation is huge. This is because the β parameter estimator is unstable for small samples, and it is undefined when the sample contains all identical cases, which would happen frequently when bootstrapping a small sample. The moral is that the bootstrap is appropriate only when the behavior of the parameter estimator under $\hat{F}$ is similar to its performance under F. This restriction is clearly violated when the sample size is tiny.

Confidence Intervals

Knowledge of the approximate variance of a parameter estimate is useful in that it tells us the degree to which we can rely on the accuracy of the estimate. As a rough rule, we can compute the standard error of the estimate as the square root of the variance and then double it to produce an approximate confidence interval. We can often be reasonably sure that the true value of the parameter is within the interval defined by the estimate plus or minus twice the standard error. This method of computing confidence intervals is quick, simple, and generally safe. However, it has several problems. For one thing, this interval is, by definition, symmetric. The upper and lower bounds are both defined by the same distance from the parameter estimate. If the parameter estimate's distribution is skewed, a symmetric confidence interval is inappropriate. Also, the distribution may have unusually light or heavy tails, in which case two standard deviations may be excessively conservative or anti-conservative. We can do better.

First, the meaning of the term *confidence interval* must be made clear. Recall that a population having unknown distribution function F has supplied us with a sample of n (possibly multivariate) cases. We are interested in estimating some unknown parameter of F, $\theta=t(F)$. We use the sample to compute an estimate of this parameter. This is the point at which many people commit a blunder that is serious in terms of strict theory but tolerable in practice. Nevertheless, understanding of this common blunder is vital to understanding confidence interval computation.

The blunder is to act as if θ is random and hence put forth an interval in which this unknown and random value of θ probably lies. For example, we may compute $\hat{\theta}=6.0$ and assert that there is a reasonable probability that θ lies within the interval (4.0, 8.0).

Strictly speaking, this chain of thought is incorrect. The problem is that θ *is not random*. It is unknown, but it is nevertheless fixed. We do not know what it really is, but it most certainly does have a value. We must rephrase the assertion.

The key is that the sample we collected from *F* is the random component, not θ. Perhaps some quirk in the random sampling caused our computed $\hat{\theta}$ to be somewhat less than θ, or perhaps greater. Certainly, large errors are less likely than small errors. So the correct question is this: Suppose θ truly equals some specified value, a presumed confidence bound. What is the probability that a sampling quirk would have caused our computed $\hat{\theta}$ to be as unusually small (or large) as we just observed? If this probability is satisfactorily large, then the specified θ is, by definition, within the confidence region.

For example, suppose we want to find a confidence interval (a lower and an upper bound) such that there is a 90 percent chance that the true θ is enclosed by this interval. Suppose also (as is the usual case) that we want to split in half the probability of θ being outside the interval. In other words, we want there to be a 5 percent chance that θ exceeds the upper limit, and the same chance that θ is below the lower limit. Assume that we collect a sample and compute, say, $\hat{\theta}$=4.5. Let us say that if θ truly equals 5.2, there is only a 5 percent chance that our sample would have been so unusual as to produce $\hat{\theta}$ as low as the 4.5 that we observed (or lower). In other words, if θ were greater than 5.2, there would be an even less than 5 percent chance of such an erroneous $\hat{\theta}$. And if θ were less than 5.2, there would be more than a 5 percent chance of $\hat{\theta}$ being this bad. So, θ=5.2 is the magic point at which the chance of $\hat{\theta}$ being observed as 4.5 or less exactly equals 5 percent. Then we say that 5.2 is the upper limit of this confidence interval. Converse logic could be applied to find the lower limit of the confidence interval. As long as we have a means of computing these probabilities, it is easy to compute confidence intervals. Sadly, we often do not have this luxury.

Suppose for the moment that we are in the enviable position taught in most beginning statistics classes: we have a plug-in estimator $\hat{\theta}$ that we know to be normally distributed with mean θ. Let $\hat{\sigma}$ be its estimated standard error, and assume that the sample is large enough that $\hat{\sigma}$ provides a decent estimate of the true standard error of $\hat{\theta}$. Following established convention, we use the symbol $\Phi(z)$ to represent the normal cumulative distribution function (CDF), and we use $\Phi^{-1}(\alpha)$ for its inverse. By definition, our estimator $\hat{\theta}$ will be less than or equal to $\theta+\hat{\sigma}\Phi^{-1}(\alpha)$ with probability α. As in the logic of the previous paragraph, $\hat{\theta}-\hat{\sigma}\Phi^{-1}(\alpha)$ is the upper endpoint of a 1–2α confidence interval for θ. Similarly, $\hat{\theta}-\hat{\sigma}\Phi^{-1}(1-\alpha)$ is the lower endpoint of the interval.

Since the normal distribution is symmetric, $\Phi^{-1}(\alpha) = -\Phi^{-1}(1-\alpha)$. Making use of this fact means that we can write the confidence interval as shown in Equation (3.18). The reason for this unconventional representation may not be apparent now, but it will be soon.

$$\left(\theta_{\mathrm{lo}}, \theta_{\mathrm{hi}}\right) = \left(\hat{\theta} + \hat{\sigma}\,\Phi^{-1}(\alpha), \hat{\theta} + \hat{\sigma}\,\Phi^{-1}(1-\alpha)\right) \tag{3.18}$$

Continue to assume that $\hat{\theta}$ has a normal distribution when it is computed from samples drawn from the parent population F. Also assume, as we must do in a bootstrap scenario, that the sample of n cases in our possession is reasonably representative of the parent population F. If we draw a bootstrap sample from the original sample and compute the plug-in estimator $\hat{\theta}^{\star}$, this quantity will have a distribution that is approximately normal, with approximate mean $\hat{\theta}$ and approximate standard deviation $\hat{\sigma}$. (As always in bootstrapping, this assumption depends on the degree to which the original sample represents the parent distribution.) This implies that $\hat{\theta}^{\star}$ will be less than $\hat{\theta} + \hat{\sigma}\,\Phi^{-1}(\alpha)$ with probability α. With this in mind and looking back at Equation (3.18), we observe something fascinating: the endpoints of confidence intervals for θ are (approximately) given by the cumulative distribution function of $\hat{\theta}^{\star}$. We can simply draw many bootstrap replications, compute the estimator for each, and use the observed quantiles of this distribution to define whatever confidence intervals we want.

The assumption of unbiased normality of $\hat{\theta}$ helps us to assume that $\hat{\theta}^{\star}$ is reasonably normal and unbiased. We also assume that the standard errors of $\hat{\theta}$ and $\hat{\theta}^{\star}$ are essentially equal. If we are willing to assume these conditions, we can make the following statement:

> If $\hat{\theta}$ follows a normal distribution with fixed variance and having expectation θ under F, then we can compute an approximate $1-2\alpha$ confidence interval for θ as the α and $1-\alpha$ empirical quantiles of $\hat{\theta}^{\star}$ under $\hat{F}$.

On first glance, this statement seems less than totally impressive. Especially, the assumption of normality is something that we are not willing to make, at least not in the context of this chapter. The whole point of the bootstrap is avoidance of such confining assumptions. We will now see how one more observation lets us generalize this technique far beyond normality.

We begin with an interesting scenario. Imagine that two experimenters, whom we will call John and Mary, are performing exactly the same experiment. They are using the same data, fitting the same model, and getting the same results. They are estimating and studying essentially the same parameter, $\theta = t(F)$, with one small difference: John's parameter is equal to ten to the power of Mary's parameter. In other words, any time they compute a parameter estimate, their estimates are related as shown in Equation (3.19).

$$\hat{\theta}_{\text{John}} = 10^{\hat{\theta}_{\text{Mary}}} \qquad\qquad (3.19)$$

Suppose $\hat{\theta}_{\text{Mary}}$ has an unbiased normal distribution with fixed variance (independent of θ). According to the previous discussion, she is entitled to perform bootstrap replications and use the α and $1-\alpha$ empirical quantiles of her $\hat{\theta}^{\star}$ as lower and upper bounds for a $1-2\alpha$ confidence interval for θ_{Mary}. This means that there is an approximately α probability that the true value of θ_{Mary} is less than her lower bound, and similarly that there is an approximately α probability that the true value of θ_{Mary} is greater than her upper bound.

What about poor John? His parameter is clearly not normally distributed. In fact, because of the exponentiation, it probably has a huge right tail and almost no left tail at all. But wait. If John and Mary are really performing exactly the same experiments, each of John's bootstrap replications is related to each of Mary's by this same ten-to-the-power rule. In particular, his ultimate α and $1-\alpha$ empirical quantiles will be ten to the power of her quantiles. Moreover, if there is probability α that the true value of θ_{Mary} is less than her computed lower bound, there must also be probability α that the true value of θ_{John} is less than ten to the power of Mary's computed lower bound, because these two events are in one-to-one correspondence. Thus, we see that John is just as entitled as Mary to use the α and $1-\alpha$ empirical quantiles of his $\hat{\theta}^{\star}$ as lower and upper bounds for a $1-2\alpha$ confidence interval for θ_{John}.

It should now be clear that this reasoning is not limited to the specific situation of one parameter being related to another by the simple power rule just employed. Any monotonic transformation preserves the ordering that defines empirical quantiles, and such a transformation also results in a one-to-one mapping of events concerning membership in confidence intervals. And as long as we can transform to unbiased normality, with the variance not depending on the parameter value, the empirical quantiles provide good confidence intervals. The key here is that *we don't even need to know the transform*, since we never explicitly perform a transform. It all happens implicitly, behind the scenes.

It gets even better: we don't even need normality. This discussion started with a normal distribution because you are probably comfortable with it, and convenient standard notation exists. But all we really need is symmetry around the parameter so that we can do the endpoint swapping shown in Equation (3.18) on page 122. Of course, we also need the shape of the parameter distribution under F to remain essentially unchanged under $\hat{F}$ so that the confidence interval has the correct width. With these things in mind, we can make the following immensely powerful statement concerning an algorithm called the *bootstrap percentile method* or the *bootstrap quantile method*:

> Let $\hat{\theta}$ be an estimator for θ. Suppose there exists a monotonic transformation h such that $h(\hat{\theta})-h(\theta)$ and $h(\hat{\theta}^{\star})-h(\hat{\theta})$ have the same distribution that is symmetric around the origin. Then we can compute a $1-2\alpha$ confidence interval for as the α and $1-\alpha$ empirical quantiles of $\hat{\theta}^{\star}$ under $\hat{F}$.

In the example of John and Mary, the transformation h was the base-ten log function, which we applied to John's bizarre parameter to get Mary's sweet normal parameter. In that example, we stipulated that Mary's parameter had a normal distribution. Now we see that this was not needed. All that we really needed from Mary's parameter was that it be symmetric around its population value (which implies that it is unbiased) and that its shape not vary as the population parameter varies. The normal distribution obviously satisfies these criteria, but a multitude of other common distributions do as well.

To be more specific about computing confidence intervals for θ, we choose the number of bootstrap replications, B, such that $\alpha(B+1)$ is an integer. Sort the resulting estimates, $\hat{\theta}^{\star i}$, in ascending order, so that $\hat{\theta}^{\star 1}$ is the smallest and $\hat{\theta}^{\star B}$ is the largest. Define the empirical quantiles as shown in Equation (3.20). The confidence interval is given by Equation (3.21).

$$\hat{G}^{-1}(p)=\hat{\theta}^{\star p(B+1)} \tag{3.20}$$

$$\text{Prob}\left\{\hat{G}^{-1}(\alpha)\leq\theta\leq\hat{G}^{-1}(1-\alpha)\right\}=1-2\alpha \tag{3.21}$$

Although this method provides a simple and often effective method for computing approximate confidence intervals for a parameter, it is not perfect. It is not terribly difficult to find distributions in which it fails to perform well because a suitable transformation does not exist, because the estimator is not unbiased, or because the variance of the estimator depends on the value of the parameter. Nevertheless, it is a firm foundation on which to build. We will soon see how improvements can be made.

Is the Interval Backward?

Before continuing, it is instructive to approach the confidence interval problem from a different direction and see how we arrive at a completely different solution. Since the method for computing confidence intervals shown in this section is often ineffective in practice, you might want to skim this section. However, you might find it interesting to see how easily intuition can lead us astray.

Recall that we are dealing with an unknown population F, and we are interested in estimating a parameter $\theta = t(F)$. We collect a sample $\hat{F}$ from this population and compute the plug-in estimator $\hat{\theta} = t(\hat{F})$ as our best guess for θ. We want to compute a $1 - 2\alpha$ confidence interval for θ.

It is natural to approach the problem by considering the distribution of the error, $\hat{\theta}_F - \theta$. (The subscript F for $\hat{\theta}$ serves as an optional reminder that $\hat{\theta}$ is a random variable under F in this context.) We want to assume nothing about the distribution of this error. But in the spirit of bootstrapping, we can draw bootstrap samples $\hat{F}^*$ from $\hat{F}$ and compute the plug-in estimator $\hat{\theta}^* = t(\hat{F}^*)$ for each. The distribution (under $\hat{F}$) of $\hat{\theta}^* - \hat{\theta}$ should resemble the distribution (under F) of the error, $\hat{\theta}_F - \theta$.

An example may make this clearer. The underlying population F in this example is e raised to the power of a normally distributed random variable with mean zero and standard deviation four. Naturally, this distribution is strictly positive and strongly skewed to the right. The simplest parameter to study is the mean, so this example will do so. One hundred cases were sampled from this distribution, and the sample mean turned out to be 602.4. One thousand bootstrap samples were taken from this original sample. Figure 3-1 is a histogram of the means of the bootstrap samples. Subtracting the original mean, 602.4, from the bootstrap sample means gives the distribution under $\hat{F}$ of the error, $\hat{\theta}^* - \hat{\theta}$. This histogram is shown in Figure 3-2. We expect this distribution to resemble the unknown distribution under F of $\hat{\theta}_F - \theta$. We'll come back to this example in a moment. But first, let's rigorously define how we will use this information, thinking about this example as we do so.

Let $\hat{H}^{-1}(\alpha)$ denote the α quantile of $\hat{\theta}^* - \hat{\theta}$, which we can estimate by bootstrapping. By assuming that the behavior of this quantity under $\hat{F}$ represents the behavior of $\hat{\theta}_F - \theta$ under F, we may assert a $1 - 2\alpha$ confidence interval for $\hat{\theta} - \theta$ as shown in Equation (3.22).

$$\text{Prob}\left\{\hat{H}^{-1}(\alpha) \leq \hat{\theta} - \theta \leq \hat{H}^{-1}(1 - \alpha)\right\} = 1 - 2\alpha \tag{3.22}$$

We really want a confidence interval for θ, not for the estimation error, so we rearrange Equation (3.22) as shown in Equation (3.23).

$$\text{Prob}\left\{\hat{\theta} - \hat{H}^{-1}(1 - \alpha) \leq \theta \leq \hat{\theta} - \hat{H}^{-1}(\alpha)\right\} = 1 - 2\alpha \qquad (3.23)$$

We could use this equation to compute the confidence intervals. However, the job is simplified if we note that $\hat{H}^{-1}(\alpha) = \hat{G}^{-1}(\alpha) - \hat{\theta}$, where $\hat{G}^{-1}(\alpha)$ was defined in Equation (3.20). This provides a simpler depiction of the confidence interval, as shown in Equation (3.24).

$$\text{Prob}\left\{2\hat{\theta} - \hat{G}^{-1}(1 - \alpha) \leq \theta \leq 2\hat{\theta} - \hat{G}^{-1}(\alpha)\right\} = 1 - 2\alpha \qquad (3.24)$$

Compare this confidence interval to that given by Equation (3.21). Observe that the left and right tails of the empirical distribution have been reversed. If the bootstrapped replications have, say, a long right tail, then Equation (3.21) will produce a confidence interval whose upper endpoint is relatively distant, while Equation (3.24) will give an interval whose left endpoint is distant. Erk!

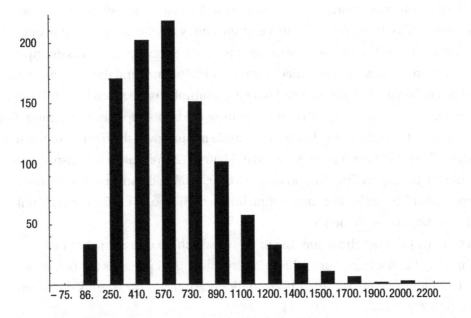

Figure 3-1. *Distribution of bootstrap sample means*

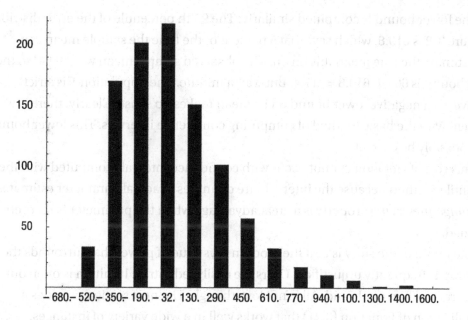

Figure 3-2. *Distribution of errors in bootstrap sample means*

So, which is correct? The answer is neither, in general. Of the two, the percentile method of Equation (3.21) is somewhat more likely to be correct in practice, especially if the parent population is skewed. In fact, despite the perfectly reasonable development of Equation (3.24), called the *basic method* or *pivot method*, this method is often problematic.

The basic method just presented has two problems. One problem is that the confidence intervals it computes are often seriously incorrect. We'll see examples of this later, although the precise reasons for its failure are beyond the scope of this text.

The other problem involves the nature of the endpoints of the interval. Refer back to Figure 3-2. This figure shows the bootstrap distribution of estimation errors $\hat{\theta}^{\star} - \hat{\theta}$ under $\hat{F}$. We assume that it reflects the unknown distribution of the errors $\hat{\theta} - \theta$ under F. To compute a 90 percent basic confidence interval for θ, the mean of the population, we need the 5'th and 95'th percentiles of this error distribution, $\hat{H}^{-1}(\alpha)$. The 5'th percentile of the error is -403.5, which says that the parameter estimate, the sample mean in this example, can be as much as 403.5 below the true mean 5 percent of the time. The mean of the observed sample was 602.4. If 5 percent of the time it can underestimate the true mean by 403.5, we conclude that the upper bound for the true mean is 602.4+403.5=1005.9, also expressed in Equation (3.23). This sounds reasonable.

The lower bound is computed similarly. The 95'th percentile of the error distribution in Figure 3-2 is 619.8, which says that 5 percent of the time the sample mean can overestimate the true mean this much. The observed sample mean was 602.4, so the lower bound is 602.4–619.8 = –17.4. But wait a minute. The population F is strictly positive, so a negative lower bound on its mean makes no sense. Clearly, there is a problem with the basic method of computing confidence intervals. This lower bound can't possibly be correct.

This sort of violation cannot occur with confidence intervals computed with the percentile method because the intervals are quantiles of actual parameter estimates. This *range-preserving* property is a great advantage when the parameter is theoretically bounded.

The moral of the story is that the modern mystique of power that surrounds the bootstrap is frequently unjustified. Users are easily led astray. Intuition is often our enemy rather than our friend. All is not lost, though. The next section provides a generalization of Equation (3.21) that works well in a wide variety of instances.

Improving the Percentile Method

In this section we explore how the percentile or quantile method previously shown can be improved so that it provides better confidence intervals in a wide variety of situations. The algorithm described here, called the *bias-corrected-accelerated* (BC_a) method, is broadly applicable and should prove effective in the majority of practical situations.

The percentile method's requirement that a fixed variance, unbiased normalizing transformation exists is fairly broad and covers a wide range of situations. Nevertheless, there are an infinite number of distributions that fail to meet this requirement. Two failure modes are especially common. First, there may exist a transformation that produces a normal (or suitable symmetric) distribution, but the transformed parameter estimator may be biased. In other words, the expectation of the transformed parameter may not be equal to the transformed value of θ. This destroys the justification of the percentile method. The second common failure mode is for the variance of the transformed estimator to depend on the value of θ. This, too, is destructive. The method of this section usually compensates for both of these problems. It must be emphasized that there are certainly other ways in which the assumptions may fail, so the confidence interval algorithm presented here must not be taken as perfect. However, theoretical studies and practical experience indicate that this method has broad applicability and can be trusted to be reasonably correct in a large number of situations.

The specific conditions under which the BC_a method is applicable are a generalization of those suitable for the percentile method. These conditions are most easily stated in terms of the normalizing transformation instead of the transformation to a symmetric distribution. We have already seen that the percentile method gives good confidence intervals when some transformation h exists (though we need not know what it is) such that the transformed parameter $h(\hat{\theta})$ has a normal distribution with mean $h(\theta)$ and fixed standard deviation σ (which we also do not need to know, but which must not depend on θ). The BC_a method provides good confidence intervals when $h(\hat{\theta})$ follows a normal distribution with mean $h(\theta)-z_0\sigma$ and standard deviation $\sigma=1+ah(\theta)$. The constants z_0 and a are estimated from the data. The former is a bias correction; it allows the expectation of $h(\hat{\theta})$ to be shifted away from $h(\theta)$. The latter, called the *acceleration*, allows the standard deviation of $h(\hat{\theta})$ z to vary with θ.

This improved algorithm is fairly easy to implement, but it is extremely complex to derive theoretically. In fact, there aren't even any good intuitive explanations for its operation. Therefore, only the algorithm and code will be shown here. Details can be found in [Efron and Tibshirani, 1993]. But be warned that the theory in that text is not for the faint of heart.

Calculation of the BC_a confidence intervals requires four steps: compute the bias correction, compute the acceleration, modify the quantile points according to these corrections, and extract the quantiles from the sorted bootstrap parameter estimates. These steps are now discussed. In the following, the symbol $\Phi(z)$ represents the normal cumulative distribution function (CDF), and $\Phi^{-1}(p)$ is its inverse.

The bias correction is easy. We count how many of the bootstrapped parameter estimates are less than the estimate for the entire sample. The bias correction term is the inverse normal CDF of the fraction of the replications less than the grand value. This is expressed in Equation (3.25).

$$\hat{z}_0 = \Phi^{-1}\left(\frac{\#\left[\hat{\theta}^{*b} < \hat{\theta} \right]}{B} \right) \tag{3.25}$$

The acceleration is somewhat more complicated to compute. We need to perform a jackknife on the parameter estimator. The data collection consists of n cases. Suppose we remove case i from the collection and compute the parameter using the remaining $n-1$ cases. Let $\hat{\theta}_{(i)}$ designate this quantity. Let $\hat{\theta}_{(.)}$ be the mean of these n jackknifed values, as shown in Equation (3.26). Then the acceleration is given by Equation (3.27).

$$\hat{\theta}_{(.)} = \frac{1}{n}\sum_{i=1}^{n}\hat{\theta}_{(i)} \tag{3.26}$$

$$\hat{a} = \frac{\sum_{i=1}^{n}\left(\hat{\theta}_{(.)} - \hat{\theta}_{(i)}\right)^3}{6\left[\sum_{i=1}^{n}\left(\hat{\theta}_{(.)} - \hat{\theta}_{(i)}\right)^2\right]^{3/2}} \tag{3.27}$$

Once the bias correction and acceleration have been computed, we need to modify the quantile points accordingly. For example, suppose we want a 90 percent confidence interval. The quantile points would be 0.05 and 0.95, assuming that we want to split the probability of failure equally above and below. A modified quantile point, α', is computed from an original α by means of Equation (3.28). This equation is applied to the upper and lower endpoints separately. Note that if the bias correction and acceleration are both zero, $\alpha'=\alpha$.

$$\alpha' = \Phi\left(\hat{z}_0 + \frac{\hat{z}_0 + \Phi^{-1}(\alpha)}{1 - \hat{a}\left(\hat{z}_0 + \Phi^{-1}(\alpha)\right)}\right) \tag{3.28}$$

The final step is to sort the B values of $\hat{\theta}^{\star b}$ into ascending order and select the appropriate pair of values from this array. The usual unbiased method for the lower bound, in which $\alpha'<0.5$, is to select element k (out of 1 through B), where k is α' times $(B+1)$, truncated down to an integer if the product is not an integer. For the upper bound, $\alpha'>0.5$, so the formula for the unbiased index is not valid. Instead, compute k as $1-\alpha'$ times $(B+1)$, truncated down to an integer if the product is not an integer. Element $B+1-k$ is the upper confidence bound.

The file BOOT_P_4.CPP on my web site contains subroutines for computing 90 percent, 80 percent, and 50 percent confidence intervals for bivariate data (like in the example shown on page 116) using the percentile and BC_a methods. The main program in this file demonstrates how the basic (pivot) intervals can be easily derived from the percentile intervals. The BC_a subroutine is listed soon, and an explanation of its operation follows. You should have no trouble modifying this routine for univariate or general multivariate cases.

The main program demonstrates and compares the three algorithms when used to find confidence intervals for the correlation coefficient of a bivariate population. Running this test program using a sample size of $n=10$ cases (small but often realistic!), a large number of bootstrap replications (B), a very large number of trials, and a theoretical correlation of 0.5, gives the results shown in the following table:

	Percentile	BC_a	Basic/Pivot
5% lower tail	10.52	6.31	20.73
5% upper tail	4.39	4.50	5.10
90% coverage	85.09	89.19	74.17
10% lower tail	17.38	11.86	24.04
10% upper tail	8.74	9.47	8.50
80% coverage	73.88	78.67	67.46
25% lower tail	33.38	27.24	32.37
25% upper tail	21.94	24.88	22.80
50% coverage	44.68	47.88	44.83

This table compares expected results with obtained results. For example, we would expect that the 5 percent lower bound for the 90 percent confidence interval would be violated 5 percent of the time. We see that for the percentile method, it was actually violated 10.52 percent of the time, but only 6.31 percent of the time for the BC_a method. The basic method scored the worst, with violations 20.73 percent of the time. The coverage is the percent of the time the true value lies within the confidence interval, and it is obtained by subtracting the sum of the failures from 100 percent. Note that coverage for the BC_a method is much better than for the percentile method, and the basic method is the grand loser.

Here is a listing of the subroutine for computing confidence intervals using the BC_a method. Since the parameter being bootstrapped is the correlation coefficient of a pair of variables, the dataset is bivariate. Note that only the code for one of the three confidence intervals is listed here because the other two are identical except for the probability constants.

```
void boot_conf_BCa (
   int n,                          // Number of cases in sample
   double *x,                      // One variable in sample
   double *y,                      // Other variable in sample
   double (*user_t) (int, double *, double *), // Compute parameter
   int nboot,                      // Number of bootstrap replications
   double *low5,                   // Output of lower 5% bound
   double *high5,                  // Output of upper 5% bound
   double *low10,                  // Output of lower 10% bound
   double *high10,                 // Output of upper 10% bound
   double *low25,                  // Output of lower 25% bound
   double *high25,                 // Output of upper 25% bound
   double *xwork,                  // Work area n long
   double *ywork,                  // Work area n long
   double *work2                   // Work area nboot long
   )
{
   int i, rep, k, z0_count;
   double param, theta_hat, theta_dot, z0, zlo, zhi, alo, ahi, temp;
   double xtemp, ytemp, xlast, ylast, diff, numer, denom, accel;

   theta_hat = user_t (n, x, y);   // Parameter for full set

   z0_count = 0;                   // Will count for computing z0 later

   for (rep=0; rep<nboot; rep++) { // Do all bootstrap reps (b from 1 to B)
     for (i=0; i<n; i++) {         // Generate the bootstrap sample
       k = (int) (unifrand() * n); // Select a case from the sample
       if (k >= n)                 // Should never happen, but be prepared
         k = n - 1;
       xwork[i] = x[k];            // Put bootstrap sample in work
       ywork[i] = y[k];
       }
```

```
  param = user_t (n, xwork, ywork);   // Param for this bootstrap rep
  work2[rep] = param;                  // Save it for CDF later
  if (param < theta_hat)               // Count how many < full set param
    ++z0_count;                        // For computing z0 later
  }

z0 = inverse_normal_cdf ((double) z0_count / (double) nboot); // Eq (3.25)
```

```
/*
  Do the jackknife for computing accel.
  Borrow xwork for storing jackknifed parameter values.
*/

  xlast = x[n-1];                    // Will replace jackknifed element with this
  ylast = y[n-1];
  theta_dot = 0.0;

  for (i=0; i<n; i++) {              // Jackknife
    xtemp = x[i];                   // Preserve case being temporarily removed
    ytemp = y[i];
    x[i] = xlast;                   // Swap in last case
    y[i] = ylast;
    param = user_t (n-1, x, y);     // Param for this jackknife, θ̂(i)
    theta_dot += param;             // Cumulate mean across jackknife
    xwork[i] = param;               // Save for computing accel later
    x[i] = xtemp;                   // Restore original case
    y[i] = ytemp;
  }

/*
  Compute accel
*/

  theta_dot /= n;                   // Equation (3.26)
  numer = denom = 0.0;
  for (i=0; i<n; i++) {             // Cumulate terms for Equation (3.27)
    diff = theta_dot - xwork[i];
```

```
    xtemp = diff * diff;
    denom += xtemp;
    numer += xtemp * diff;
    }

  denom = sqrt (denom);
  denom = denom * denom * denom;
  accel = numer / (6.0 * denom);        // Equation (3.27)

/*
  Compute the outputs
*/

  qsortd (0, nboot-1, work2);            // Sort ascending

  zlo = inverse_normal_cdf (0.05);
  zhi = inverse_normal_cdf (0.95);
  alo = normal_cdf (z0 + (z0 + zlo) / (1.0 - accel * (z0 + zlo)));  // Equation (3.28)
  ahi = normal_cdf (z0 + (z0 + zhi) / (1.0 - accel * (z0 + zhi)));  // Also do upper bound

  k = (int) (alo * (nboot + 1)) - 1;     // Unbiased quantile estimator for lower bound
  if (k < 0)                             // (Subtract 1 because origin 0 here, 1 in text)
    k = 0;
  *low5 = work2[k];

  k = (int) ((1.0-ahi) * (nboot + 1)) - 1; // Ditto for upper bound
  if (k < 0)
    k = 0;
  *high5 = work2[nboot-1-k];
... Other outputs computed here
}
```

The subroutine on my web site also processes two more confidence intervals. These are omitted in the text because they are redundant. Let us now follow the code just shown.

The first step is to compute $\hat{\theta}$ (theta_hat in the code) for the full dataset. This quantity is needed during the bootstrap loop where we count (in z0_count) how many times the bootstrap sample's parameter is less than the full sample's parameter. This count will ultimately be used to compute z0 using Equation (3.25).

The next step is to perform the jackknife. The last data point (in x[n-1] and y[n-1]) is saved because this point will replace point i for the i'th jackknife evaluation. The value of the parameter is computed and saved in xwork, then the i'th point is replaced. Also, the mean of the parameters is cumulated so that we can compute $\hat{\theta}_{(.)}$ (called theta_dot in the code) using Equation (3.26) later. After the jackknife is complete, Equation (3.27) is used to compute the acceleration.

The hard work is now complete. The final step is to use Equation (3.28) to correct the lower and upper confidence probabilities and select the corresponding elements from the sorted array of bootstrapped parameters. Note that we must protect against the pathological condition of the tail probability being too small relative to the number of bootstrap replications, resulting in the unbiased position being outside the sorted array. Also note that the explanation of the selection in the text assumes that the elements of the array are numbered 1 through B, while the indexing in the C code starts at zero.

Hypothesis Tests for Parameter Values

We saw on page 70 how hypothesis tests could be used to compute confidence in classification decisions. They could also be used to establish thresholds for making classification decisions. This section discusses how to test hypotheses concerning the value of population parameters. Because much of this material is similar to what was presented starting on page 70, it would be good to review that material before proceeding.

As was the case for classification, the usual procedure is to specify two hypotheses that are mutually exclusive and exhaustive. The *null hypothesis* represents the status quo, the uninteresting condition. The *alternative hypothesis* usually represents what we want to believe. In any case, the alternative hypothesis is what we will attempt to assert by indirect means. The logic is straightforward:

1) Assume that the null hypothesis is true. In the context of this section, this means that we assume that the parameter in question, θ, has a value specified by the null hypothesis. Often this value will be zero.

2) Compute the value of the parameter estimate, $\hat{\theta}$.

3) Compute the probability that, if the null hypothesis were true, we could have observed a value as extreme as the $\hat{\theta}$ we did observe.

4) If this probability is small, we then assert that the null hypothesis is probably not true and thereby assert that the alternative hypothesis is probably true.

Sometimes the alternative hypothesis is *two-sided*, meaning that the experimenter is interested in asserting that the true parameter is either above or below the null hypothesis value. But most often the alternative hypothesis is *one-sided*, meaning that only one direction is interesting. For example, a person developing a system for trading financial markets will probably have a null hypothesis that the true return is zero, and an alternative hypothesis that the true return is positive. A negative return garners no interest. In one-sided situations, it is customary to lump the other side in with the null hypothesis so that the null and alternative hypotheses cover all possibilities. The uninteresting side can then be conveniently and safely ignored. Thus, in the market-trading example, the actual null hypothesis is that the proposed trading rule is either worthless or a loser, but we proceed with the test as if the null hypothesis were simply that the trading system has a true return of exactly zero.

Why can we get away with this? Recall that the foundation of the procedure is that we presume the null hypothesis to be true and then compute the probability of observing a return as good as or better than what we obtained. Of the members of the null hypothesis family, a presumed true return of zero is closest to a positive return, and hence most difficult to distinguish from a true positive return. It provides the most stringent test. If the true return were less than zero, a positive observed return would be even less likely than if the true return were zero. The probability computed from assuming a true return of zero is the largest we could obtain among the null hypothesis family. So if this probability is satisfactorily small, we can be assured that no other member of the null family would produce a larger probability and thereby torpedo our logic.

Having collected a sample and computed $\hat{\theta}$, how do we decide if we should reject the null hypothesis that θ equals some specified value? We already know how to do this: compute a confidence interval and see if the hypothesized value of θ lies outside the interval. The discussion on page 120 showed that the lower bound of a confidence interval is the value of θ at which we have the desired probability of obtaining a $\hat{\theta}$ as large as or larger than the observed value. The converse logic defines the upper bound.

In the example of evaluating the performance of a trading system, the null hypothesis is that the mean return is zero. To perform a hypothesis test with a significance threshold of 5 percent, we compute the lower 5 percent bound of a confidence interval for the true mean return. If this lower bound is less than zero, a hypothetical return of zero implies greater than 5 percent probability of having obtained a return as high as what we observed, so we do not reject the null hypothesis. We go back to the drawing board for a new trading system. But if the lower bound of the confidence interval happens to exactly equal zero, we know that there is only a 5 percent chance that a truly worthless trading system would have given us a return as large as or larger than what we obtained, so we reject the null hypothesis and assert that our system may be a money-maker. Finally, if the lower bound of the confidence interval is greater than zero, we know that a truly worthless system would have even less than 5 percent probability of having given results this good. This is because the positive lower bound marks the exact 5 percent threshold, so the smaller value of zero would make a decent return still less likely. If desired, we could compute the probability associated with a true return of zero by finding the tail area beyond zero.

Bootstrapping Ratio Statistics

We saw in Chapter 1 that statistics based on the ratio of success to failure have valuable theoretical and practical properties. In fact, designers of market trading systems are fond of the venerable Sharpe Ratio as a measure of risk/reward, and especially the profit factor (ratio of total wins to total losses) due to its ability to generalize well. Unfortunately, the Sharpe Ratio, the profit factor, and many other performance measures involving ratios require special consideration when they are bootstrapped. This section explores two of the most dangerous pitfalls.

First, whenever a ratio is involved, we must be wary of the danger of dividing by zero. The user must ensure that the data supplied to any automatic computation routine is safe in the sense that the probability of a degenerate bootstrap sample is negligibly small. For example, if the collection of gains/losses includes almost no losses, the denominator of a success factor such as the profit factor may be zero for some bootstrap samples, those that happen to include only gains. Also, if the dataset contains numerous duplicate trade profits, the denominator of the Sharpe Ratio (or any performance statistic defined as the ratio of mean gain to variation of gains) may occasionally be zero or dangerously tiny. Even if the automatic routine takes precautions against dividing by zero, results will probably be so distorted as to be worthless.

The second potential problem involves the choice of the confidence interval or hypothesis test method. In the section that begins on page 125 we saw that the *basic method* of computing confidence intervals could be seriously problematic compared to the *percentile method* and the BC_a *method*. This consideration also applies to hypothesis tests, since confidence intervals and hypothesis tests are intimately related. The graphs that begin with Figure 3-3 on page 139 illustrate the performance of these three methods for confidence bounds, along with the implied hypothesis tests. All of these examples are based on hypothetical market trading systems, although the conclusions reached apply to any application in which the performance statistic being bootstrapped is a ratio of success to failure or a ratio of mean gain to variation of gain.

These figures show the violation rate for the lower and upper confidence interval endpoints of worthless systems having 50 trades. The statistic being tested is named in each figure's caption. In all graphs, the basic method is depicted with a solid line, the percentile method is a dashed line, and the BC_a method is dotted. The theoretically correct value is also shown as a straight solid line running diagonally through the figure.

To understand the import of these graphs, recall that an ideal algorithm for computing confidence intervals would provide endpoints that are violated with probability exactly equal to their intended values. For example, suppose we desire a 90 percent symmetric confidence interval. Ideally, the computed lower bound would exceed the true population value 5 percent of the time, and the computed upper bound would be less than the true value 5 percent of the time. Departures from this ideal represent shortcomings in the algorithm. If the confidence bound is violated less often than it is supposed to be, the bound is overly conservative and hence pessimistic. The more serious error is if the bound is violated more often than it is supposed to be, as then our application can fail more often than we expect.

Figure 3-3 and those following show the degree to which each of the three competing algorithms approaches ideal performance. Violation rates for ten different tail probabilities ranging from 0.01 through 0.10 are plotted, with the rate computed by the bootstrap along the horizontal axis and the actual violation rate along the vertical axis. In all cases, the trade data is totally random, implying worthless systems.

It bears repeating that if the confidence interval algorithms were perfect, their violation rates would fall exactly on the diagonal line. In other words, if the bootstrap algorithm identifies a particular value as being the lower 5 percent bound (position 0.05 on the horizontal axis), then we should in fact find that this bound is indeed violated 5 percent of the time (0.05 on the vertical axis). This is true for all confidence probabilities, so we should observe a plot of a straight line at 45 degrees.

Unfortunately, this is not at all what happens. Observe that in all cases the basic method (solid line) is the worst performer in that it is furthest from the perfect diagonal. The BC_a method is usually the most accurate, although it occasionally is slightly less accurate than the percentile method. On the other hand, the percentile method has a tendency to exceed the correct reject rate. This is usually considered to be a more serious error than the converse, as it means that the computed confidence bound is too tight (optimistic). In most applications an excessively wide interval is better than one that is excessively narrow.

Figures 3-3 and 3-4 show the violation rates for Sharpe Ratio lower and upper bounds, respectively. Note that relatively speaking, performance of the basic method is worst in the extreme tails. At a tail probability of 0.01, the observed violation rate is zero for both the lower and upper bounds. This implies conservatism taken to an extreme.

Figures 3-5 and 3-6 demonstrate that the situation for the profit factor is incredibly worse. At a desired reject rate of 0.1, the basic method's lower bound is violated with probability 0.006, and its upper bound is violated with probability 0.17. The confidence interval is obviously off-center in that the lower bound is almost never violated and the upper bound is violated almost twice as often than it should be.

Finally, we see in Figures 3-7 and 3-8 that working with the log of the profit factor instead of the raw profit factor helps the situation a lot, although the cure is not complete.

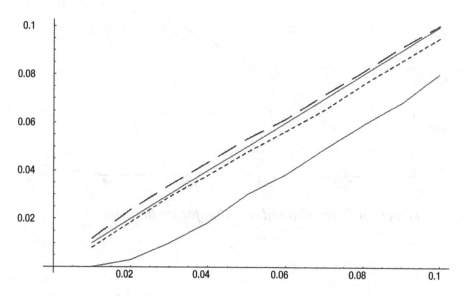

Figure 3-3. *Lower confidence bound rejection for Sharpe Ratio*

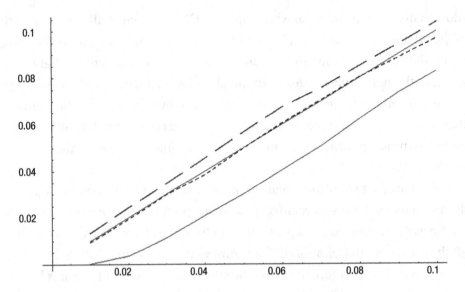

Figure 3-4. *Upper confidence bound rejection for Sharpe Ratio*

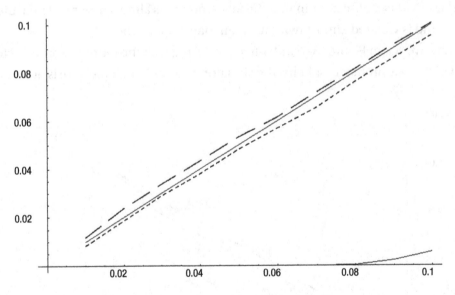

Figure 3-5. *Lower confidence bound rejection for profit factor*

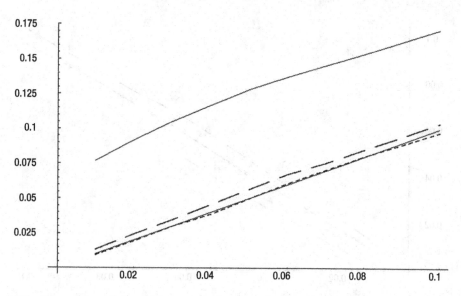

Figure 3-6. *Upper confidence bound rejection for profit factor*

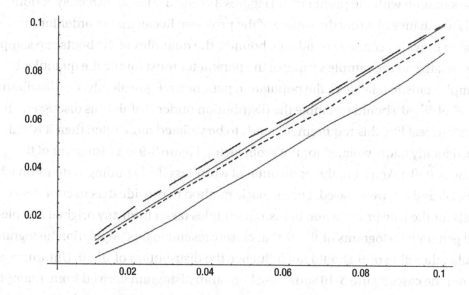

Figure 3-7. *Lower confidence bound rejection for log profit factor*

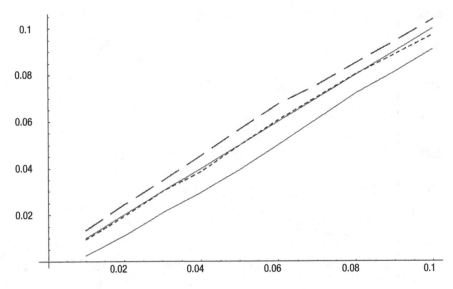

Figure 3-8. *Upper confidence bound rejection for log profit factor*

The situation with the profit factor (Figures 3-5 and 3-6) is so extremely serious that it is worth investigating the source of the problem. Recall that in order for the basic method to provide accurate confidence bounds, the quantiles of the bootstrap sample statistic relative to the sample's value of the parameter must mimic the quantiles of the sample statistic relative to the population parameter. Equivalently, the distribution under $\hat{F}$ of $\hat{\theta}^{\star} - \hat{\theta}$ should resemble the distribution under F of $\hat{\theta} - \theta$ as discussed on page 125. In real life, this requirement tends to be violated more often than it is satisfied. It is particularly badly violated for the profit factor. Figure 3-9 is a histogram of the differences $\hat{\theta} - \theta$ taken from the population of worthless ($\theta = 1$) trading systems on which Figures 3-5 and 3-6 were based. For the basic method to provide decent confidence intervals for the true profit factor, bootstrap samples taken from any original sample should generate histograms of $\hat{\theta}^{\star} - \hat{\theta}$ that closely resemble this population histogram. In particular, the value of $\hat{\theta}$ should not influence the distribution of $\hat{\theta}^{\star} - \hat{\theta}$. Unfortunately, this is not the case. Figure 3-10 shows the bootstrap histogram derived from a sample having a profit factor of 0.39, and Figure 3-11 shows that from a sample having a profit factor of 2.29. Neither of these histograms comes even close to representing the population distribution of the error.

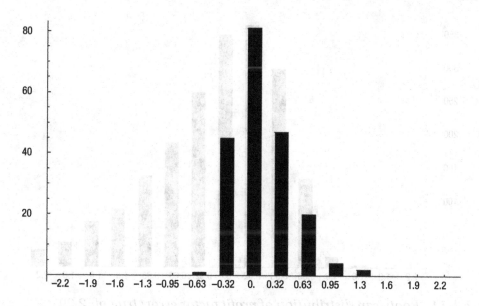

Figure 3-9. *Population distribution of profit factor error with n=50*

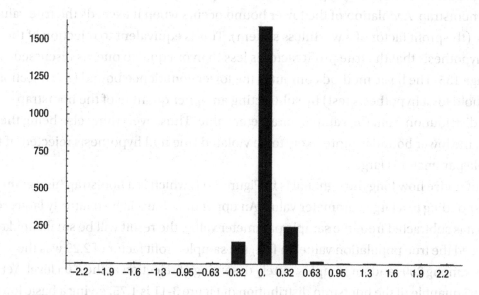

Figure 3-10. *Bootstrap distribution of profit factor error; true pf=0.39*

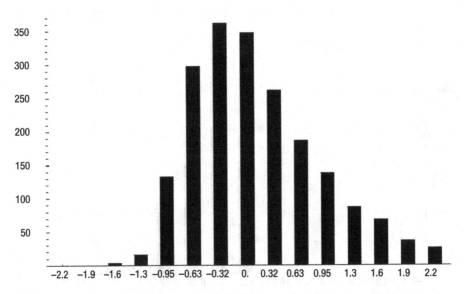

Figure 3-11. *Bootstrap distribution of profit factor error; true pf=2.29*

It is now easy to see why Figures 3-5 and 3-6 portray such peculiar results for the basic bootstrap. A violation of the lower bound occurs when it exceeds the true value of one (the profit factor of a worthless system). This is equivalent to rejection of the null hypothesis that the true profit factor is less than or equal to one, as discussed on page 135. The basic method computes the lower confidence bound (the rejection threshold for a hypothesis test) by subtracting an upper quantile of the bootstrap error distribution from the sample parameter value. Thus, everything else being the same, the lower bound is more likely to be violated (the null hypothesis rejected) if the sample parameter is large.

But notice how long the right tail is in Figure 3-11, which is a bootstrap histogram corresponding to a large parameter value. An upper quantile will be relatively large, so when it is subtracted from the sample parameter value, the result will be small, unlikely to exceed the true population value. In fact, the sample profit factor of 2.29 was the largest among 200 random samples, surely deserving of rejection at the 0.05 level. Yet the 0.95 quantile of the bootstrap distribution of Figure 3-11 is 1.75, giving a basic lower bound of 2.29–1.75=0.54, which is not even close to exceeding the true value of 1.0. The situation conspires against violation of the lower bound. Samples that have a large profit factor, a value that should inspire rejection of the null hypothesis, also have a long right tail, which results in an unnaturally small lower bound. This makes it all the more difficult to reject the null hypothesis.

The opposite effect occurs with the upper confidence bound. The basic method computes this bound by subtracting a low quantile of the bootstrap error distribution (a negative number) from the sample parameter estimate. For the upper bound to be violated, the sample parameter value must be significantly less than the true population value. But as Figure 3-10 demonstrates, the left tail of the bootstrap distribution is extremely short when the sample profit factor is small. The result is that the upper confidence bound is too close to the sample profit factor and hence too easily violated, as shown in Figure 3-6.

We conclude this section with a look at how the three confidence interval algorithms behave when the trading system is profitable. Figures 3-12 through 3-17 portray systems having 50 trades with a true profit factor of 2.33. Observe that in general the differences between the methods are even more profound than they were for worthless systems!

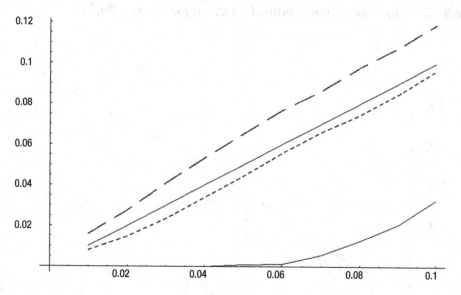

Figure 3-12. *Lower confidence bound rejection for Sharpe Ratio*

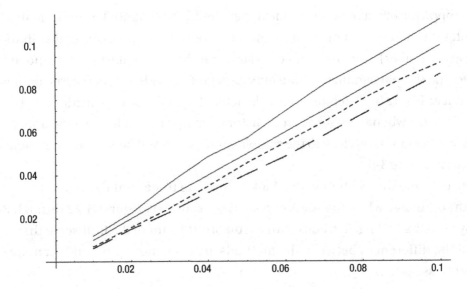

Figure 3-13. *Upper confidence bound rejection for Sharpe Ratio*

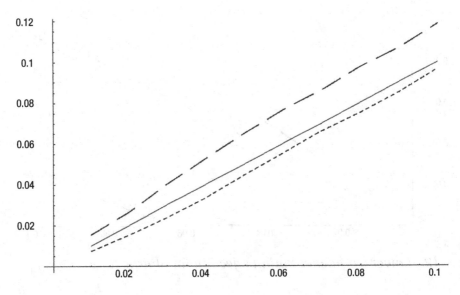

Figure 3-14. *Lower confidence bound rejection for profit factor*

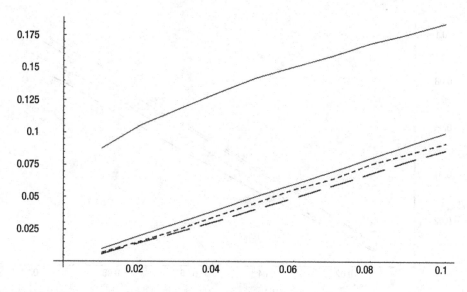

Figure 3-15. *Upper confidence bound rejection for profit factor*

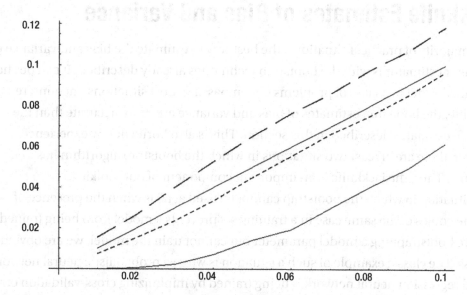

Figure 3-16. *Lower confidence bound rejection for log profit factor*

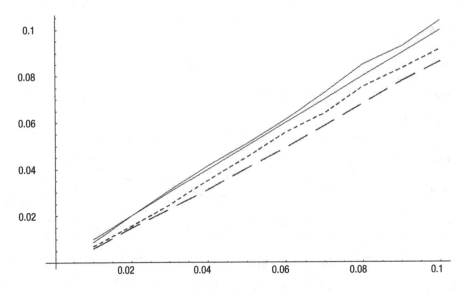

Figure 3-17. *Upper confidence bound rejection for log profit factor*

Jackknife Estimates of Bias and Variance

In the majority of practical situations, the best way to estimate the bias and variance of a parameter estimator is with the bootstrap techniques already described. It can be shown theoretically that in a class of problems encompassing most situations encountered in real life, the bootstrap estimates of bias and variance are more accurate than the jackknife estimates described in this section. This is also borne out by experience. However, there are at least two situations in which the bootstrap algorithm has problems. Thus, the jackknife is an important component of our toolkit.

A situation in which the bootstrap cannot be used at all is when the presence of multiple copies of the same case in a training set prevents a model from being trained. If we are bootstrapping a model parameter but cannot train the model, we are obviously helpless. The classic example of such a situation is when a probabilistic neural network or general regression neural network is being trained by minimizing cross-validation error. It is important that for each holdout case, its twin not be present in the remaining kernel set. A bootstrap training set will have many such duplications, preventing successful optimization. Support vector machines can also suffer when multiple copies of training cases are present.

A more general and common situation in which the bootstrap is at a disadvantage is when the presence of multiple copies of a case leads to instability in the parameter estimate. Recall the fundamental assumption of the bootstrap: the behavior of the parameter under practical sampling from the parent distribution F is reflected in the behavior of the parameter under bootstrap sampling from $\hat{F}$. If the fact that $\hat{F}$ is necessarily a finite (and often small) discrete distribution leads to unusual behavior in the parameter calculated from bootstrap samples, the bootstrap algorithm is degraded or worthless. We will explore this situation in an example later.

The explanation of how the jackknife achieves its goals is not terribly difficult, but it is quite long and not especially important in the current context. Therefore, its theoretical justification is omitted in this text. Detailed explanations can be found in [Efron and Tibshirani, 1993] and [Efron, 1982]. Slightly less detailed but more intuitive discussions of the jackknife can be found in [Masters, 1993]. Here, we consider only how to compute bias and variance estimates using the jackknife.

We start by repeating a definition seen earlier. If we remove case i from the collection and compute the parameter using the remaining $n-1$ cases, we let $\hat{\theta}_{(i)}$ designate this quantity. Define $\hat{\theta}_{(.)}$ to be the mean of these n jackknifed values, as shown in Equation (3.26) on page 130. As usual, let $\hat{\theta}$ be the parameter computed using the entire set of n cases. Then the jackknife estimate of bias is given by Equation (3.29), and the estimate of variance is given by Equation (3.30).

$$\text{bias} = (n-1)\left(\hat{\theta}_{(.)} - \hat{\theta}\right) \tag{3.29}$$

$$\text{var} = \frac{n-1}{n}\sum_{i=1}^{n}\left(\hat{\theta}_{(i)} - \hat{\theta}_{(.)}\right)^2 \tag{3.30}$$

A subroutine for computing jackknife estimates of bias and variance can be found in the file BOOT_P_5.CPP on my web site. This subroutine is now listed, and an explanation follows:

```
void jack_bias_var (
    int n,                          // Number of cases in sample
    double *data,                   // The sample
    double (*user_t) (int, double *, double *), // Compute param
    double *rawstat,                // Raw statistic of sample, theta-hat
    double *bias,                   // Output of bias estimate
    double *var,                    // Output of variance estimate
    double *work                    // Work area n long
)
```

```
{
    int i;
    double last, temp, param, diff, theta_dot;

    last = data[n-1];
    theta_dot = 0.0;

    for (i=0; i<n; i++) {
        temp = data[i];             // Preserve case being temporarily removed
        data[i] = last;             // Swap in last case
        param = user_t (n-1, data, NULL);  // Param for this jackknife
        theta_dot += param;         // Cumulate mean across jackknife
        work[i] = param;            // Save for computing variance later
        data[i] = temp;             // Restore original case
        }

    theta_dot /= n;                 // Equation (3.26)

    *rawstat = user_t (n, data, NULL);
    *bias = (n - 1) * (theta_dot - *rawstat); // Equation (3.29)

    *var = 0.0;
    for (i=0; i<n; i++) {
        diff = work[i] - theta_dot;
        *var += diff * diff;
        }

    *var *= (n - 1.0) / (double) n; // Equation (3.30)
}
```

The jackknife implementation is straightforward. The last case in the dataset is preserved because it will take the place of each temporarily removed case. The parameter for each omission is computed and saved, and the sum is cumulated to compute the mean later. Equation (3.29) supplies the bias estimate, and Equation (3.30) supplies the variance.

The file BOOT_P_5.CPP also contains a demonstration program for comparing the jackknife bias and variance estimates with those of the bootstrap in a special but not rare application. The *profit factor* of a financial trading system is defined as the ratio of the sum of wins over the sum of losses. A completely random system will be expected to have a profit factor of 1.0 because on average its total wins will equal its total losses. High-quality systems have profit factors significantly exceeding 1.0, while losing systems have profit factors less than 1.0.

The fact that profit factor is a ratio can be problematic for the bootstrap algorithm. If the sample is not very large, it is possible for some of the bootstrap samples to contain few losses, resulting in an unusually small denominator, which in turn produces an unusually large ratio. In fact, it is even possible for a random bootstrap selection to contain no losses at all, resulting in an infinite ratio. This is especially likely if the original sample happens to contain relatively few losses. On the other hand, it would require an extreme situation for the jackknife's removal of a single case to result in a huge or infinite profit factor. Thus, we would expect the jackknife to be superior to the bootstrap in this application, at least for small samples.

To demonstrate this effect, the test program was run with 10,000 tries of a sample of size $n=50$. The relatively large amount of 2,000 bootstrap replications was employed to ensure fairness. The program fixes the standard deviation of the profits at $1,000 and lets the user specify a mean profit per trade. This test let the mean be $100, producing a modestly profitable system with a true profit factor of about 1.29 and a mean estimated profit factor of about 1.39, indicating a true bias of about 0.10. The bootstrap estimate of bias was 0.12, while the jackknife's was 0.11, slightly better. Also, the jackknife's bias estimate had a slightly lower standard deviation than the bootstrap's. Across the 10,000 tries, the observed parameter variance was 0.287. The mean variance estimate from the jackknife was 0.329, a reasonable approximation to the true value. The bootstrap's mean variance estimate was 0.555, almost twice the correct value. This magnitude of error is a direct consequence of the fact that the bootstrap replications do not fairly represent sampling from the original distribution. However, be aware that this is an unusual situation. In the majority of applications, the bootstrap is to be preferred to the jackknife.

Bootstrapping Dependent Data

All of the bootstrap algorithms presented so far assume that the cases that make up the original sample are independent. In simple terms, this means that the observed value of each case is not influenced in any way by the value of any other case or cases in the sample. This assumption is not just a theoretical nicety; it is crucial to the bootstrap. Even small violations can have a severe impact on results. This section presents several methods for handling dependent data.

When the data is a time series, we should always be suspicious that it may contain serial dependencies. For example, profits in a trend-following market trading system applied to a trendy market may experience unusually long periods of success followed by long periods of failure. If the bootstrap is used to study this sort of data, the results will at best be compromised and at worst be worthless.

The most common method for detecting dependency is to study the *autocorrelation* of the series. This is the sequence of correlations at lags of one, two, three, etc. It measures the degree to which the value of a case is linearly influenced by its recent predecessors. But be warned that autocorrelation is not the only form of dependency. There are many forms of serious dependency that will induce no autocorrelation at all and hence be undetectable by the usual methods. On the other hand, extensive experience indicates that autocorrelation is by far the most common manifestation of dependency. Moreover, methods for detecting other forms of dependency are cumbersome, especially if the experimenter has no prior knowledge of the form the dependency may take. For this reason, most people rely on autocorrelation as their only test for dependency. Canned software packages for assessing autocorrelation are widely available and are a useful tool for deciding if a dependent bootstrap is warranted.

This section presents two very different modifications of the bootstrap. Neither is perfect, and each has its own strengths and weaknesses. The *stationary bootstrap* [Politis and Romano, 1994] is the easier one to implement, and it has broad applicability. But results from the stationary bootstrap tend to have relatively high error variance. Its chief competitor, the *tapered block bootstrap* [Paparoditis and Politis, 2002], is much more complicated, and it suffers from significant limitations to its applicability. But when it can be used, it generally provides the most reliable (lowest error) results of all common dependent bootstrap algorithms.

Estimating the Extent of Autocorrelation

Most canned statistical packages that compute and display autocorrelation include significance tests that help the user decide if the observed correlation is real, as opposed to being due to random fluctuations that can be expected in uncorrelated data. If the correlation appears serious and the user decides to use a stationary bootstrap, the extent of the correlation must be assessed. Visual inspection of plots is vital. In addition, [Politis and White, 2004] provides an automated test that is a useful adjunct to (but not a substitute for!) more conventional tests. This section presents their method for estimating the maximum lag distance at which correlations are large enough to require remediation.

Let X_i for $i \in Z$ be a strictly stationary real-valued series having (usually unknown) mean $\mu = E(X_i)$. Its *autocovariance function* at lag k is defined in Equation (3.31).

$$R(k) = E\left[\left(X_i - \mu\right)\left(X_{i+|k|} - \mu\right)\right] \tag{3.31}$$

When we observe a sample $x_1, \ldots, x_n$, the standard way to estimate the autocovariance of the parent population is shown in Equation (3.32), with the sample mean being defined in Equation (3.33) and the *autocorrelation* function in Equation (3.34).

$$\hat{R}(k) = \frac{1}{n}\sum_{i=|k|}^{n}(x_i - \bar{x})\left(x_{i-|k|} - \bar{x}\right) \tag{3.32}$$

$$\bar{x} = \frac{1}{n}\sum_{i=1}^{n} x_i \tag{3.33}$$

$$\hat{r}(k) = \hat{R}(k)/\hat{R}(0) \tag{3.34}$$

The autocorrelation is similar to an ordinary correlation in that it ranges from a minimum of –1 when the deviation of the current point from the series' mean is exactly opposite the deviation at the indicated lag, to a maximum of +1 when the deviations are equal. An autocorrelation of zero means that there is no linear relationship between the current point and the point at the indicated lag.

The [Politis and White, 2004] method for assessing the extent of autocorrelation finds the smallest integer lag m such that Equation (3.35) is satisfied.

$$\left|\hat{r}(m+k)\right| < 2\sqrt{\log_{10} N / N} \text{ for } k = 1, \ldots \max\left[5, \sqrt{\log_{10} N}\right] \tag{3.35}$$

The value of m that satisfies this equation is assumed to be the lag after which autocorrelations are negligible. Despite the theoretical foundation underlying this automatically generated number, it always pays to examine a plot (or at least a numerical list) of autocorrelations to visually confirm m's value. Later, when the two dependent bootstraps are discussed, m will play a role in the selection of an appropriate block size for bootstrap sampling. A subroutine that computes the autocorrelation function and m can be found in the file DEP_BOOT.CPP on my web site. Here is a listing of the subroutine, with an explanation of its operation following:

153

```
int correlation_extent (
   int n,                           // Length of series
   int maxlag,                      // Max lag to test
   double *x,                       // The series
   double *autocov                  // Output of autocovariance, maxlag+1 long
   )
{
   int i, lag, kn, nsmall, m;
   double mean, sum, thresh;


   // Compute the autocovariance of the sample.


   mean = 0.0;
   for (i=0; i<n; i++)                          // Equation (3.33)
      mean += x[i];
   mean /= n;


   for (lag=0; lag<=maxlag; lag++) {
      sum = 0.0;
      for (i=lag; i<n; i++)                      // Equation (3.32)
         sum += (x[i] - mean) * (x[i-lag] - mean);
      autocov[lag] = sum / n;
      }

   // Use the Politis and White [2004] heuristics

   kn = (int) (sqrt(log10(n)) + 0.5);
   if (kn < 5)
      kn = 5;

   thresh = 2.0 * sqrt (log10 (n) / n);

   nsmall = 0;
   m = maxlag;

   for (i=1; i<=maxlag; i++) {                    // Equation (3.35)
      if (fabs (autocov[i] / autocov[0])< thresh) // Equation (3.34)
         ++nsmall;
```

```
  else
    nsmall = 0;

  if (nsmall >= kn) {
    m = i - kn;
    break;
    }
  }

if (m < 1)
  m = 1;

return m;
}
```

This subroutine begins by computing the autocovariance for lags through a user-specified upper limit. This limit should be a conservative upper bound on the extent of autocorrelation. The autocorrelation is not precomputed. Rather, it is computed as needed using Equation (3.34). Equation (3.35) is checked in a loop that starts at a lag of i=1 and continues through the maximum lag computed. As long as the condition in Equation (3.35) is satisfied, the counter nsmall is incremented. Any time the condition fails, the counter is reset. When the counter reaches its required limit, m is found.

The Stationary Bootstrap

The first dependent bootstrap algorithm to be presented is the *stationary bootstrap* of [Politis and Romano, 1994]. Its original claim to fame was its ability to produce bootstrap samples that, when taken from a weakly dependent stationary series, themselves form a stationary series. This is useful when dealing with multivariate parameters of the entire series. Earlier dependent bootstraps designed for handling multivariate parameters were either limited to parameters of fixed multivariate dimension or, if not so limited, did not produce a stationary bootstrap series. Although stationarity is not often strictly required, it is a nice property to have. The stationary bootstrap was initially shunned because there was widespread belief that it was seriously inefficient compared to nonstationary competitors. However, it has recently been discovered that these purported inefficiencies were based on erroneous calculations. It is actually quite competitive, although as we will see, it is somewhat less efficient than the other dependent bootstrap that we will present. For the purposes of this text, its primary benefits are that it is both easy to compute and broadly applicable. Stationarity of the bootstrap samples is a nice bonus, and its slightly substandard efficiency is a small price to pay. The stationary bootstrap is therefore an essential tool.

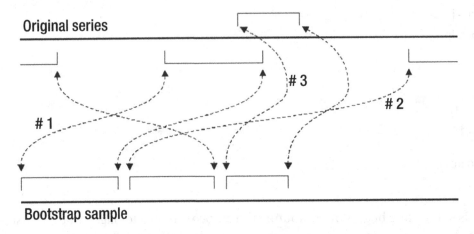

Figure 3-18. *Assembling a stationary bootstrap sample*

Figure 3-18 illustrates how a stationary bootstrap sample is constructed. The complete sample, which has the same length (number of observations) as the original series, is composed of a set of blocks taken from the original series. Each block has a random starting point and a random length. If a block runs past the end of the original series, it wraps around to the beginning. The final block will usually need to be truncated to prevent it from extending past the end of the bootstrap sample series.

In Figure 3-18, the first (randomly chosen) bootstrap block begins about a third of the way into the original series. The second block starts near the end of the original series and wraps to the beginning. The third block partially overlaps the first block in the original series. Because the starting points and block lengths are random, many points in the original series will be duplicated, while many others will not appear in the bootstrap sample.

The starting point for each block is uniformly distributed across the original series. In other words, all original cases are equally likely to be chosen to begin each block. The length of each block follows a geometric distribution. For a specified parameter p in (0, 1), the probability that any given block will contain k contiguous observations is given by Equation (3.36).

$$P(k) = p(1-p)^{k-1} \tag{3.36}$$

The mean of a geometric distribution is $1/p$, so if we want the mean block length to be b, we set $p=1/b$.

There is an easy way to define blocks whose lengths follow a geometric distribution. Pick a random starting point for the block. Then, with probability p declare the block complete and start a new block. Conversely, append another point to the current block with probability $1-p$. In other words, after picking a starting point, we repeatedly flip a biased coin. As long as the event having probability $1-p$ occurs, keep extending the block, wrapping to the beginning of the original series if necessary. As soon as the event having probability p occurs, start a new block from a new random starting point.

Code for drawing a stationary bootstrap sample can be found in the file DEP_BOOT. CPP on my web site. Here is a subroutine that performs this task:

```
void SBsample (int n, int blocksize, double *x, double *bootsamp)
{
  int i, pos;
  double p;

  p = 1.0 / blocksize;           // Parameter for geometric distribution

  pos = (int) (unifrand() * n);  // Pick a random starting point
  if (pos >= n)                  // Should never happen
    pos = n - 1;                 // But avoid disaster

  for (i=0; i<n; i++) {          // Build the bootstrap sample
    bootsamp[i] = x[pos];        // Get a case
    if (unifrand() < p) {        // Implement the geometric distribution
      pos = (int) (unifrand() * n);  // We may choose a new random position
      if (pos >= n)              // (protected as before)
        pos = n - 1;
    }
    else
      pos = (pos + 1) % n;       // Or we may simply advance circularly
  }
}
```

Choosing a Block Size for the Stationary Bootstrap

There is no universally excellent algorithm for computing the optimal mean block length for the stationary bootstrap (or any other dependent bootstrap, for that matter). However, [Politis and White, 2004] provide an algorithm that has a solid theoretical foundation and that seems to usually work well in practice. The theory behind the algorithm is far beyond the scope of this text. All we will do here is state the algorithm in terms of equations and source code.

A function that will be needed regularly is the *flat-top lag window* defined by Equation (3.37). This function has domain [-1, 1]. Its middle is uniformly equal to one, and it tapers to zero at both ends.

$$\lambda(t) = \begin{cases} 1 & \text{if } |t| \in [0, 1/2] \\ 2(1 - |t|) & \text{if } |t| \in [1/2, 1] \\ 0 & \text{otherwise} \end{cases} \tag{3.37}$$

The quantities given by Equations (3.38) and (3.39) are defined in terms of this lag window and the sample autocovariance function of Equation (3.32). In these equations, we let $M = 2m$, where m is the autocorrelation extent given by Equation (3.35) on page 153.

$$\hat{G} = \sum_{k=-M}^{M} \lambda(k/M) \, |k| \, \hat{R}(k) \tag{3.38}$$

$$\hat{g}(w) = \sum_{k=-M}^{M} \lambda(k/M) \, \hat{R}(k) \cos(wk) \tag{3.39}$$

Finally, we need the quantity given by Equation (3.40). The integral makes this a bit tricky.

$$\hat{D}_{SB} = 4\hat{g}^2(0) + \frac{2}{\pi} \int_{-\pi}^{\pi} \left[1 + \cos(w) \right] \hat{g}^2(w) \, dw \tag{3.40}$$

Once all of these quantities have been found, the (theoretically) optimal mean block size for the stationary bootstrap is given by Equation (3.41). Recall that N is the number of cases in the sample.

$$b_{SB} = \left[\frac{2\hat{G}^2}{\hat{D}_{SB}} \right]^{1/3} N^{1/3} \tag{3.41}$$

Code for performing all of these calculations can be found in the file DEP_BOOT.CPP on my web site. We will now work through this code, one part at a time.

The quantity $\hat{g}(w)$ defined in Equation (3.39) is needed several times, so we compute it with a separate subroutine. The summation is symmetric in k, so we sum one side only, double it, and add in the center term $k=0$. The $k=M$ term is zero, so there is no point in including it in the summation. As will be the case throughout this code, the flat-top lag window of Equation (3.37) is embedded in the code, rather than being computed separately.

```
static double gw (double w, int M, double *autocov)
{
   int k;
   double lambda, sum;

   sum = 0.0;
   for (k=1; k<M; k++) {
     lambda = (double) k / M;
     lambda = (lambda < 0.5) ? 1.0 : (2.0 * (1.0 - lambda));
     sum += lambda * autocov[k] * cos (k * w );
     }

   // The summation is symmetric, so double one side and add center (k=0) term
   return 2.0 * sum + autocov[0];
}
```

The integrand in Equation (3.40) is best made a separate function to facilitate integration. Unfortunately, it involves two nuisance parameters, the autocovariance vector and M. It would be easy enough to pass them in the integrand's parameter list here. However, you may want to substitute a fast, highly accurate numerical integration routine for the relatively crude (but effective!) method employed here. For this reason, these two nuisance parameters are declared as statics and passed as module globals. This way the user can substitute a canned integration routine.

```
static int this_M;              // Facilitates future use of canned integration
static double *this_autocov;
```

```
static double integrand (double w)
{
   double temp;
   temp = gw (w, this_M, this_autocov);
   return (1.0 + cos (w)) * temp * temp;
}
```

The main routine begins here. The first few of its steps are straightforward. Subroutine correlation_extent() (presented on page 154) computes m as defined in Equation (3.35). Then, Equation (3.38) is evaluated. Once again, note that the summation is symmetric in k, so we need to evaluate only one side. Also, the center term is zero.

```
int optimal_SB_size (
   int n,                  // Number of cases in sample
   double *x,              // Variable in sample
   double *autocov         // Work area n long (Actually, not all n used)
   )
{
   int i, k, maxlag, m, nint;
   double sum, prior_sum, ghat, lambda, dhat, term, space;

/*
   Compute m as the minimum integer after which correlation appears negligible.
   Also compute the autocovariance.
*/

   maxlag = n / 4;         // This is very conservative
   if (maxlag > 50)        // And this is fairly conservative
      maxlag = 50;

   m = correlation_extent (n, maxlag, x, autocov);

/*
   Compute G-hat per page 7 of Politis and White (2004).
   Note that the sum is symmetric, so we only need to evaluate one side,
   double it, and add in the center (k=0). But we don't even need to worry
   about the center, because this term is zero.
   Also note that lambda(1)=0 so we do not need to do k=M.
*/
```

160

```
ghat = 0.0;
for (k=1; k<2*m; k++) {              // Equation (3.38)
   lambda = (double) k / (2.0 * m);
   lambda = (lambda < 0.5) ? 1.0 : (2.0 * (1.0 - lambda));
   ghat += lambda * k * autocov[k];
   }
ghat *= 2.0;
```

The next section of the code handles the integration in Equation (3.40). We use the crude method of naive trapezoidal subdivision. It is not as fast as more intelligent methods, but it is more than accurate enough for this application, and it enjoys the blessing of (relative) simplicity.

The first step is to set the two static nuisance parameters that avoid the need for passing them in the integrand's parameter list. Initialize the mean of trapezoidal heights (in sum) to be the mean of the integrand at its endpoints. Also initialize space to be the distance between the endpoints. Since the integrand is symmetric in w, we integrate from zero to pi and double the result.

The integration is performed in the next loop. The mean height from the prior iteration is preserved in prior_sum, the spacing is halved, and the mean of the integrand is computed for the refined spacing. The two quantities are combined to give the new, more accurate result. Lower and upper limits are imposed on the number of subdivisions to prevent pathological premature escape and endless looping, respectively. Both of these events are extremely unlikely, but the insurance is cheap. The ultimate convergence indicator is the absolute or relative change in the computed integral.

Finally, the mean height of the integrand is multiplied by pi to give the actual integral, doubled because we summed only one side, and multiplied by $2/\pi$ as per Equation (3.40).

```
this_M = 2 * m;              // Lets us avoid passing nuisance parameters
this_autocov = autocov;      // In case we want to use a canned package later
nint = 1;                    // Number of new integration points
sum = 0.5 * (integrand (0.0) + integrand (PI)); // O riginal interval
space = PI;                  // Spacing for that original interval
```

```
for (;;) {                          // Endless loop waits for convergence or give up
   prior_sum = sum;                 // Convergence indicator also holds old estimate
   space *= 0.5;                    // Spacing for the upcoming subinterval
   sum = 0.0;                       // Will cumulate subdivision here
   for (i=0; i<nint; i++)           // Sum the refinement term
      sum += integrand (space + 2 * i * space); // Subdivide
   sum /= nint;                     // Refinement term
   sum = 0.5 * (prior_sum + sum);   // This is the refined estimate

   nint *= 2;                       // Number of terms in next refinement subdivision
   if (nint < 64)                   // Avoid convergence check until stable
      continue;

   if (nint >= 4096)                // Unlikely, but avoid huge time
      break;
   term = fabs (sum - prior_sum);   // This is how much we just changed
   if ((term < 1.e-10) || (term / (1.e-60 + fabs (sum)) < 1.e-10))
      break;                        // This is the real (incredibly conservative) convergence test
   }

sum *= 4;                           // Mean height * PI * 2 * (2 / PI)
```

The last few lines of the routine are trivial. Complete the evaluation of Equation (3.40) and then evaluate Equation (3.41). Round up to find the optimal mean block size.

```
term = gw (0.0, this_M, autocov);
dhat = 4.0 * term * term + sum;     // Equation (3.40)

term = 2.0 * ghat * ghat / dhat;    // Equation (3.41)
term = exp (log (term * n) / 3.0);  // Raise ratio to the 1/3 power
return (int) (term + 0.999999999);  // Round up
}
```

The Tapered Block Bootstrap

The *tapered block bootstrap* of [Paparoditis and Politis, 2001] is similar to the stationary bootstrap in that it randomly samples blocks of points in order to construct a bootstrap sample that captures dependencies in the data. However, it is different in some significant ways:

- The statistic being estimated must be *approximately linear*. This makes the tapered block bootstrap more restrictive than the stationary bootstrap. Only the mean will be considered in this text. See [Paparoditis and Politis, 2001] for a rigorous definition of an approximately linear statistic and examples of such statistics.

- The points that go into the bootstrap sample are not taken from the original raw data but rather are taken from a series derived from the *influence function* of the statistic. This will be discussed soon.

- All of the blocks are the same size. The stationary bootstrap uses random lengths.

- Blocks never wrap around the end of the series as they did in the stationary bootstrap.

- An irksome implication of the previous two points is that if the length of the original series is not an integer multiple of the block length, the bootstrap sample will be shorter than the original series, necessitating corrections to performance estimates.

- Rather than being a continuous series, each bootstrap sample is a collection of wide pulses of data, each of which tapers smoothly to zero at its beginning and its end. This accounts for the name *tapered block bootstrap*.

The *influence function* corresponding to a statistic is, roughly, a measure of the degree to which a change in the probability of a value in the population domain impacts the statistic. Rigorous definitions can be found in many advanced statistics texts. We will keep to an intuitive level.

Suppose, for example, that you have taken a course and had three exams on which you scored grades of 85, 86, and 87. Your grade average so far is 86. Only your term paper remains. If you score, say, 86 on this paper, your average will not be impacted at all. It will still be 86. So you could say that a score of 86 would have zero influence on your mean grade. On the other hand, suppose your dog ate the term paper and you received a grade of zero. Your average would then plummet. The influence of zero would be very negative. Similarly, a score of, say, 90 would have a small positive influence on your mean grade.

The influence function for the mean is trivial. It is computed simply by subtracting the sample mean. In particular, let x_i, $i=1, ..., n$ be the original sample, and let our statistic of interest be the mean of the population from which the sample is drawn. Then the collection of influence function values yi corresponding to the original sample is obtained by centering the data as shown in Equation (3.42).

$$y_i = x_i - \frac{1}{n}\sum_{k=1}^{n} x_k \quad \text{for } i=1,...,n \tag{3.42}$$

The taper window is similar to the flat-top lag window used in the optimal block length computation in that it is flat in the middle and it tapers to zero at both ends. Keep in mind, however, that these two windows serve entirely different purposes. The stationary bootstrap did not taper the data. It used the lag window to taper the impact of the autocorrelation function at distant lags. Here, in the tapered block bootstrap, the taper window is applied to each block individually. This is illustrated in Figures 3-19 through 3-21 on page 166. Figure 3-19 shows a typical taper window. Figure 3-20 is a correlated time series to which we want to apply the bootstrap. Figure 3-21 shows a tapered block bootstrap sample from this series. Notice how the effect of the taper window on the four blocks is visible in this figure. Of course, in practice we would use more blocks of shorter length. The window effect is more easily illustrated when we use few blocks.

A useful generic taper window is defined by Equation (3.43). The developers of this algorithm recommend a value of $c=0.43$.

$$h(t)=\begin{cases} t/c & \text{if } t \in [0,c] \\ 1 & \text{if } t \in [c,1-c] \\ (1-t)/c & \text{if } t \in [1-c,1] \end{cases} \tag{3.43}$$

After the taper window is computed, it needs to be normalized to have squared length equal to b, the block size. This increases the weight in the center to compensate for the lower weight in the ends. This normalization, as well as the discrete expression of the continuous general taper window, is shown in Equation (3.44).

$$w_i = \frac{\sqrt{b}\, h\left(\dfrac{i-0.5}{b}\right)}{\sqrt{\displaystyle\sum_{j=1}^{b} h^2\left(\dfrac{j-0.5}{b}\right)}} \quad \text{for } i=1,\ldots,b \tag{3.44}$$

Once the influence function has been computed for each sample point using Equation (3.42) and the taper window weights have been computed using Equations (3.43) and (3.44), the bootstrap sample can be constructed. Ideally, the number of cases in the sample, n, is an integer multiple of the block size, b. If it is not, the length of the bootstrap sample will be less than n, as shown in Equation (3.45). In this equation, $\lfloor\ \rfloor$ is the floor function, the largest integer less than or equal to its argument. In other words, you divide the original sample length by the block size to get the number of blocks that will be in the bootstrap sample, throw away any remainder because incomplete blocks are not included, and multiply this result by the block size to get the length of the bootstrap sample.

$$m = b \lfloor n/b \rfloor \tag{3.45}$$

The starting point of each block within the source sample (the set of influence function values corresponding to the original data points) is chosen randomly in the range one through n–b+1, making every possible block equally likely. The chosen block of b points is multiplied point-by-point by the b taper window points of Equation (3.44). This is repeated until the required number of blocks have been assembled into a single bootstrap sample.

Recall that this text considers only one test statistic: the mean. Thus, the mean of each bootstrap sample is computed to build the collection of bootstrapped values of the statistic. If a different test statistic is studied, its influence function must replace that of the mean in Equation (3.42). But before doing so, you should become familiar with [Paparoditis and Politis, 2001] to avoid using a statistic that does not satisfy the requirements of the tapered block bootstrap.

If the number of original sample points is not an integer multiple of the block size, the bootstrap sample will be shorter than the original sample. The implication is that the standard deviation of the bootstrapped test statistic will be inflated by a factor of sqrt(n/m) where m is the length of the bootstrap sample as shown in Equation (3.45). Therefore, computed performance measures like the standard error of the statistic or quantiles must be multiplied by sqrt(m/n). Of course, if the sizes of the original and bootstrapped samples are equal, this factor will be one.

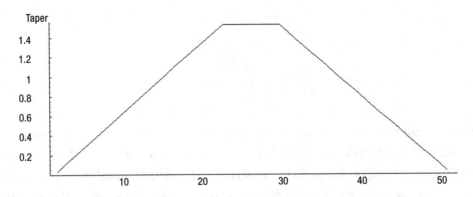

Figure 3-19. *A normalized taper window with c=0.43*

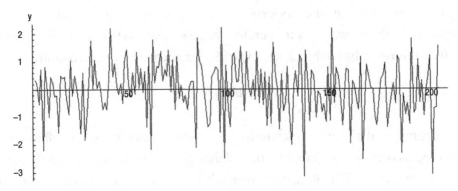

Figure 3-20. *An autocorrelated series*

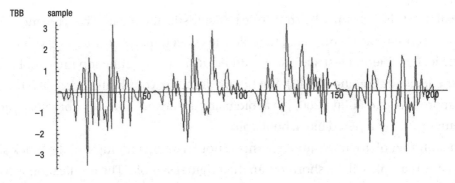

Figure 3-21. *A tapered block bootstrap sample*

166

The complete source code for the tapered block bootstrap can be found in the file DEP_BOOT.CPP on my web site. We examine this code now. The following subroutine computes the weights for the taper window using Equations (3.43) and (3.44):

```
void make_taper (int blocksize, double *window)
{
  int i, low, high;
  double w, sum;

  low = 0;
  high = blocksize - 1;

  for (;;) {
    w = (low + 0.5) / (double) blocksize;    // Position in block

    if (w < 0.43) {                          // If near edge
      w /= 0.43;                             // Taper upwards
      window[low++] = w;                     // Insert the taper
      window[high--] = w;                    // It is symmetric
    }
    else {                                   // Come here when well away from edge
      while (low <= high)                    // Fill in the center
        window[low++] = 1.0;                 // With full value
      break;                                 // Done
    }
  }

// Normalize the length of the window

  sum = 0.0;   // Will cumulate squared length here
  for (i=0; i<blocksize; i++)
    sum += window[i] * window[i];

  w = sqrt (blocksize / sum);
  for (i=0; i<blocksize; i++)
    window[i] *= w;
}
```

The taper window should be precomputed. The following subroutine builds a single bootstrap sample from the influence function values in x:

```
void TBBsample (int n, int blocksize, double *window,
          double *x, double *bootsamp)
{
  int i, j, k, pos;

  j = 0;                                      // Will index bootsamp
  k = (int) (n / blocksize);                  // Number of blocks

  while (k--) {                               // Count blocks done
    pos = (int) (unifrand() * (n-blocksize+1)); // Pick a random starting point
    if (pos > (n - blocksize))                // Should never happen
      pos = n - blocksize;                    // But avoid disaster
    for (i=0; i<blocksize; i++)               // Build the bootstrap sample
      bootsamp[j++] = x[pos+i] * window[i];   // Get a windowed case
  }
```

It can be useful to estimate quantiles of the distribution of the deviation of the sample mean from the population mean. In the conditions under which the tapered block bootstrap is most likely to be effective, the previously disparaged *basic method* of computing confidence intervals (page 125) will often outperform the *percentile method* and sometimes even the BC_a method. Thus, when the test statistic is the sample mean, as is the case here, you can subtract bootstrap quantiles from the original sample mean to find a confidence interval for the population mean. Similarly, you can test the null hypothesis that the population mean is zero by comparing the original sample mean to the appropriate bootstrap quantile.

For example, you could reject the null hypothesis at the 0.05 level if the sample mean equals or exceeds the 0.95 quantile. You should confirm in your own minds that this is equivalent to computing the lower bound of a 90 percent *basic* confidence interval and seeing if this lower bound equals or exceeds zero. More generally, the p-level associated with a given sample mean is one minus the cumulative distribution function of the bootstrap collection. Also note that *percentile method* confidence limits are obtained by adding the bootstrap quantiles to the sample mean instead of subtracting them (pivoting around the mean). The corresponding hypothesis test for a zero population mean is performed by adding the sample mean to a low quantile (such as the 0.05 quantile) of the bootstrapped differences and seeing if the sum exceeds zero. It may be beneficial to perform both tests and confirm that they provide similar results. If the results are highly different, which would happen if the bootstrap distribution is seriously asymmetric, the situation should be investigated.

The following subroutine, included in the DEP_BOOT.CPP source file on my web site, computes quantile estimates of the distribution of the sample mean around its population value using the tapered block bootstrap. Note that since the test statistic is the sample mean and its influence function is computed simply by centering the data, it makes little sense to call an outside function to accomplish this task. However, we do so here to clearly illustrate the tapered block bootstrap algorithm. Also note that the returned value must be normalized as discussed on page 165.

```
double QuantileMeanTBB (
    int n,                  // Number of cases in sample
    double *x,              // The sample
    int blocksize,          // Block size
    int nboot,              // Number of bootstrap replications to do
    double q,               // Desired quantile, 0-1
    double *xinf,           // Work area n long for influence function values
    double *bs,             // Work area n long for bootstrap sample
    double *reps,           // Work area nboot long for replications
    double *window          // Work area blocksize long for window
    )
{
    int i, k, iboot, subscript;
    double mean;

    if (q <= 0.5)           // Formula for unbiased subscript only works if q<=.5
        subscript = (int) (q * (nboot + 1)) - 1;
    else
        subscript = nboot - (int) ((1.0 - q) * (nboot + 1));

    if (subscript < 0)      // Ensure that a silly user didn't put us outside bounds
        subscript = 0;
```

```
else if (subscript >= nboot)
   subscript = nboot - 1;
influence_mean (n, x, xinf);          // Compute influence function for each case
make_taper (blocksize, window);       // Compute the tapered window
k = blocksize * (int) (n / blocksize);  // Length of TBB sample (<=n)

for (iboot=0; iboot<nboot; iboot++) {
   TBBsample (n, blocksize, window, xinf, bs);
   mean = 0.0;                         // Compute mean of bootstrap sample
   for (i=0; i<k; i++)                 // (Which is the test statistic)
      mean += bs[i];
   mean /= k,
   reps[iboot] = mean;                 // Save the bootstrapped values
   }

qsortd (0, nboot-1, reps);            // Sort ascending for cumulative dist func

return sqrt ((double) k / (double) n)* reps[subscript];   // Page 165 normalization
}
```

Choosing a Block Size for the Tapered Block Bootstrap

[Paparoditis and Politis, 2001] provides an algorithm for automatically choosing a block size for the tapered block bootstrap. It is similar to that for the stationary bootstrap, although it is slightly simpler because no integration is involved.

This algorithm uses the same flat-top lag window given by Equation (3.37) on page 158. It also uses the same autocovariance estimate shown in Equation (3.32) on page 153. We compute the following quantities:

$$\Gamma = -5.25 \sum_{k=-M}^{M} \lambda(k/M)k^2 \hat{R}(k) \tag{3.46}$$

$$\Delta = 1.1 \left[\sum_{k=-M}^{M} \lambda(k/M)\hat{R}(k) \right]^2 \tag{3.47}$$

The constants 5.45 and 1.1 in the prior two equations are a function of the shape of the tapering window, and they are valid only for the window specified in Equations (3.43) and (3.44) with c=0.43 as recommended by the authors.

The theoretically optimal block length for the tapered block bootstrap is then given by Equation (3.38).

$$b_{TBB} = \left[\frac{4\Gamma^2}{\Delta} \right]^{1/5} N^{1/5} \tag{3.48}$$

The code for computing this quantity can be found in the file DEP_BOOT.CPP on my web site. Here is the subroutine:

```
int optimal_TBB_size (
   int n,                // Number of cases in sample
   double *x,            // Variable in sample
   double *autocov       // Work area n long (Actually, not all n used)
   )
{
   int k, maxlag, m;
   double gamma, lambda, delta, term;

/*
   Compute m as the minimum integer after which correlation appears negligible.
   Also compute the autocovariance.
*/

   maxlag = n / 4;       // This is very conservative
   if (maxlag > 50)      // And this is fairly conservative
      maxlag = 50;

   m = correlation_extent (n, maxlag, x, autocov);

/*
   Compute Gamma per page 114 of Paparoditis and Politis.
   Note that the sum is symmetric, so we only need to evaluate one side,
   double it, and add in the center (k=0). But we don't even need to worry
   about the center, because this term is zero.
   Also note that lambda(1)=0 so we do not need to do k=M.
*/
```

```
   gamma = 0.0;
  for (k=1; k<2*m; k++) {                    // Equation (3.46)
     lambda = (double) k / (2.0 * m);
     lambda = (lambda < 0.5) ? 1.0 : (2.0 * (1.0 - lambda));
     gamma += lambda * k * k * autocov[k];
     }
   gamma *= -10.9;

/*
   Compute Delta. This, too, is symmetric, but the center term is not zero.
*/

   delta = 0.0;
   for (k=1; k<2*m; k++) {                    // Equation (3.47)
     lambda = (double) k / (2.0 * m);
     lambda = (lambda < 0.5) ? 1.0 : (2.0 * (1.0 - lambda));
     delta += lambda * autocov[k];
     }
   delta = 2.0 * delta + autocov[0]; // 2 * side + center
   delta = 1.1 * delta * delta;

/*
   Compute the optimal block size
*/

   term = 4.0 * gamma * gamma / delta;
   term = exp (log (term * n) / 5.0 );      // Equation (3.48)
   return (int) (term + 0.999999999);       // Round up
}
```

What If the Block Size Is Wrong?

Actually, an incorrect block size encompasses two larger issues: What if our data contains dependencies, but we use an ordinary bootstrap? Or, what if our data is perfectly independent, but we use a dependent bootstrap anyway? These concepts fall under the heading of incorrect block size because both the stationary bootstrap and the tapered block bootstrap devolve into an ordinary bootstrap when the block size is one. This section explores the price that we pay for using an inappropriate block size.

All of these studies employ an autoregressive time series having a single weight at a lag of one. This is commonly abbreviated as *AR(1)*. The population mean of the series is zero, and the error is normally distributed with unit variance. In other words, the series is defined by Equation (3.49), in which ε is normally distributed with mean zero and variance one.

$$X_i = cX_{i-1} + \varepsilon \tag{3.49}$$

Three values of c are used: 0.0, which produces a truly independent series, suitable for an ordinary bootstrap; 0.5; and 0.9, which produces an extremely correlated series, far beyond what one normally encounters in practice. For $c=0.0$ and 0.5, a sample size of $n=100$ is used. When $c=0.9$, the sample size is set to 500. Block sizes (the horizontal axis) range from $b=1$ (an ordinary bootstrap) to $b=25$ (which is very large when $n=100$).

The test statistic in all cases is the sample mean. Two different aspects of this test statistic are studied. Figures 3-22 (on page 174) through 3-27 pertain to estimating the standard error of the mean. Figures 3-28 through 3-33 pertain to estimating the 0.1 quantile of the sample mean relative to the population mean. This quantity is typically used for computing confidence intervals for the population mean, as well as for performing hypothesis tests on the value of the population mean.

Within each of these two sets of tests and for each of the three values of the AR(1) weight c, two figures are presented. The first in each pair is the average bias of the quantity being estimated (this quantity being the standard error of the mean or the 0.1 quantile of the mean). This bias is the estimated value minus the true value, which, of course, is known analytically. The second figure in each pair is the root-mean-square (RMS) error of the estimate.

Each figure contains two graphs. The dotted line pertains to the stationary bootstrap, and the solid line pertains to the tapered block bootstrap. The following points are worthy of note:

- When estimating the standard error of the mean, the tapered block bootstrap usually outperforms the stationary bootstrap in terms of both bias and RMS error.

- The relative performance of the two bootstraps in this situation depends of the nature of the block size error. When the block size is near its best value, the tapered block bootstrap is superior. This continues to hold true if the block size is too large. But if the block size is too small, the stationary bootstrap can be slightly superior.

- The relationship just described for the standard error of the mean does not hold for the 0.1 quantile of the mean. In fact, at the best block size, the two algorithms perform nearly identically in terms of RMS error. And if the block size is too large, the stationary bootstrap can significantly outperform the tapered block bootstrap in terms of RMS error.

- If the data is truly independent but a dependent bootstrap is still used, as long as the block size is small, the price paid in RMS error is not terribly serious. The bias is significantly increased, but this is probably less important than the RMS error.

- If the data has serial dependence, using even a small, very suboptimal block size is tremendously better than using an ordinary bootstrap (block size 1). The falloff in both bias and RMS error for both the standard error and the 0.1 quantile is rapid as the block size increases from 1. And *overspecifying the block size is much less harmful than underspecifying it.*

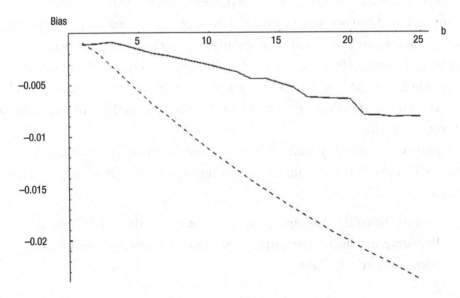

Figure 3-22. *Bias of standard error with c=0.0*

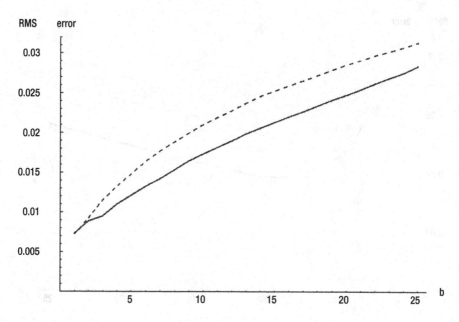

Figure 3-23. *RMS error of standard error with c=0.0*

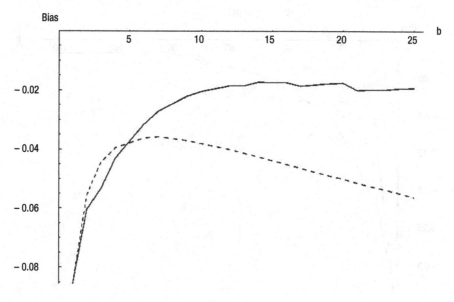

Figure 3-24. *Bias of standard error with c=0.5*

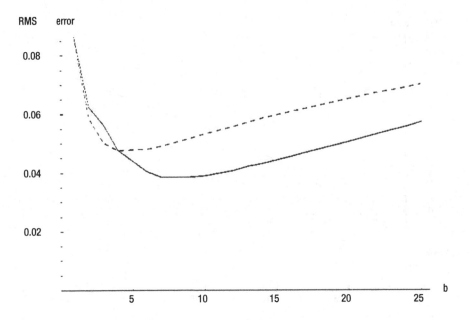

Figure 3-25. *RMS error of standard error with c=0.5*

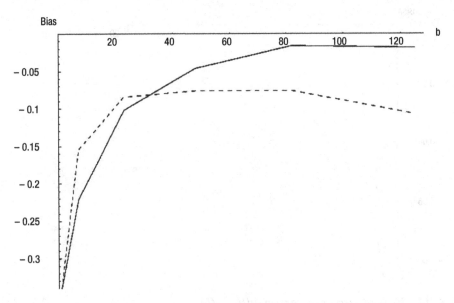

Figure 3-26. *Bias of standard error with c=0.9*

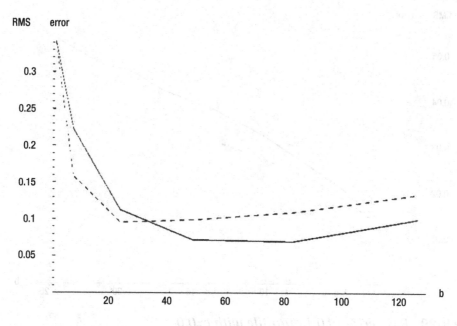

Figure 3-27. *RMS error of standard error with c=0.9*

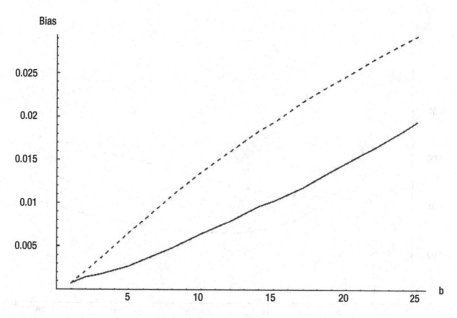

Figure 3-28. *Bias of 0.1 quantile with c=0.0*

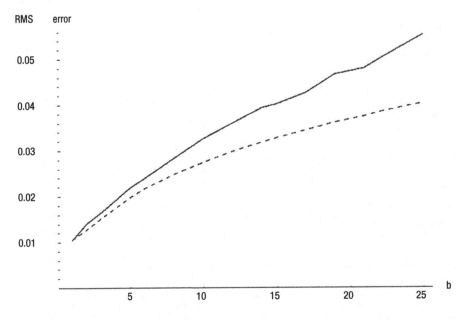

Figure 3-29. *RMS error of 0.1 quantile with c=0.0*

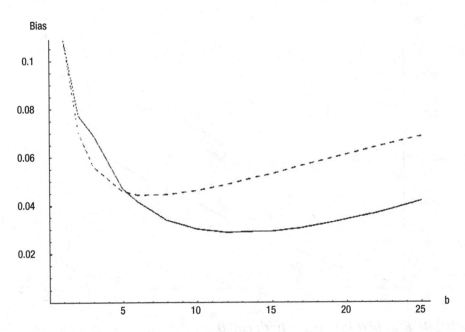

Figure 3-30. *Bias of 0.1 quantile with c=0.5*

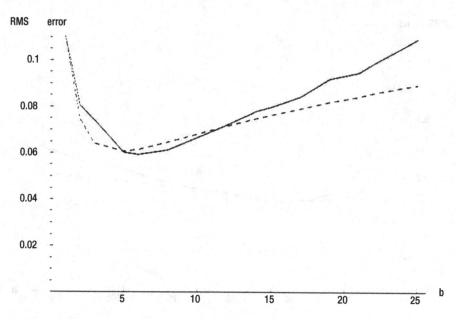

Figure 3-31. *RMS error of 0.1 quantile with c=0.5*

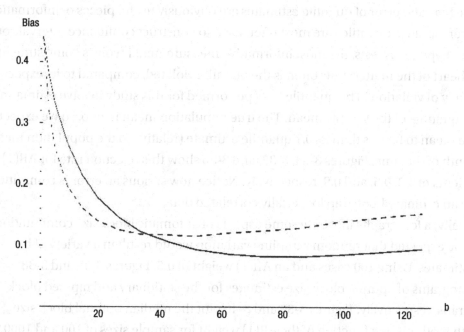

Figure 3-32. *Bias of 0.1 quantile with c=0.9*

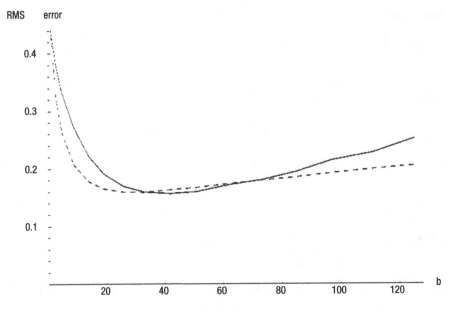

Figure 3-33. *RMS error of 0.1 quantile with c=0.9*

The bias and error of quantile estimates are obviously useful pieces of information. However, because quantiles are most often used to construct confidence intervals or perform hypothesis tests, the most informative measurement involves going straight to the heart of the matter: how often is the quantile violated, compared to its expected frequency of violation? The quantile tests performed for this study involve estimating the 0.1 quantile of the sample mean. The true population mean is zero, so we expect the sample mean to be less than its 0.1 quantile estimate (relative to the population mean) one-tenth of the time. Figures 3-34, 3-35, and 3-36 show this rejection rate for AR(1) coefficients of 0.0, 0.5, and 0.9, respectively. Notice how serious an error is committed if we use an ordinary bootstrap for serially correlated data!

Finally, a few graphs illustrate some aspects of automatic block size computation. It is to be expected that random sampling variation would result in a variety of block size estimates. Using 100 cases and an AR(1) weight of 0.5, Figures 3-37 and 3-38 are histograms of optimal block size estimates for the stationary and tapered block bootstraps, respectively. Figures 3-39 and 3-40 plot the median optimal block size for both methods as a function of the AR(1) weight for sample sizes of 100 and 1000, respectively.

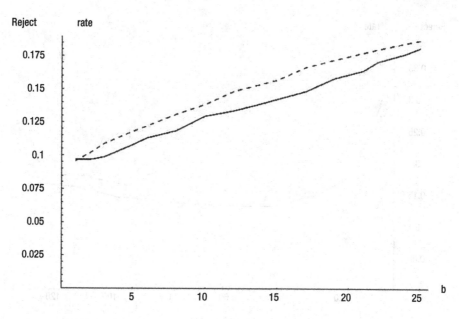

Figure 3-34. *Rate of 0.1 rejection with c=0.0*

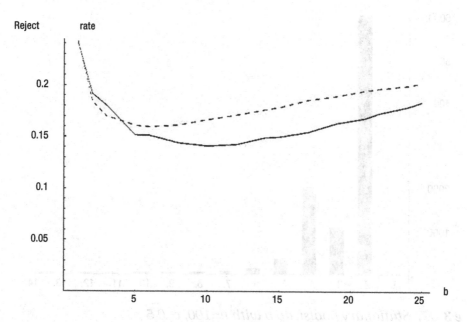

Figure 3-35. *Rate of 0.1 rejection with c=0.5*

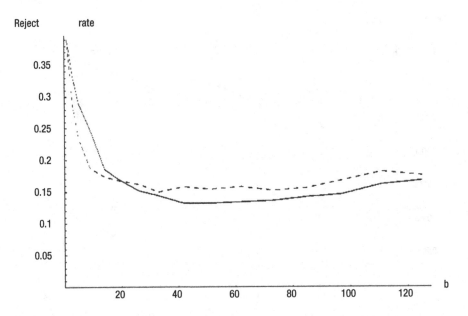

Figure 3-36. *Rate of 0.1 rejection with c=0.9*

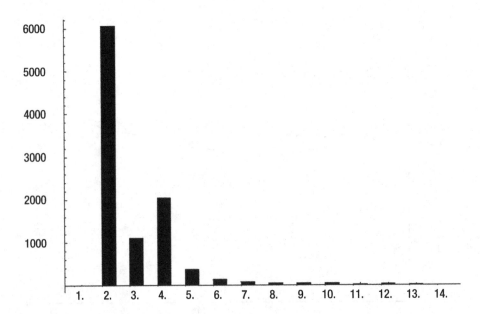

Figure 3-37. *Stationary bootstrap b with n=100, c=0.5*

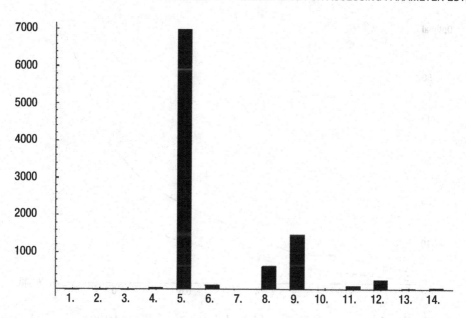

Figure 3-38. *Tapered block bootstrap b with n=100, c=0.5*

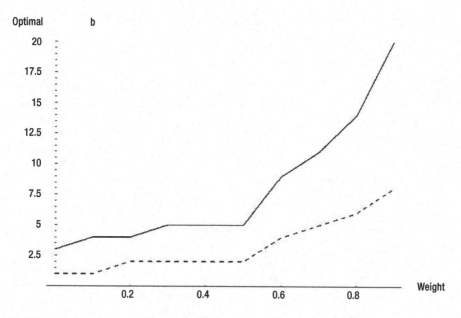

Figure 3-39. *Stationary and tapered block bootstrap b, n=100*

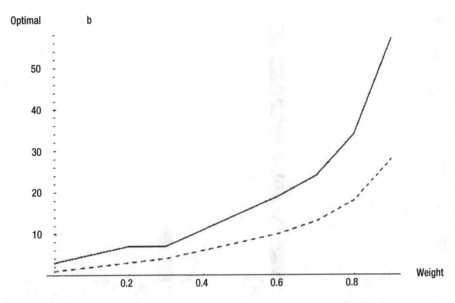

Figure 3-40. *Stationary and tapered block bootstrap b, n=1000*

CHAPTER 4

Resampling for Assessing Prediction and Classification

- Partitioning the Error

- Cross Validation

- Bootstrap Estimation of Population Error

- Efron's E0 and E632 Estimates of Error

- Comparing the Error Estimators for Prediction and Classification

The most common procedure for assessing the performance of a prediction or classification model is to split the data collection into two subsets. The model is trained with one dataset, the *training set*, and then tested with a completely independent dataset, the *test set* or *validation set* or *out-of-sample set*. (The choice of term is usually personal preference.) A performance measure, such as mean squared error, median absolute error, profit factor, cost, or any other custom figure, is computed for the independent set. This performance measure provides a quality judgment for the model. Naive researchers stop here. But enlightened researchers go one step further. They use the methods described in Chapter 3 to evaluate the reliability of the performance measure.

© Timothy Masters 2018
T. Masters, *Assessing and Improving Prediction and Classification*,
https://doi.org/10.1007/978-1-4842-3336-8_4

This chapter presents an entirely different way to use resampling to assess the performance of a prediction or classification model. The problem with the methods shown earlier is that in real-life applications it may be economically or logistically infeasible to collect two large datasets, one for training and one for testing. Perhaps we have been handed the only dataset we will ever see for this project, and it is so small that splitting it into training and validation sets is out of the question. What do we do? There are sophisticated techniques for using a single dataset as both a training set and a validation set. Almost miraculously, we can compute performance measurements that are nearly as reliable as those produced by more straightforward methods. We do pay a huge price in computation time and a modest price in complexity. But if we can afford the cost, these methods are valuable. This chapter describes resampling algorithms for model quality assessment.

Partitioning the Error

To describe the ideas behind the algorithms of this chapter, some notation is necessary. This notation is largely concerned with partitioning the model's error into several components. Unfortunately, there is little notational consistency in the existing literature. The notation chosen for this text is generally followed by at least one other authority, and it follows conventions common to related fields. It is quite different from the notation used in some of the leading texts on this subject, but the ideas are the same. They are simply expressed with different symbols.

We collect a dataset from a population having unknown distribution F. This dataset consists of n cases. Each case comprises a predictor variable x, which may be a scalar or vector, and a predicted variable y, which will always be a scalar. Sometimes it will be convenient to emphasize the fact that the predictor and predicted are inseparable by using the letter t (for *training*) to represent them both. Case i in this dataset is depicted by $t_i=(x_i, y_i)$. Note that although this notation implies numerical prediction, classification can be handled equally well by letting y_i identify the class membership of case i.

The entire dataset is represented by T. When we train our model using T as a training set, the trained model is called η_T. When this trained model is applied to a value x of the predictor variable in order to estimate the corresponding value of the predicted variable, the model's prediction is called $\eta_T(x)$. Sometimes for economy we will abbreviate a generic prediction of a model simply by η.

When the model makes a prediction, we may want to compare the prediction η with the true value y of the predicted variable and measure the error in some manner. Let $Q[y, \eta]$ be the error measure. A traditional error measure for models is mean squared error, in which case the squared difference shown in Equation (4.1) would be employed. For classification, we might define $Q[y, \eta]$ to be zero if $y=\eta$ and one otherwise.

$$Q[y, \eta] = (y-\eta)^2 \tag{4.1}$$

Suppose the trained model is applied to each member of the dataset used to train it. The mean error (as defined by whatever Q happens to be) across all of these training cases is called the *apparent error*. The most common procedure is to train the model by minimizing this mean error. However, there is nothing in our development that requires this to be so. We may optimize any measure we want and then judge the quality of the trained model using a different criterion. The apparent error is defined in terms of the model's performance when it is applied to the training cases and the Q function is applied to each prediction. The method of training is irrelevant. The apparent error is defined in Equation (4.2).

$$\text{err}_{\text{App}} = \frac{1}{n} \sum_{i=1}^{n} Q(y_i, \eta_T(x_i)) \tag{4.2}$$

The apparent error is well known to be optimistically small (called *training bias* in prior chapters) because the model η_T is being tested on the very data on which its training was based. When sufficient data is available, we deal with this situation by testing the trained model on a new dataset. This lets us estimate the expected error of the trained model when it is put to use in the general population. The expected error of a model that has already been trained on the training set is called the *prediction error*. The key here is that we have a particular trained model in hand and are speaking of its expected future performance. The prediction error is expressed in Equation (4.3). This equation use capital X and Y out of deference to the usual mathematical convention that capitals are used to designate hypothetical random variables as opposed to actually observed values. The expectation is over the population F of (X, Y) pairs. The model η_T is defined by the fixed training set T.

$$\text{err}_{\text{Pred}} = E_F\left[Q\left[Y, \eta_T(X) \right] \right] \tag{4.3}$$

There is another nearly identical type of error that is different in a subtle but important way. The prediction error just shown is based on a realized model. What if we expand the expectation to include all possible training sets? In other words, T consists of n cases drawn from F. We can speak of the prediction error for the model trained on this T, and we could even approximate this prediction error by means of an independent validation set. But now suppose we draw an entirely new training set and repeat the procedure. We would surely obtain a slightly different prediction error. We need to be able to consider the expected prediction error across all possible training sets. The *population error* is defined in Equation (4.4).

$$\text{err}_{\text{Pop}} = E_F\left[\text{err}_{\text{Pred}}\right] \tag{4.4}$$

We already know that for any training set, the apparent error will almost certainly be optimistically less than the prediction error. This is because the model will attempt to accommodate idiosyncrasies of the training set, to the detriment of its performance in the general population. The degree to which the prediction error exceeds the apparent error is called the *excess error*. (In earlier chapters this effect has been called training bias, an ill-defined term. Here we must become rigorous.) This is shown in Equation (4.5).

$$\text{err}_{\text{Excess}} = \text{err}_{\text{Pred}} - \text{err}_{\text{App}} \tag{4.5}$$

The most important type of error for this chapter is the *expected excess error*. Recall that the prediction error and the apparent error are both dependent on a particular training set T. Just as we did for the population error, we want to consider the expected value of the excess error across all possible training sets. This is shown in Equation (4.6).

$$\text{err}_{\text{ExpExs}} = E_F\left[\text{err}_{\text{Pred}} - \text{err}_{\text{App}}\right] = E_F\left[\text{err}_{\text{Pred}}\right] - E_F\left[\text{err}_{\text{App}}\right] \tag{4.6}$$

This equation illustrates the fact that we can think of expected excess error in either of two ways. We can consider individual training sets and look at the excess error of each, computing the expectation of this difference across all possible training sets. This is probably the most straightforward concept. Alternatively, we can think of the expected apparent error, subtracting this from the population error. You should carefully ponder both interpretations.

Cross Validation

Cross validation is widely known in the research community. Of the algorithms presented in this chapter, cross validation is the simplest to understand and implement, and it is often the fastest. It has the desirable property of being almost completely unbiased in its estimate of the model's expected future performance. And it has very wide applicability, while some of the best algorithms cannot be used in conjunction with some important model training algorithms. But it has one serious disadvantage: its variance is large, often unacceptably so. In other words, it is excessively susceptible to the vagaries inherent in the random selection of the original dataset. The experimenter collects a sample, trains a model using this sample, and then uses cross validation to assess the expected future performance of the model. But if the experimenter collects a second independent sample and performs the same procedures, the figure produced by cross validation this time may be dramatically different from the figure produced the first time. The fact that this figure is nearly unbiased (on average it will be correct) is of little comfort in the face of such large variance. And the worst part of all is that most experimenters don't know this is the case. Cross validation is widely used and trusted by people who do not know that there are superior methods of doing essentially the same thing. Nonetheless, because cross validation is easy and sometimes our only choice, this section describes the procedure.

The concept behind cross validation is almost trivially simple. Split the dataset into two parts, using one part for training the model and the other part for independent testing of the model, exactly as we would do if we had a surfeit of data. But rather than making the validation set large enough for a reliable performance estimate, make it tiny, allocating the vast majority of the cases to the training set. In fact, it is common to use just one case in the validation set. Then, after training and testing the model with this split, merge the validation set back into the dataset and remove a different validation set. Repeat this procedure until all cases have taken part in testing once. The mean test error across all replications is the cross validation error estimate.

The most common and effective version of cross validation withholds one case at a time. This is the method that will be described here. It should be obvious how to modify the algorithms for withholding multiple cases. We already have the notation that η_T is the model after training with the complete dataset T. Let $\eta_{T(i)}$ represent the model after it has been trained using a dataset consisting of T without case i. Then the cross validation estimate of the model's expected future error is given by Equation (4.7).

$$\text{err}_{\text{CV}} = \frac{1}{n} \sum_{i=1}^{n} Q\left[y_i - \eta_{T(i)} \right]$$

(4.7)

The code for cross validation may be found in the file BOOT_C_1.CPP on my web site. This file also contains code for all of the algorithms in this chapter, as well as a test program. Here is the cross validation code. An explanation follows. Some comparative test results appear at the end of this chapter.

```
void cross_validation (
    int n,                          // Number of cases in sample
    int npred,                      // Number of predictor variables
    double *data,                   // N by npred+1 set of predictors followed by predicted
    void (*tt) (                    // Train and test model
        int ntrain,                 // Number of training cases
        int ntest,                  // Number of test cases
        int npred,                  // Number of predictor variables
        double *train,              // Ntrain by npred+1 matrix of predictors followed by predicted
        double *test,               // Above is training set; This is validation set
        double *predicted),         // Output of validation set predictions
    double *mean_err                // Output of mean error estimate
    )
{
    int i, j;
    double temp, *excluded, *test, train_err, test_err;

    *mean_err = 0.0;                        // Will cumulate mean error here
    test = data + (n - 1) * (npred + 1);    // Last case in dataset

    for (i=0; i<n; i++) {                    // Exclude one case at a time
        excluded = data + i * (npred + 1);   // Excluded case
        for (j=0; j<=npred; j++) {            // Swap excluded case to test spot at end
            temp = test[j];
            test[j] = excluded[j];
            excluded[j] = temp;
        }
```

190

```
  tt (n-1, 1, npred, data, test, &temp);
  err = q (test[npred], temp);        // User's error measure (true, predicted)
  *mean_err += err;                   // Cumulate for mean error

  for (j=0; j<=npred; j++) {          // Swap to restore original order
    temp = test[j];
    test[j] = excluded[j];
    excluded[j] = temp;
    }
  }
 *mean_err /= n;
}
```

The dataset being cross validated is assembled into one array. Each case occupies npred+1 positions because the predictor variables come first, with the predicted variable last. The user must supply a subroutine called tt that fits a model to a supplied training set and computes the predicted values for a validation set.

The last case's position in the data array will always be used as the test spot. The cross validation loop swaps other cases into this spot one at a time. The mean error is cumulated and returned to the user. One reason for the popularity of cross validation is apparent: it is almost trivially easy to implement.

Note, by the way, that although cumulating the mean error across validation folds is the most common procedure, it is not mandatory. Sometimes it can be useful to preserve the predictions in a single vector and then compute a grand performance measure for the vectors of predicted and true values. This facilitates sophisticated error measures such as quantiles of absolute differences.

Bootstrap Estimation of Population Error

This section describes how a straightforward bootstrap algorithm can be used to estimate population error. This algorithm is *not* recommended for general use; the E0 and E632 algorithms described in the next two sections are almost always better. However, the direct bootstrap described here is the foundation on which the more advanced algorithms are built. Understanding its operation is essential to understanding of its successors.

It would be useful to review the bootstrap estimator of bias described on page 106, as the underlying concepts are similar to those studied here. In our current situation, we have a population F from which we randomly draw the cases that make up our training set T. We train the model and test its performance on T, giving its apparent error. This error is optimistic, and the population error can be expected to exceed the apparent error by what has been called the *excess error*. Unfortunately, without an independent validation set, we do not know either the excess error or the population error, which are the quantities that we really want to know. All we know is the optimistic apparent error. However, we can use the bootstrap to estimate the excess error, which can then be added to the apparent error to *roughly* estimate the population error.

How do we use the bootstrap to estimate the excess error? We do it almost the same way we estimated the bias of a parameter: let the empirical distribution function $\hat{F}$ play the part of the unknown $\hat{F}$, and take numerous bootstrap samples from $\hat{F}$. For each bootstrap collection, train the model. The model's error on the bootstrap sample is the apparent error for that sample. The model's error on the complete dataset is the prediction error, because $\hat{F}$ is the entire population. Subtracting gives the excess error for this bootstrap sample. The average of this quantity over a large number of bootstrap replications is the expected excess error.

We can make the algorithm more rigorous by means of a few definitions and equations. Let $T^\star$ be a training set created by bootstrap sampling from T. In other words, we take n cases from T, randomly selected with replacement. Train the model with $T^\star$. The prediction error of this model is computed by averaging its error across the cases in the population from which the bootstrap was drawn, the original dataset. This is expressed in Equation (4.8).

$$\mathrm{err}_{\mathrm{Pred}}\left(T^\star\right) = \frac{1}{n}\sum_{i=1}^{n} Q\left(y_i, \eta_{T^\star}\left(x_i\right)\right) \tag{4.8}$$

Let k_i be the number of times original case i appears in $T^\star$. The apparent error associated with $T^\star$ is computed by averaging the model's error across the cases in $T^\star$. This is shown in Equation (4.9).

$$\mathrm{err}_{\mathrm{App}}\left(T^\star\right) = \frac{1}{n}\sum_{i=1}^{n} k_i\, Q\left(y_i, \eta_{T^\star}\left(x_i\right)\right) \tag{4.9}$$

The excess error associated with $T^\star$ is the difference between its predicted and apparent errors. It would be wasteful to compute both Equations (4.8) and (4.9) and subtract, because the same error terms appear in both. Factoring out this common quantity gives us Equation (4.10).

$$\text{err}_{\text{Excess}}\left(T^\star\right)=\frac{1}{n}\sum_{i=1}^{n}\left(1-k_i\right)Q\left(y_i,\,\eta_{T^\star}\left(x_i\right)\right) \tag{4.10}$$

Equation (4.10) must be evaluated for each of many (several hundred to several thousand) bootstrap replications. The average excess error over these replications provides the estimated expected excess error, which is added to the apparent error for the full sample to *roughly* estimate the population error.

The file BOOT_C_1.CPP on my web site contains a subroutine for computing the bootstrap estimate just described. This file also contains subroutines for all of the algorithms in this chapter, along with a test program that compares their performance. Here is a listing of the bootstrap routine:

```
void bootstrap (
   int n,                    // Number of cases in sample
   int npred,                // Number of predictor variables
   double *data,             // The n by npred+1 dataset of predictors followed by predicted
   int nboot,                // Number of bootstrap replications
   void (*tt) (              // Train and test model
      int ntrain,            // Number of training cases
      int ntest,             // Number of test cases
      int npred,             // Number of predictor variables
      double *train,         // Ntrain by npred+1 matrix of predictors, predicted
      double *test,          // Above is training set; This is validation set
      double *predicted),    // Output of validation set predictions
   double *mean_err,         // Output of error estimate
   double *bootsamp,         // Work area n * (npred+1) long
   double *predicted,        // Work area n long
   int *count                // Work area n long
   )
{
   int i, rep, k;
   double err, apparent, excess, *tptr;
```

```
   excess = 0.0;                       // Cumulates excess error

 for (rep=0; rep<nboot; rep++) {  // Do all bootstrap reps (b from 1 to B)
   memset (count, 0, n * sizeof(int));

   for (i=0; i<n; i++) {                // Generate the bootstrap sample
     k = (int) (unifrand() * n);        // Select a case from the sample
     if (k >= n)                        // Should never happen, but be prepared
       k = n - 1;

     memcpy (bootsamp + i * (npred+1),  // Put case in bootstrap sample
           data + k * (npred+1), (npred+1) * sizeof(double));
     ++count[k];                        // Count inclusion of this case
     }

   tt (n, n, npred, bootsamp, data, predicted); // Train and predict

   for (i=0; i<n; i++) {                // Compute mean error.
     tptr = data + i * (npred+1);  // This case is here
     err = q (tptr[npred], predicted[i]);
     excess += (1.0 - count[i]) * err; // Equation (4.10)
     }
   } // For all bootstrap replications

 excess /= n * nboot; // Computed excess error is grand mean: Equation (4.10)
/*
 Compute apparent error. Add it to excess to get population error estimate.
*/
 tt (n, n, npred, data, data, predicted); // Train and predict

 apparent = 0.0;
 for (i=0; i<n; i++) {
   tptr = data + i * (npred+1);           // This case is here
   err = q (tptr[npred], predicted[i]);
   apparent += err;
   }

 apparent /= n;

 *mean_err = apparent + excess;
}
```
194

Efron's E0 Estimate of Population Error

The second problem with duplication of test cases in the model's training set is that some models are rendered inoperative by such duplication. The standard example is probabilistic and general regression neural networks. Unless the smoothing constant is very large, a test case that is identical to a training case will produce an essentially perfect prediction. This takes the problem of optimism to an extreme.

Bradley Efron proposed a straightforward way of circumventing the problem: prevent it. Choose the bootstrap training sets as usual. But for each bootstrap replication, test only those original cases that are not present in the training set. Compute the mean error of these cases and use this as the estimate of the population error. This is called the *E0* estimate of population error.

The literature contains two methods for computing E0. The original method, described in [Efron, 1983], sums all errors and divides by the total number of cases tested. This is the method that will be presented here. More recent research into the theoretical underpinnings of E0 described in [Efron and Tibshirani, 1993] leads to an algorithm that splits the averaging into two steps. First, for each original case, the errors for all bootstrap replications not containing that case are summed and then divided by the number of such replications, giving a mean error for that case. Then, this mean is summed across all cases and divided by the number of cases to get a grand mean. This latter method is more computationally complex than the former. However, they are asymptotically identical, and their performance in both simulations and actual applications is essentially identical. Therefore, the former is chosen here. If you're interested, you should have no problem modifying the supplied code to split the averaging as just described.

Some new notation is needed in order to rigorously describe the algorithm. As usual, T is the original dataset and $T^{\star}$ is a bootstrap sample. Let $T^{\triangledown}$ designate the set of cases in T that are not in $T^{\star}$. Let $\#(T^{\triangledown})$ be the cardinality of (number of cases in) $T^{\triangledown}$. Then Efron's original E0 estimate of population error is given by Equation (4.11).

$$E0 = \frac{\sum\limits_{b=1}^{B}\sum\limits_{i \in T_b^{\triangledown}} Q\left(y_i, \eta_{T_b^{\star}}(x_i)\right)}{\sum\limits_{b=1}^{B} \#\left(T_b^{\triangledown}\right)} \tag{4.11}$$

The file BOOT_C_1.CPP on my web site contains a subroutine to compute E0. Here is a listing of the E0 routine:

```
void E0 (
    int n,                      // Number of cases in sample
    int npred,                  // Number of predictor variables
    double *data,               // The n by npred+1 dataset of predictors followed by predicted
    int nboot,                  // Number of bootstrap replications
    void (*tt) (                    // Train and test model
        int ntrain,                 // Number of training cases
        int ntest,                  // Number of test cases
        int npred,                  // Number of predictor variables
        double *train,              // Ntrain by npred+1 matrix of predictors, predicted
        double *test,               // Above is training set; This is validation set
        double *predicted),         // Output of validation set predictions
    double *mean_err,       // Output of error estimate
    double *bootsamp,       // Work area n * (npred+1) long
    double *predicted,      // Work area n long
    int *count              // Work area n long
    )
{
    int i, rep, k, ntot;
    double err, *tptr;

    *mean_err = 0.0;            // Cumulates excess error
    ntot = 0;

    for (rep=0; rep<nboot; rep++) { // Do all bootstrap reps (b from 1 to B)

        memset (count, 0, n * sizeof(int));

        for (i=0; i<n; i++) {               // Generate the bootstrap sample
            k = (int) (unifrand() * n);     // Select a case from the sample
            if (k >= n)                     // Should never happen, but be prepared
                k = n - 1;
```

```
    memcpy (bootsamp + i * (npred+1),    // Put case in bootstrap sample
        data + k * (npred+1), (npred+1) * sizeof(double));
    ++count[k];                          // Count inclusion of this case
    }

tt (n, n, npred, bootsamp, data, predicted); // Train and predict

    for (i=0; i<n; i++) {                // Compute mean error.
        if (count[i])                    // If this case was in training set
            continue;                    // Do not test it
        tptr = data + i * (npred+1);     // This case is here
        err = q (tptr[npred], predicted[i]);  // Error of true, predicted
        *mean_err += err;                // Cumulate numerator of Equation (4.11)
        ++ntot;                          // And denominator
        }
    } // For all bootstrap replications

if (ntot)                                // Extremely unlikely to fail in practice
    *mean_err /= ntot;                   // But avoid division by zero
}
```

This code is similar to the previous routine. The bootstrap samples are created and used for training in the same way. The count vector is used only as a true/false flag here, as opposed to an actual counter in the previous routine. You might observe a small inefficiency in this implementation of Equation (4.11). For each bootstrap replication, the prediction needs to be computed only for cases not in the training set. But this routine computes the prediction for all cases, going on to compute and sum the error only for the appropriate cases. This inefficiency allows use of a general training/ predicting routine like tt(), which simplifies coding. In the vast majority of modeling situations, the training time is so much greater than the prediction time, often by orders of magnitude, that eliminating unnecessary predictions would produce no discernable increase in speed. However, for those applications in which prediction is unusually slow, it may be better to send a flag vector to tt() or use some other method to limit predictions.

Efron's E632 Estimate of Population Error

The E0 estimator described in the previous section is excellent and can definitely be recommended for general use. It never tests a case that is also in the training set, meaning that it can be used for any model. And its variance is reasonably small, about as good as can be expected with the current state of the art. It has only one trait that some may consider problematic in some applications: it is moderately conservatively biased. On average, it tends to overestimate the true population error of the model. Keep in mind that as long as this characteristic is not taken to an extreme, conservative bias is not generally regarded as a serious problem. Underestimation of the population error, as happens with the ordinary bootstrap, is far more dangerous because it leads to false optimism. When the E0 method is used, we can usually be confident that the actual population error is more likely to be less than the value we compute, rather than greater. This should be a comfort. Nevertheless, it may be worthwhile to attempt to reduce the inherent conservative bias of E0 to reduce the possibility that we may erroneously reject a model based on unjust pessimism. The E632 estimator attempts to do this.

Consider for a moment the problem with using apparent error (the error resulting from testing the training set) as an estimate of population error. The testing procedure is not fair because only cases used for training are tested. This set is not representative of the entire population because it is too similar to the training set, resulting in optimism. Now consider E0. Exactly the opposite happens. Training cases are deliberately excluded from testing, meaning that the test set is not representative of the entire population because its cases are too dissimilar to the training set. In real life, cases identical to or almost identical to training cases will appear. E0 discourages this, so E0 is pessimistic.

The E632 algorithm seeks a compromise between these two extremes. A fair approach might be to test the model by sampling inside and outside the training set with probabilities that reflect how this would happen in real life. More easily, we could adjust for sampling after the fact. As the sample size grows, the probability that any given case appears in a given bootstrap sample converges to $1-e^{-1}\approx0.632$. The heuristic suggested by Efron is to estimate the population error as a weighted sum of E0 and the apparent error, using weights determined by the sampling probabilities. His estimator, called E632, is given by Equation (4.12).

$$E632 = 0.632 \ E0 + 0.368 \ Err_{App} \tag{4.12}$$

There is no need to list code for computing E632, as it is straightforward. A suitable subroutine can be found in the file BOOT_C_1.CPP on my web site.

Comparing the Error Estimators for Prediction

This chapter has presented a variety of methods for computing estimates of the population error of a model. The following points should be remembered:

- Cross validation is the easiest to implement and often the fastest to execute. It works for any model, and it is almost perfectly unbiased. But it often has dangerously high variance, especially if the training is unstable. For this reason, cross validation is not generally recommended unless there are no alternatives. It is good but not the best.

- Straight bootstrapping is probably the worst possible choice. It is not applicable to models that cannot handle duplicate training cases, nor can it be used with models that are compromised by test cases that are also in the training set. And it is dangerously biased toward underestimating the true population error. It really has nothing to recommend it.

- E0 is almost always the best choice when the training procedure allows duplicate cases. It is applicable to many models because it never tests training cases. It is not terribly sensitive to unstable learning situations, such as models whose training depends heavily on stochastic methods. In practice, its variance approximately ties cross validation in stable learning environments, and it significantly outperforms cross validation in unstable environments. Its only disadvantage is that it tends to overestimate the true population error. This is annoying and occasionally problematic, but conscientious researchers should not usually be troubled by this.

- E632 is a suitable alternative to E0 when the pessimistic bias of E0 is worrisome. It is almost always applicable to any model because even models like probabilistic nets that disallow general testing of cases that are also in the training set, allow this testing when it is known in advance that the training set in its entirety is being tested. This means that the apparent error, needed for E632, can be computed along with the E0 that can always be computed. But keep in mind that researchers who care will prefer the conservative nature of E0 over E632's honesty that is subject to variance.

The file BOOT_C_1.CPP on my web site contains subroutines for all of these algorithms, along with a test program. This program generates artificial data that follows the model shown in Equation (4.13). In this model, x_1 and x_2 have a standard normal distribution. The error term ε is normally distributed with mean zero and variance specified by the user on the command line.

$$y = x_1 - x_2 + \varepsilon \qquad (4.13)$$

An ordinary linear model is fit to the dataset, and then all of the algorithms of this chapter are used to estimate the population mean squared error. The fit model is also tested with independent test data to ascertain its true error. This process is repeated for many tries, as specified by the user. Averages across all of these tries are printed, along with the standard deviation of the various estimators. Using an error variance of 1.0, a sample size of 15 cases, and 1,000 bootstrap replications, the mean population error was 1.260. The results are summarized in the following table:

	Mean Estimated Error	Standard Deviation
Cross validation	1.285	0.547
Bootstrap	1.207	0.494
E0	1.622	0.679
E632	1.321	0.546

As expected, the straight bootstrap on average underestimated the true error, giving 1.207 rather than 1.260. At the other extreme, E0 significantly overestimated the error, averaging 1.622. Is this a problem? Conservatives will think not, even though E0's standard deviation of the estimate was also the highest at 0.679. As will be seen in the next table, this degree of overestimation is primarily due to the unusually small sample size, just 15 cases. This is pushing any statistical procedure. But you must decide for yourself. Cross validation and E632 performed almost identically, with cross validation having slightly less bias and E632 having slightly less standard deviation. This closeness is because this model and dataset are extremely stable. Cross validation is subverted primarily by instability. In the current situation, either could be appropriate if low bias is important.

This experiment was repeated with a more realistic sample size of 100 cases. The actual mean error was 1.031. The results are summarized in the following table:

	Mean Estimated Error	Standard Deviation
Cross validation	1.029	0.146
Bootstrap	1.026	0.146
E0	1.060	0.151
E632	1.026	0.146

In this experiment, all four methods performed satisfactorily. Again, E0 overestimated the error, but only by a small amount, less than 3 percent. The other three algorithms underestimated the error by a trivial amount. Again, E0 had the highest standard deviation, but only by about 3 percent. When overall quality is this high, the small overestimation of error produced by E0 is likely to be preferred to the trivial underestimation of the others.

Comparing the Error Estimators for Classification

The previous example showed some differences between the various methods for estimating population error. But it left the incorrect impression that cross validation is comparable to the others in effectiveness. This is because the smooth error function used in that example, mean squared error of a simple model, creates a stable learning environment. Classification is generally not so stable. Small changes in the data can result in sudden large jumps in the error. The example in this section explores this situation.

The file BOOT_C_2.CPP on my web site is practically identical to the file BOOT_C_1. CPP already explored. The algorithms for estimating population error are identical to what we have already seen. The main differences are in the function that defines the prediction error, and the main program that does the comparative tests. The test program generates bivariate data having moderate positive correlation. A scatterplot of the data for a class would show an ellipse whose main axis points up and to the right. Two classes are generated, with their data shifted roughly perpendicular to the main axis by an amount specified by the user on the command line. A linear model is fit to the data, with the predicted variable having the value –1.0 for one class and +1.0 for the other. The prediction error is defined to be 0.0 if the true and predicted values have the same sign, and 1.0 if they have opposite signs.

Two tests were run. The first uses a sample size of 15 cases, 1,000 bootstrap repetitions, 10,000 trials, and a separation of zero. In other words, this tests a hopeless task, as the two classes have identical distributions. As expected, the mean observed error is 0.500. The results of this test are shown in the following table:

	Mean Estimated Error	Standard Deviation
Cross validation	0.499	0.176
Bootstrap	0.442	0.125
E0	0.500	0.110
E632	0.434	0.102

Cross validation again demonstrates that it is essentially unbiased. This should not be surprising in this test because since the model is worthless, there is a 50-50 chance that any given case will be misclassified. The same logic applies to E0. We normally expect that E0 will be pessimistically biased. But this happens only when the model has some effectiveness. In this case, E0's forced exclusion of test cases from the training set is innocuous. Unfortunately, this carries over to E632. Because one of E632's components is truly unbiased and the other is strongly optimistically biased, E632 itself exhibits significant optimistic bias. Be warned.

This test also clearly demonstrates the principal criticism of cross validation: its large variance. The standard deviation of the cross validation estimator here is 0.176, very much inferior to the 0.110 of E0. As is often the case, E632 has the lowest standard deviation of all, 0.102. But this does us little good when it has such strong optimistic bias.

A second test, in which the model has some predictive power, was also run. In this test, the mean observed error was 0.124. The results of this test are shown in the following table:

	Mean Estimated Error	Standard Deviation
Cross validation	0.127	0.092
Bootstrap	0.115	0.086
E0	0.156	0.087
E632	0.125	0.077

This test shows E632 to be the best estimator. It is almost completely unbiased, and it has the lowest standard deviation. However, E0 has only moderately greater standard deviation, and it exhibits its usual comforting pessimistic bias. Considering the situation with the prior test, this should be taken into consideration when a decision between using E0 and E632 is made. Once again, cross validation has the highest standard deviation, and the straight bootstrap has dangerous optimistic bias.

Summary

You might peruse this chapter and be left with the impression that resampling for estimation of a model's population error is interesting but overly complex and not really useful for their projects. This is too bad because the fact is that the resampling methods shown here are almost miraculous and should be taken extremely seriously. Think about it a moment. The straightforward approach to testing a model involves collecting a whole new independent dataset, which is troublesome at best and impossible at worst. The methods shown here avoid this task. The entire data collection can be put to work, simultaneously training the model for future use as well as providing a good estimate of its future effectiveness. Such power is not to be brushed off carelessly.

The chief impediment to resampling for model evaluation is computer time. These methods require training the model dozens, hundreds, or perhaps even thousands of times. Many experts consider 50 bootstrap replications to be sufficient for many tasks, although I am not in this camp. Several hundred replications should do for all but the most demanding applications. When training a model even once taxes available computational resources, training it enough times for a reliable error estimate may be prohibitively expensive. On the other hand, how expensive are test cases? And how expensive would unexpected failure be? These issues are all too frequently overlooked, with finger crossing taking their place. As computer power grows, limited resources should become less and less of a problem. Resampling for error estimation should ideally become the normal development procedure, not a rare and special process.

CHAPTER 5

Miscellaneous Resampling Techniques

- Bagging
- AdaBoost
- Comparing the Boosting Algorithms
- Permutation Training and Testing

Previous chapters explored some uses for resampling techniques. In Chapter 3 we saw that the bias and variance of parameter estimators could themselves be estimated. This included model parameters as well as performance measures based on independent test data. Then in Chapter 4 we saw that performance measures for a model could be safely obtained from the very same data that was used to train the model. In this chapter we will explore assorted methods for using resampling to improve the performance of models. In particular, we will witness a marvelous phenomenon: it will be shown how a model whose performance is only slightly better than random guessing can be used to create a super-model whose performance is markedly better than the original. These techniques are extremely expensive in terms of computational requirements. However, in situations in which performance is more important than cost, resampling methods for model building are priceless.

The essential idea behind most of the algorithms in this chapter is that it is difficult for a single model to learn how to effectively accommodate every possible pattern in the grand population. Instead, we need to train a set of specialist models, each of which focuses on a subset of the training set. Then, when an unknown case is to be processed, all of the specialists are consulted, and their consensus opinion is obtained. The means by which the training set is partitioned, and the method for obtaining a consensus, are what differentiate the algorithms. But the central idea is the same in all.

© Timothy Masters 2018
T. Masters, *Assessing and Improving Prediction and Classification*,
https://doi.org/10.1007/978-1-4842-3336-8_5

Bagging

Sometimes the simplest approach to a problem is nevertheless useful. This is surely the case with *bagging* (short for *bootstrap aggregation*) as described in [Breiman, 1994]. This algorithm is about as simple as it gets, yet in some applications it can substantially reduce error rates. All bagging does is repeatedly take bootstrap samples from the training set and train a copy of the model for each sample. If the model makes a numerical prediction, the bagging prediction is the mean prediction of all of the bootstrap models. If the model is a classifier, then the bagging classification is decided by majority vote of the bootstrap models. As long as at least a few dozen bootstrap replications are done, the bagging result can be good.

The chief criterion for determining if bagging will improve a classifier is the stability of the underlying model. Bagging (and, in fact, most resampling methods) is most useful when the model is unstable. A stable model is one for which small changes in the training set cause only small changes in the model's predictions. For example, ordinary linear regression on data that does not have inordinately heavy tails is generally stable because adding or removing a few training points has small impact on the final regression equation. Also, nearest neighbor classifiers are usually stable because each region of the population space typically has sufficient representation that addition or deletion of a few cases makes no difference; another point is probably lying around to take the place of one that is removed. These models will probably not be aided by bagging. On the other hand, if a model is constructed by stepwise variable selection and the candidate predictors have similar predictive power, the resulting model may be unstable. Replacing just a few cases could cause a different subset of predictors to be chosen, with the result that the final models will be quite different. Models constructed by tree building, or those trained by stochastic processes, also tend to be unstable. Such models are often significantly improved by bagging.

A Quasi-theoretical Justification

Like many resampling techniques, bagging is based more on practical observation that it works than on rigorous theoretical explanations of how it works. The worst ramification of this fact is that it is often difficult to tell in advance whether bagging will help a given application. Nonetheless, it is possible to get a feel for its operation by studying some expressions of error and examining how these errors are interrelated. If you are uncomfortable with statistical theory, you may safely skip this section, as it is not directly related to implementing bagging algorithms.

We have a population F from which we sample cases (x, y), where x is a predictor variable (scalar or vector) and y is the scalar variable that we want to be able to predict. A collection of n of these cases serves as a training set that we shall call T. This training set is used to train a model called η_T. When this model is given a predictor x, its prediction is designated $\eta_T(x)$. For a given fixed training set, we define the model's error as the expectation over (X, Y) of the squared difference between the true and predicted values. In other words, we use ordinary mean squared error.

Right now, we are more interested in the expected value of this error over all possible training sets. Loosely stated, this is the average error we could expect if we collect a training set and train a model, obviously an important quantity. This error is defined in Equation (5.1).

$$e = E_T E_{X,Y} \left(Y - \eta_T \left(X \right) \right)^2 \tag{5.1}$$

Suppose we were able to collect an infinite number of training sets from F and train a model for each. We could define a super-model by aggregating all of these individual models. The prediction of this aggregated model would be the mean of the predictions of the component models. This prediction is given by Equation (5.2). The expected error of this aggregated model is given by Equation (5.3).

$$\eta_A \left(x \right) = E_T \eta_T \left(x \right) \tag{5.2}$$

$$e_A = E_{X,Y} \left(Y - \eta_A \left(X \right) \right)^2 \tag{5.3}$$

Even though we can obviously never realize η_A because it requires infinite sampling from F, it is still useful to compare its expected error, e_A, to that of a single realization of a model, e. The first step toward this goal is to expand Equation (5.1) as shown in Equation (5.4). Then remove the extraneous E_T, substitute the definition of η_A, and interchange the order of expectation, as shown in Equation (5.5).

$$e = E_T E_{X,Y} Y^2 - 2E_T E_{X,Y} Y\eta_T(X) + E_T E_{X,Y} \eta_T^2(X) \tag{5.4}$$

$$e = E_{X,Y} Y^2 - 2E_{X,Y} Y\eta_A(X) + E_{X,Y} E_T \eta_T^2(X) \tag{5.5}$$

Similar expansions can be done for e_A, the hypothetical population bagging error defined in Equation (5.3). Squaring and distributing the expectation gives Equation (5.6). Substituting the definition of η_A gives Equation (5.7).

$$e_A = E_{X,Y} Y^2 - 2E_{X,Y} Y\eta_A(X) + E_{X,Y} \eta_A^2(X) \tag{5.6}$$

$$e_A = E_{X,Y} Y^2 - 2E_{X,Y} Y\eta_A(X) + E_{X,Y} \left(E_T \eta_T(X)\right)^2 \tag{5.7}$$

It can be seen that e_A and e are identical except for the rightmost term in the $E_{X,Y}$ expectation. The squared expectation of a random variable is less than or equal to the expectation of the square of that random variable. This tells us that e_A cannot exceed e due to the inequality shown in Equation (5.8).

$$\left[E_T \eta_T(X)\right]^2 \le E_T \eta_T^2(X) \tag{5.8}$$

The degree to which this inequality is unequal determines the degree to which this (impossible) infinite bagging procedure would improve on a single model. If the training set size is so large and the model so stable that the model changes little as the training set is changed, then the inequality will be nearly equal and bagging will not be much help. But if the various training sets produce highly variable models, the inequality will be large and bagging will be useful.

Unfortunately, we cannot take direct advantage of this property because η_A requires infinite sampling from F. All we can do is collect bootstrap samples from the empirical distribution function $\hat{F}$. (The empirical distribution function is discussed on page 103.) Each bootstrap sample is clearly inferior to the entire training set when it comes to training a model, because numerous cases are ignored. Will the benefit of collecting many such inferior samples, training a model for each, and averaging their predictions overcome the damage due to the individual inferiorities? The answer is that it depends on the model. Bagging is caught in a tug-of-war. On the side of good is the inequality in Equation (5.8), and on the side of evil is the fact that each bootstrapped model is inferior to the model computed from the entire dataset. If the grand model is nearly optimal and stable, evil will win, and the bagged model will actually be slightly inferior to a single model. But if the model is not so excellent, good will win, and the bagged model will be superior. If possible, test both methods before committing to one or the other. A few rounds of large-block cross validation is often revealing. But be comforted that even when bagging proves detrimental, the damage is almost always small. Also be comforted that we will soon see methods that are considerably more effective than bagging, though at the price of increased complexity. It is almost always a price worth paying.

The Component Models

The bagging code that appears in the next section, as well as most of the remaining algorithms in this chapter, requires a set of pointers to Model objects. The file MLFN.CPP on my web site contains a simple implementation of a multiple-layer feedforward network that serves as the test model for these algorithms. The user is free to use any model desired, as long as a few reasonable conventions are followed.

The user provides an array of models ready to be trained. The demonstration program in ARCING_M.CPP makes this a global array and constructs the models in the main module. The convention employed in this chapter, which is the standard in the literature, is that when classifying, each model is able to be trained to predict a numeric value in the range [−1, 1]. There will be one model (output variable) for each class. For a given case, the output corresponding to the case's class will be trained to predict +1, and all other outputs will be trained to predict −1. The code supplied here assumes that the user's model has no inherent limitations that would prevent such training. This code employs a separate model for each predicted output. It should be easy for you to modify this code to use a single model that predicts all outputs, if this is desired. However, most researchers agree that the use of separate models improves performance. Also, some algorithms that appear later in this chapter require separate models for each output. Employing this convention for bagging simplifies the comparative studies that will be performed after all algorithms have been presented.

The bagging algorithm as implemented here will call neither the constructor nor the destructor for any model. It assumes that the models have already been constructed and are globally visible. The following member functions will be called:

Model::reset (): Remove any existing training data and prepare for receiving a new training set.

Model::add_case (double *case): Add a case to the training set. All predictor variables appear first in the vector, followed by the single predicted variable.

Model::add_case (double *case, double prob): This is identical to the prior function except that an importance weight for the case is supplied. Bagging will not use this function, but some of the later algorithms will. The model must not assume that this sums to one.

Model::train (): Train the model.

Model::predict (double *input, double *output): Given an input vector, compute the scalar predicted output value.

If you write your own model, only these member functions are required by the algorithms in this chapter. However, note that neither the sample model in MLFN.CPP nor any of the algorithms in this chapter provide for error checking of any sort. Cautious readers will include whatever degree of protection seems appropriate for the application.

Code for Bagging

My web site contains two versions of the bagging algorithm. The file ARCING.CPP contains a relatively simple version limited to prediction of one variable and strictly binary classification. The file ARCING_M.CPP contains a more general version that is capable of predicting any number of variables simultaneously, as well as classification of any number of classes. The latter version is presented here because it illustrates several important points discussed later. Bagging is implemented as a class in which the constructor is responsible for training the models. Separate member functions for numerical prediction and classification are employed. For clarity, no means of user interruption are provided, and no checks for insufficient memory are performed. Readers who use these algorithms in their own programs should provide for both of these needs. Here is the class declaration:

```
class Bagging {

public:
  Bagging (int n, int nin, int nout, double *ts et, int nmods);
  ~Bagging ();
  void numeric_predict (double *input, double *output);
  int class_predict (double *input);

private:
  int nmodels;          // Number of models (nmods in constructor call)
  int nout;             // Number of outputs
  double *work;         // Work area for holding output and training case
  int *count;           // Classification voting counter
};
```

The constructor for the bagging class trains all of the component models and prepares the bagging model for use as a predictor or classifier. Here is the code for the constructor. An explanation of its operation follows.

```
Bagging::Bagging (
  int n,                // Number of training cases
  int nin,              // Number of inputs
  int nouts,            // Number of outputs
  double *tset,         // Training cases, ntrain by (nin+nout)
  int nmods             // Number of bootstrap replications
  )
{
  int i, k, iboot, iout;
  double *tptr;

  nout = nouts;
  nmodels = nmods;
  work = (double *) malloc ((nin + nout) * sizeof(double));
  count = (int *) malloc (nout * sizeof(int));
/*
  Build the bootstrap training sets and train each model
*/
```

```
for (iboot=0; iboot<nmodels; iboot++) {

   for (iout=0; iout<nout; iout++)
      models[iboot*nout+iout]->reset();  // Prepares the reusable model

   for (i=0; i<n; i++) {                    // Build this bootstrap training set
      k = (int) (unifrand() * n);           // Select a case from the sample
      if (k >= n)                           // Should never happen, but be prepared
        k = n - 1;
      tptr = tset + k * (nin + nout);       // Point to this case
      memcpy (work, tptr, nin * sizeof(double));  // This case's input
      for (iout=0; iout<nout; iout++) {    // Each output handled by separate model
         work[nin] = tptr[nin+iout];        // This output
         models[iboot*nout+iout]->add_case (work);   // Add it to the model's training set
         }
      }

   for (iout=0; iout<nout; iout++)
      models[iboot*nout+iout]->train();   // Train the model
   } // For the nmodels bootstrap replications
}
```

The user supplies a training set to the constructor. Each case consists of the nin predictors followed by the nout predicted values. A work area of sufficient length to hold a case is allocated. An integer work vector that will be used only for classification is also allocated.

The user specifies via nmodels the number of bootstrap models to create. This is typically anywhere from several dozen to several hundred. In actuality, nmodels*nout models will be used, because each output has its own personal model. However, it is convenient to think of models in both ways, according to whichever is clearer at the time. Within a bootstrap replication, the first step is to call reset() for each of the nout models. Since the model objects are reusable, this clears previous training data. Then each model's training set is cumulated. Each model's input vector is the same, so the randomly selected training case has its inputs copied to the work vector. One at a time, each output value is copied to the last position in the work vector, and this vector is appended to the corresponding model's training set via add_case(). After all training sets are complete, the models for this bootstrap replication are all trained. This is all that the constructor needs to do.

Separate member functions are used for numeric prediction and for classification. The numeric routine simply averages the predictions of the component models. Its code is as follows:

```
void Bagging::numeric_predict (double *input, double *output)
{
   int i, imodel;
   double out;

   for (i=0; i<nout; i++)
     output[i] = 0.0;

   for (imodel=0; imodel<nmodels; imodel++) {
     for (i=0; i<nout; i++) {
       models[imodel*nout+i]->predict (input, &out);
       if (out > 1.0)      // This hard limiting aids stability
         out = 1.0;        // Design your model so that [-1,1] is its
       if (out < -1.0)     // natural range, so this has minimal impact
         out = -1.0;
       output[i] += out;
       }
     }

   for (i=0; i<nout; i++)
     output[i] /= nmodels;
}
```

The numeric prediction code shown previously contains one feature that might better be removed for some applications. The predictions made by the component models are hard limited at +/−1 before contributing to the mean prediction. This is because the demonstration application in ARCING_M.CPP, like many real-life applications, is aided by hard limiting when the model is capable of occasional predictions wildly beyond its natural trained range. If the user's application happens to have a different natural range or if meaningless wild predictions from the component models are not possible, this limiting should be changed or eliminated.

The classification member function assumes (as is fairly standard) that training cases are such that the output corresponding to the correct class is +1 and all other outputs are −1. Thus, the classifier chooses the output having maximum value as the correct class. A vote is taken across all bootstrap models, and the winner is returned. This implementation does not attempt to break ties. Here is the code:

```
int Bagging::class_predict (double *input)
{
  int i, imodel, ibest, bestcount;
  double best, out;

  for (i=0; i<nout; i++)      // An output for each class
    count[i] = 0;             // Counts votes for each class

  for (imodel=0; imodel<nmodels; imodel++) {      // All bootstrap models
    for (i=0; i<nout; i++) {                       // Try each class
      models[imodel*nout+i]->predict (input, &out);
      if ((i == 0) || (out > best)) {              // Keep track of max output
        best = out;
        ibest = i;                                 // Winning class so far
        }
      }
    ++count[ibest];                                // Count winners
    }

/*
  At this time, count[i] is the number of times output i was the greatest
*/

  for (i=0; i<nout; i++) {
    if ((i == 0) || (count[i] > bestcount)) {
      bestcount = count[i];
      ibest = i;
      }
    }

  return ibest;
}
```

The files ARCING.CPP and ARCING_M.CPP on my web site contain demonstration programs that compare bagging and the other algorithms of this chapter with each other and with a single model. Detailed experimental results from these programs will appear near the end of this chapter.

AdaBoost

This section describes an algorithm that is similar to bagging in that it attempts to improve the power of a model by training multiple copies of the model on different subsets of the training set. But the means by which this algorithm selects the training cases, and the method for combining the models' results, are much more sophisticated than bagging. It is not always more effective, mind you. But it certainly is a lot more sophisticated than bagging, and the potential for superior performance is there.

AdaBoost is a term first coined by Yoav Freund and Robert Schapire in their classic paper [Freund and Schapire, 1997]. *Boosting* is a generic term describing a technique for combining multiple versions of a weak classifier to produce a better classifier. (It is actually a rigorously defined term from learning theory, but we will avoid needless details here.) The specific technique proposed in the paper just cited is adaptive in that it changes its training procedures to adapt to the dataset and model. This led to the name *AdaBoost*.

There is no longer a single algorithm called AdaBoost. Numerous variations have appeared, and more appear regularly. Most work on this algorithm has been in the area of classification, and only classification will be considered in this text. Numerical prediction is far more complex and less understood at this time. If you're interested in using AdaBoost for numerical prediction, you are advised to search the current literature for the latest methods.

Binary AdaBoost for Pure Classification Models

The simplest version of AdaBoost performs binary (two-class) classification and works with models that provide class decisions only. Later we will see how models that provide confidence indicators along with their class decisions can be used to improve the AdaBoost algorithm. Also, two versions of AdaBoost that can handle any number of classes will be presented later. However, these generalizations are considerably more complex and are not always needed. We begin with the easy case.

One impediment to the general use of AdaBoost is that it performs best when the training algorithm for the underlying model is capable of accepting different degrees of importance to different cases in the training set. AdaBoost will tell the training routine that it is particularly important to focus on being able to correctly classify certain cases, while performance on other cases is less critical. The vast majority of classification models in existence can be modified in this way, so people writing their own custom software should have no trouble implementing AdaBoost. However, if the user is forced to link with a proprietary library that does not have this capability, operation will be somewhat compromised. Nevertheless, AdaBoost can still be used. This topic will be discussed in the next section. For now, we assume that the model is capable of being trained with arbitrary weights assigned to each case.

The AdaBoost algorithm begins by assuming that all training cases are equally important. The model is trained, and the training set is classified using the trained model. The model's overall error is recorded. Then each training case is examined individually. If the model correctly classifies a case, this case's importance for future training is reduced. Conversely, cases that are misclassified are made more important. Another instance of the model is trained using this modified importance distribution. This model will primarily focus on cases that were previously misclassified, paying less attention to cases that the previous model correctly classified. The overall performance of this model is recorded, and the importance distribution is adjusted again. This chain of events is repeated anywhere from a dozen times to hundreds of times. When finished (a usually arbitrary decision at this level of development), we have in our possession a collection of models, each of which focuses on a different weighting of the training cases. We also have a record of the overall performance of each of these models. To classify a test case, all of the models are invoked. A consensus decision is reached by combining all of the classification decisions, assigning importance to each model according to its previously recorded overall performance. Models that did well during training are given more consideration than models that did poorly. The result is a classification decision that is usually superior to what could be made by a single copy of the model trained on the unweighted training set.

To make the binary AdaBoost algorithm more rigorous, let us denote the n cases in the training set by $\{(x_1, y_1), ..., (x_n, y_n)\}$. Each x_i is a predictor, a scalar, or a vector. Each y_i is either +1 or −1, according to the class membership of the case. Decide in advance that we will train a total of T models that will be indexed by $t=1, ..., T$. The binary AdaBoost algorithm proceeds as follows:

Let $t=1$.

Initialize the importance distribution for equal weighting of all cases:

$$D_1(i) = 1/n \quad for\ i=1, ... n \tag{5.9}$$

1) Train model h_t using the current importance distribution, D_t. Each model is trained to make a prediction of +1 or −1, the binary class decision.

2) Compute the error of this model h_t as the total weight of all misclassified cases. One rough interpretation of this quantity is the probability of an incorrect decision. The double brackets, ⟦ ⟧, indicate the value 1.0 when the expression they enclose is true, and 0.0 when false.

$$\varepsilon_t = \sum_{i=1}^{n} \llbracket h_t(x_i) \neq y_i \rrbracket D_t(i) \tag{5.10}$$

3) If the model is perfect ($\varepsilon_t=0$) or worthless ($\varepsilon_t>0.5$), quit iterating, even if t has not yet reached T. Keep this model in the final model set if it is perfect. Discard this model if it is worthless.

4) Compute the reweighting coefficient:

$$\beta_t = \frac{\varepsilon_t}{1-\varepsilon_t} \tag{5.11}$$

5) Update the importance distribution by multiplying the importance of each correctly classified case by β_t. Then normalize D so that it sums to 1.0, allowing us to think of it as a probability distribution. This equation is not nearly as fierce as it looks on first sight. The sum of the two terms in parentheses evaluates to β_t for correct cases and 1.0 for incorrect cases. The denominator normalizes D to sum to one.

$$D_{t+1}(i) = \frac{D_t(i)\left(\llbracket h_t(x_i)=y_i \rrbracket \beta_t + \llbracket h_t(x_i) \neq y_i \rrbracket\right)}{\sum_{i=1}^{n} D_t(i)\left(\llbracket h_t(x_i)=y_i \rrbracket \beta_t + \llbracket h_t(x_i) \neq y_i \rrbracket\right)} \tag{5.12}$$

6) If $t<T$, let $t=t+1$ and go to step 1. Otherwise, we are finished.

You might note that it is not really necessary to update D after training the last model. A trivial amount of work could be saved by breaking out of the training loop after step 3 when $t=T$. However, because the final distribution is sometimes of interest (it identifies the hardest remaining cases) and represents a negligibly small amount of computation, the algorithm is usually implemented as shown previously.

When the AdaBoost classifier is used on a test case, all T component models are evaluated. Each model is given a weight equal to 0.5 times the log of the reciprocal of its β weight. Thus, models having lower error are given a higher weight. The weighted sum of the models' predictions is computed. The sign of this sum defines the AdaBoost classification decision. This is expressed in Equation (5.13). Note that the factor of 0.5 is superfluous, as multiplication by any constant will have no effect on the sign of the sum. It is included here out of deference to some theoretical results not quoted in this text. If you are interested, you can see, for example, [Freund and Schapire, 1997].

$$h_{AdaBoost}(x) = \text{sign}\left[\sum_{t=1}^{T} .5 \log\left(\frac{1-\varepsilon_t}{\varepsilon_t}\right) h_t(x)\right] \tag{5.13}$$

The code for this algorithm can be found in the file ARCING.CPP on my web site. Here is the class declaration for the class that implements binary AdaBoost for models without confidence information:

```
class AdaBoostBinaryNoConf {

public:
   AdaBoostBinaryNoConf (int n, int nin, double *tset, int nmods);
   ~AdaBoostBinaryNoConf ();
   int class_predict (double *input);

private:
   int nmodels;        // Number of models (nmods in constructor call)
   double *alpha;      // Nmods long alpha constant for each model
   double *dist;       // N long probability distribution
   double *h;          // N long work area for saving model's predictions
};
```

The constructor trains all component models and makes ready for AdaBoost classification. Here is a listing of the constructor. An explanation of its operation follows. It may be useful to review the Model class member functions, listed on page 210.

```
AdaBoostBinaryNoConf::AdaBoostBinaryNoConf (
   int n,            // Number of training cases
   int nin,          // Number of inputs
   double *tset,   // Training cases, ntrain by (nin+nout, where nout=1)
   int nmods         // Number of models (T in the text)
   )
{
   int i, imodel;
   double *tptr, temp, eps, beta, out;

   nmodels = nmods;
   alpha = (double *) malloc (nmodels * sizeof(double));      // Will be log(1/beta)
   dist = (double *) malloc (n * sizeof(double));             // D, importance distribution
   h = (double *) malloc (n * sizeof(double));                // Work area

/*
   Initialize distribution to be uniform
*/

   temp = 1.0 / n;
   for (i=0; i<n; i++)                    // Equation (5.9)
      dist[i] = temp;

/*
   Main training loop trains sequence of models
*/

   for (imodel=0; imodel<nmodels; imodel++) { // t from 1 to T

      models[imodel]->reset();        // Prepares the reusable model

      for (i=0; i<n; i++) {              // Build this training set
         tptr = tset + i * (nin + 1);    // Point to this case
         models[imodel]->add_case (tptr, dist[i]); // Add it to the model's training set
         }
```

```
    models[imodel]->train ();                    // Train this model (algorithm step 1)
```

```
/*
    Compute eps as the probability of error. Use this to compute optimal alpha weight.
    The 'if' statement in this loop must be true when the actual class (as stored in the training set) is not
    the predicted class (as defined by the model's prediction). This particular implementation uses a
    model having positive prediction for the first class, negative for the second.
*/
```

```
    eps = 0.0;
    for (i=0; i<n; i++) {                        // Equation (5.10)
        tptr = tset + i * (nin + 1);             // Point to this case
        models[imodel]->predict (tptr, &out);    // Evaluate the model

        if (out > 0.0)                           // If it predicts first class
            h[i] = 1.0;                          // Flag it this way
        else                                     // But if it predicts second class
            h[i] = -1.0;                         // Other flag value

        if ((tptr[nin] * h[i]) < 0.0)            // Error test: true & predicted same sign?
            eps += dist[i];                      // Model erred if signs opposite
    }

    if (eps <= 0.0) {                            // Unusual situation of model being perfect
        nmodels = imodel + 1;                    // No sense going any further
        alpha[imodel] = 0.5 * log ((double) n);  // Arbitrary large weight
        break;                                   // Exit training loop early
    }

    if (eps > 0.5) {                             // Unusual situation of model being worthless
        nmodels = imodel;                        // No sense going any further
        break;                                   // So exit training loop early
    }
```

```
    alpha[imodel] = 0.5 * log ((1.0 - eps) / eps);   // Weight for Equation (5.13)
    beta = eps / (1.0 - eps);                        // Equation (5.11)

/*
    Adjust the relative weighting, then rescale to make it a distribution
*/

    temp = 0.0;
    for (i=0; i<n; i++) {                             // Equation (5.12)
        tptr = tset + i * (nin + 1);                 // Point to this case
        if ((tptr[nin] * h[i]) > 0.0)                // If this case correctly classified
            dist[i] *= beta;                         // Reduce its probability weighting
        temp += dist[i];                             // Cumulate total probability for rescale
    }

    for (i=0; i<n; i++)                              // A probability distribution must sum to 1.0
        dist[i] /= temp;                            // Make it so

    } // For all models
    free (dist);
    free (h);

}
```

The constructor first allocates three work areas. The weights used in Equation (5.13) will be saved in alpha, the importance probability distribution D will be maintained in dist, and h serves to hold the model's predictions from where they are computed and first used to compute the error, to where they are used a second time during the distribution update. This avoids the need to have the model make the same predictions twice. Note that as implemented here, it would be cleaner to let h and dist be automatic variables in the constructor instead of making them be private class members. However, more general implementations of the class may make these quantities available for inspection later. In this case, they should be freed by the destructor instead of at the end of the constructor.

The distribution *D* is initialized to uniformity, and then the main loop begins. The first step in the loop is to reset the model to prepare it to receive a new training set. The training set is copied to the model, and the model is trained. Then each case in the training set is processed by the trained model. By the definition of this version of the algorithm, the model is unable to provide confidence information. It only makes a class decision. This is codified by checking the sign of the model's prediction and explicitly setting the class prediction to +1 or −1 accordingly. In a later algorithm we will see that this potentially useful information does not need to be discarded so casually. It can be put to use. But here we are assuming that the numerical value is meaningless, which is the situation for many models, so we explicitly discard it. If the case is misclassified, we cumulate in eps the importance of this case.

It is possible (though not very likely) that the model is able to classify the training set perfectly. In this happy situation, there is no need to continue. We assign a large weight to the model and break out of the main loop, noting how many models exist. In the distinctly less happy situation that the model is more likely to err than to be correct, we also break out of the loop in dismay. The worthless model is discarded.

This is a convenient time to compute and save (in alpha) the model's scoring weight that will be needed in Equation (5.13) when the AdaBoost algorithm is called upon to evaluate a test case.

The last step in each pass through the main loop is to adjust the importance distribution. The adjustment factor β is computed with Equation (5.11), and then Equation (5.12) is used to decrease the importance of each case that is correctly classified. The model's predictions were saved in h so that they do not need to be computed again at this time. The denominator of Equation (5.12) is cumulated as the reweighting is done, and the normalization to unit sum is performed.

The final action is to free the work areas that are no longer needed. The destructor will free alpha, which is needed for classification. As noted earlier, if a more general version of the class keeps h and dist available for inspection, they should be freed in the destructor.

The member function that performs an AdaBoost classification is shown next. It may be that the very first model (*t*=1) proved worthless, in which case no classifier exists. This rare event should be flagged to the user. Otherwise, Equation (5.13) is summed, and its sign is checked. For consistency with multiple-class versions of AdaBoost, this function returns the integer 0 for the first class, and 1 for the second class. The code for this function is as follows:

```
int AdaBoostBinaryNoConf::class_predict (double *input)
{
    int i;
    double sum, out;

    if (nmodels == 0)      // Abnormal condition of no decent models
        return -1;         // Return an error flag

    sum = 0.0;
    for (i=0; i<nmodels; i++) {
        models[i]->predict (input, &out);
        if (out > 0.0)              // If it predicts first class
            sum += alpha[i];
        else if (out < 0.0)     // But if it predicts second class
            sum -= alpha[i];
    }

    if (sum > 0.0)
        return 0;   // Flags first class
    else
        return 1;   // Flags second class
}
```

Probabilistic Sampling for Inflexible Models

The AdaBoost algorithm just presented, as well as the AdaBoost algorithms yet to
be shown in this chapter, require the underlying model to have the ability to accept
an importance for each training case. This may not always be possible. The training
algorithm itself may preclude differential weighting. Or the user may be linking to a
precompiled proprietary library without source code. In this situation, AdaBoost can still
be used with only a modest performance penalty. This section shows how the algorithm
of the previous section can be suitably modified. The modification shown here can be
applied to any AdaBoost algorithm.

Only the part of the code involved in collecting the models' training sets needs to be modified. All other aspects of the AdaBoost algorithm remain unchanged. The difference is that instead of copying every training case to the model and differentially weighting each, we randomly sample (with replacement) from the original training set in order to construct the models' training sets. The probability of selecting each case is set equal to the relative importance distribution value for the case. (Recall that the importance distribution usually sums to one by design, so we can treat it as an actual probability distribution. If not, we must normalize it to sum to one.) The only potential pitfall to this method is that some models are unable to tolerate multiple copies of the same training case in the training set. The sampling method described cannot be used with such models.

The sampling algorithm requires two work vectors beyond those required by AdaBoost itself. These are shown next. Notice that we need to define a *resolution factor*, set to five here. This is an arbitrary multiplier, and the sampling will operate correctly as long as this is set to any positive value. The trade-off is that larger multiplier values require more memory and result in faster operation. A value of five is quite large, and greater values are almost certainly overkill.

```
m = 5 * n;   // Resolution factor = 5;
cdf = (double *) malloc (n * sizeof(double));
idist = (int *) malloc (m * sizeof(int));
```

We now skip directly to the model-building loop. This code is shown next, and an explanation follows:

```
for (imodel=0; imodel<nmodels; imodel++) {

    models[imodel]->reset();        // Prepares the reusable model

/*
   Build the table for sampling from tset with distribution 'dist'
*/

    cdf[0] = dist[0];                     // Compute cumulative distribution function
    for (i=1; i<n; i++)
       cdf[i] = cdf[i-1] + dist[i];
    cdf[n-1] = 1.0 + 1.e-8;               // Avoid fpt roundoff error causing overrun later
```

```
   j = -1;                              // Will be a starting subscript
   temp = 0.0;
   for (i=0; i<m; i++) {                // Build table of starting subscripts
     while (temp <= (double) i) {       // Must always be true for i=0
       ++j;                             // So that j advances beyond -1
       temp = m * cdf[j];
       }
     idist[i] = j;                      // Place starting subscript in table
     }

/*
   Use this table to build this training set by randomly sampling from tset
*/

   for (i=0; i<n; i++) {                // Build this training set
     temp = unifrand ();                // Uniform [0,1)
     j = (int) (m * temp + 0.999999) - 1; // Index into starting subscript table
     if (j < 0)                         // Happens very rarely
       j = 0;                           // So must be prepared
     j = idist[j];                      // Table guaranteed <= true index
     while (temp > cdf[j])              // This refinement loop should almost
       ++j;                             // always end the first time or quickly
     tptr = tset + j * (nin + 1);       // Point to this case
     models[imodel]->add_case (tptr);   // Add it to the model's training set
     }

   models[imodel]->train (); // Train this model
```

Consider for the moment a primitive method for randomly sampling integers with preassigned probabilities. For any given value between zero and one, a uniform [0, 1] random number has probability equal to that given value of being less than or equal to it. For example, there is probability 0.27 that a uniform random number will be less than or equal to 0.27. So we could cumulate the individual probabilities, finding the *cumulative distribution function* (CDF). We then generate a uniform random number and search upward through the CDF table, stopping when the random number no longer exceeds the CDF entry. We will land at or below that bin with probability equal to the value in the bin, and we will land at or below the next bin down with probability equal to the value in the lower bin. Thus, we will land in the given bin with probability equal to the difference between its value and the value just below. Because of the definition of the CDF, this is simply the desired probability.

The first action taken in the previous code is to build the CDF table. The final entry, 1.0 in theory, is explicitly set to a slightly larger value to prevent cumulated round-off error from causing search overruns later. We could then use the search technique just described to conduct our random sampling.

The problem with this primitive sampling method is that if the CDF table is large, an inordinate amount of time may be spent searching for the correct bin. It is inefficient to have to start at the bottom each time. The algorithm can be considerably improved by calculating a better starting point. The idea is that we need a guarantee that the starting point is not already above the bin being sought, but it should nevertheless be as large as possible to minimize the search time. The solution is to build a table of starting subscripts. When the uniform random number is generated, it is multiplied by the length of the starting subscript table to quickly index an entry in that table. This entry is the starting subscript for the CDF search. Naturally, it will be impossible to guarantee that we will always start at precisely the subscript being sought, because probabilities can always cluster in arbitrary ways. But by using a large index table, we can achieve high enough resolution that extensive searches are rare. This is what is done by the previous code.

Binary AdaBoost When the Model Provides Confidence

It is frequently the case that not only does a model provide a classification decision, but it also provides a numerical value that can be interpreted as a confidence level for the decision. There is no necessity for this numerical value to be a probability. In fact, it does not even need to be always monotonically related to some rigorously defined confidence measure. If we can assume nothing more than a statement such as "large numerical values usually imply higher confidence in the decision than small numerical values," we can make use of this information. This section shows how to incorporate a model's confidence level into the binary AdaBoost algorithm already described. Theoretical details concerning this algorithm can be found in [Schapire and Singer, 1998].

The underlying model should be similar to the model assumed by the previous AdaBoost algorithm. It is a binary classifier, with the first class being signified by +1 in the training set, and the second by –1. The model should attempt to make similar predictions. The model is presumed to have predicted the first class if its prediction is positive, and the second class if negative. So far, this is identical to the previous algorithm's model characteristics. But now, the magnitude of the prediction is interpreted as confidence. Predictions having large absolute value are interpreted as being more confident than predictions having small absolute value. For compatibility with the model used in the demonstration program, as well as the published literature, the model's output value is hard limited here to +/–1. It is good if this is the natural range of your model so that the hard limiting has little or no impact. This limiting aids stability when the underlying model has a chance of occasionally producing wild output values. Also, having a known natural range helps keep this AdaBoost algorithm efficient in a way that will be seen later.

The overall structure and operation of this AdaBoost algorithm is similar to the no-confidence version: a component model is trained, its error is computed, and the probability vector conveying each case's importance is updated in preparation for the next round of training. The final decision is based on the sign of the weighted sum of the models' predictions. There are only two differences. The first difference is that in the no-confidence version of AdaBoost, the model's grand error is computed by tallying individual misclassifications. But when the model does convey confidence in its decision, a more generalized error figure is computed by considering the magnitude in addition to the sign of each prediction. Individual errors involving predictions with large magnitude have a greater impact on the model's grand error than those involving small magnitude. The other difference involves computation of the alpha weights used indirectly to update the probability vector and used directly to perform the final classification of a test case. The previous algorithm computed this value as the log of a ratio involving the misclassification error. It is not so easy here. This algorithm must numerically minimize a function of alpha. Luckily, this function is well behaved and easily minimized, as will be seen later in the chapter.

We now present the binary AdaBoost algorithm for models providing confidence. Again, denote the training set by $\{(x_1, y_1), ..., (x_n, y_n)\}$. Each x_i is a predictor, a scalar, or a vector. Each y_i is either +1 or –1, according to the class membership of the case. Decide in advance that we will train a total of T models that will be indexed by $t=1, ..., T$. The algorithm proceeds as follows:

Let $t=1$.

Initialize the importance distribution for equal weighting of all cases:

$$D_1(i) = 1/n \quad for \ i = 1, \dots n \tag{5.14}$$

1) Train model h_t using the current importance distribution, D_t. Each model is trained to make a prediction at least approximately in the interval $[-1, 1]$.

2) Compute the prediction-quality vector as the product of each case's true and predicted values. A positive product implies a correct decision, and a negative product implies an incorrect decision. The magnitude of the product reflects the import of this quality/failure measure.

$$u_t(i) = y_i h_t(x_i) \tag{5.15}$$

3) If $u_t(i)$ is never negative for any training case i, we have a rare degenerate condition in that the model never makes an incorrect decision. Quit here, keeping this final model.

4) If $u_t(i)$ is never positive for any training case i, we have a rare degenerate condition in that the model never makes a correct decision. Discard this model and quit here.

5) Compute the optimal alpha by minimizing the quantity shown in Equation (5.16). It should be apparent that as long as at least one $u_t(i)$ is positive and one is negative, a minimum exists. This is because the sum will increase without bound as alpha becomes very positive or negative. If you're interested, you can differentiate this quantity twice to prove that exactly one minimum exists.

$$Z_t = \sum_{i=1}^{n} D_t(i) \exp(-\alpha_t u_t(i)) \tag{5.16}$$

6) Update the importance distribution as shown in Equation (5.17). Alpha is almost always positive. Observe that in this situation, cases having a large positive $u_t(i)$ (confidently correct) will have

their importance significantly decreased. Cases having a large negative $u_t(i)$ (confidently incorrect) will have their importance significantly increased. And cases whose $u_t(i)$ is near zero will have their importance left almost unchanged. The denominator normalizes D to sum to one.

$$D_{t+1}(i) = \frac{D_t(i)\, e^{-\alpha_t u_t(i)}}{Z_t} \tag{5.17}$$

7) If $t<T$, let $t=t+1$ and go to step 1. Otherwise, we are finished.

When the AdaBoost classifier is used on a test case, all T component models are evaluated. Each model is given a weight equal to its alpha as computed in step 5. The weighted sum of the models' predictions is found. The sign of this sum defines the AdaBoost classification decision. This is expressed in Equation (5.18).

$$h_{AdaBoost}(x) = \text{sign}\left[\sum_{t=1}^{T} \alpha_t h_t(x) \right] \tag{5.18}$$

The code for this algorithm can be found in the file ARCING.CPP on my web site. Here is the class declaration for the class that implements binary AdaBoost for models that supply confidence information:

```
class AdaBoostBinary {

public:

  AdaBoostBinary (int n, int nin, double *tset, int nmods);
  ~AdaBoostBinary ();
  int class_predict (double *input);

private:
  int nmodels;        // Number of models (nmods in constructor call)
  double *alpha;      // Nmods long alpha constant for each model
  double *dist;       // N long probability distribution
  double *u;          // N long work area for saving model's error products
};
```

The constructor is responsible for training all of the component models and preparing for AdaBoost classification. The constructor code appears here, and an explanation follows:

```
AdaBoostBinary::AdaBoostBinary (
   int n,                       // Number of training cases
   int nin,                     // Number of inputs
   double *tset,                // Training cases, ntrain by (nin+nout, where nout=1)
   int nmods                    // Number of models
   )

{
   int i, imodel, ngood, nbad;
   double *tptr, temp, sum, h;
   double x1, y1, x2, y2, x3, y3;

   nmodels = nmods;
   alpha = (double *) malloc (nmodels * sizeof(double));
   dist = (double *) malloc (n * sizeof(double));
   u = (double *) malloc (n * sizeof(double));

/*
   Initialize distribution to be uniform
*/

   temp = 1.0 / n;
   for (i=0; i<n; i++)
      dist[i] = temp;

/*
   Main training loop trains sequence of models
*/

   for (imodel=0; imodel<nmodels; imodel++) {
      models[imodel]->reset();     // Prepares the reusable model
```

```
  for (i=0; i<n; i++) {                               // Build this training set
    tptr = tset + i * (nin + 1);                      // Point to this case
    models[imodel]->add_case (tptr, dist[i]); // Add it to the model's training set
    }


  models[imodel]->train ();                           // Train this model

/*
  Compute the u vector.
*/

  sum = 0.0;
  ngood = nbad = 0;                                   // Degenerate if all cases good or bad

  for (i=0; i<n; i++) {
    tptr = tset + i * (nin + 1);                      // Point to this case
    models[imodel]->predict (tptr, &h);               // Make a prediction
    if (h > 1.0)                                       // Hard limiting for a potentially wild model
      h = 1.0;                                         // like a neural net helps stability
    if (h < -1.0)
      h = -1.0;

    u[i] = h * tptr[nin];                              // Error indicator is predicted times true

    if (u[i] > 0.0)                                    // Class prediction is correct
      ++ngood;                                         // This lets us detect degenerate situation

    if (u[i] < 0.0)                                    // Class prediction is incorrect
      ++nbad;                                          // If degenerate, optimal alpha does not exist
    }

  if (nbad == 0) {                                     // Unusual situation of model never failing
    nmodels = imodel + 1;                             // No sense going any further
    alpha[imodel] = 0.5 * log ((double) n);           // Heuristic big value
    break;                                            // So exit training loop early
    }
```

231

```
      if (ngood == 0) {                    // Unusual situation of model being worthless
         nmodels = imodel;                 // No sense going any further
         break;                            // So exit training loop early
         }
```

```
/*
   Compute the optimal alpha.
*/
```

```
      local_n = n;                         // Static copies of these things
      local_dist = dist;                   // Avoids messy parameter passing
      local_u = u;                         // For function being minimized
      glob_min (-1.0, 1.0, 3, 0, 0.0, alpha_crit, &x1, &y1, &x2, &y2, &x3, &y3);
      brentmin (20, 0.0, 1.e-6, 1.e-4, alpha_c rit, &x1, &x2, &x3, y2);
      alpha[imodel] = x2;                  // Optimal alpha left in x2 by brentmin()
```

```
/*
   Adjust the relative weighting, then rescale to make it a distribution
*/
```

```
      sum = 0.0;

      for (i=0; i<n; i++) {                // Equation (5.16)
         dist[i] *= exp (-alpha[imodel] * u[i]);
         sum += dist[i];    // Cumulate total probability for rescale
         }

      for (i=0; i<n; i++)                   // A probability distribution must sum to 1.0
         dist[i] /= sum;                    // M ake it so
      } // For all models

   free (dist);
   free (u);
}
```

Like its no-confidence predecessor, this constructor starts by allocating work areas and initializing the importance probability distribution to uniformity. The main loop begins as before also, copying the training set to the model (along with each case's weight) and training the model.

Then it changes. For each case in the training set, the model is invoked. In this implementation, the model's output is hard limited to $[-1, 1]$. This is because the neural net employed by the test program is occasionally capable of producing a wild prediction, which would skew subsequent calculations. The u vector is computed and saved, and a count is kept of how many positive and negative terms are encountered. The only purpose of these counters is to detect degenerate conditions. We need at least one positive and one negative product to proceed. Degeneracy is rare in practice.

The quantity Z_t defined in Equation (5.16) must now be minimized to find the optimal alpha. This is done in two steps. First, glob_min() is called to find a trio of points having the property that the central point is lower than its neighbors. This bounds the minimum. Then brentmin() is called to quickly and efficiently refine this minimum. The code for these two subroutines can be found in the file MINIMIZE.CPP on my web site. To avoid the need for passing nuisance parameters, pointers to the information required by the criterion function are kept in static variables visible to this function. The criterion function that evaluates Equation (5.16) is as follows:

```
static double alpha_crit (double trial_alpha)
{
   int i;
   double sum;

   sum = 0.0;

   for (i=0; i<local_n; i++)
      sum += local_dist[i] * exp (-trial_alpha * local_u[i]);
   return sum;
}
```

After all models have been trained, AdaBoost classification is performed as shown in Equation (5.18) on page 229. Every model is invoked on the test case, their predictions are weighted by their alphas, and the sum is cumulated. The sign of this sum determines the classification decision. This code is shown here:

```
int AdaBoostBinary::class_predict (double *input)
{
    int i;
    double h, sum;

    if (nmodels == 0)        // Abnormal condition of no decent models
        return -1;           // Return an error flag

    sum = 0.0;

    for (i=0; i<nmodels; i++) {

        models[i]->predict (input, &h);

        if (h > 1.0)         // Hard limiting for a potentially wild model
            h = 1.0;         // like a neural net helps stability

        if (h < -1.0)
            h = -1.0;

        sum += alpha[i] * h;  // Equation (5.18)
    }

    if (sum > 0.0)
        return 0;            // Flags first class
    else
        return 1;            // Flags second class
}
```

AdaBoost.MH for More Than Two Classes

AdaBoost becomes considerably more complex when there is a need to choose from among more than two classes. Several quite different (and differently behaved) versions are in use as of this writing. This section describes the version known as AdaBoost.MH, which is derived by directly extending many aspects of the binary version. The next section will present an interesting alternative.

Even though AdaBoost.MH handles any number of classes, the component models upon which it relies are univariate in output. Each of the T AdaBoost component models is composed of a set of submodels, one for each class and each predicting a scalar value.

Let K be the number of classes. One could implement AdaBoost.MH by using T models, each of which predicts K outputs simultaneously. However, this approach is more difficult to program and usually has inferior performance in real applications. The better approach, which is used here, is to employ KT models, each of which predicts a single scalar variable.

AdaBoost.MH is designed to be able to take advantage of confidence information provided by the component models. If such information is not available, the user simply sets the model's output to +/−1 as determined by the sign of the prediction. The remainder of this section will treat the component models as if they provide meaningful confidence determined by the magnitudes of their predictions. If you have a situation in which this is not the case, it is understood that suitable (and trivial) modifications will be made to constrain the output to +1 or −1.

The AdaBoost.MH algorithm is similar to the binary AdaBoost algorithm shown in the previous section. An importance/probability distribution is initialized to be uniform before the main loop commences. Within the main loop, model set t (of T altogether) is trained based on this distribution. The training set is processed, and a grand error figure is computed. This error is stored for use by the classifier later. The distribution is modified to emphasize difficulties and downplay successes.

This set of operations is repeated T times. However, there are several important differences. Model set t is not a single model, but rather a set of K models, where K is the number of classes. Also, the importance distribution is an n by K array, where n is the number of training cases. AdaBoost.MH focuses not only on cases that are difficult to classify but on classes that are difficult to separate.

Denote the training set by $\{(x_1, y_1), ..., (x_n, y_n)\}$. Each x_i is a predictor, a scalar or a vector. Each y_i is a vector K long. The element of this vector corresponding to the class of this case is +1, and all other elements are −1. Decide in advance that we will train a total of T model sets that will be indexed by $t=1, ..., T$. The algorithm proceeds as follows:

Let $t=1$.

Initialize the importance distribution for equal weighting of all cases and classes:

$$D_1(i, k) = 1/nK \text{ for } i = 1, ... n; \ k = 1, ... K \tag{5.19}$$

1) Train model h_t (actually a set of K single-output models) using the current importance distribution, D_t. Each model is trained to make a prediction at least approximately in the interval [−1, 1].

2) Compute the prediction-quality array as the product of each case's true and predicted values for each output (class flag). A positive product implies a correct decision, and a negative product implies an incorrect decision. The magnitude of the product reflects the import of this quality/failure measure. In Equation (5.20), $y_{i,k}$ is the k'th element of the output vector for case i, and $h_t[k](x_i)$ is the output of the k'th submodel when presented with training set input x_i.

$$u_t(i,\ k) = y_{i,k} h_t[k](x_i)$$ (5.20)

3) If $u_t(i,k)$ is never negative for any i and k, we have a rare degenerate condition in that the model never makes an incorrect decision. Quit here, keeping this model.

4) If $u_t(i,k)$ is never positive for any i and k, we have a rare degenerate condition in that the model never makes a correct decision. Discard this model and quit here.

5) Compute the optimal alpha by minimizing the quantity shown in Equation (5.21). As discussed in the previous section, a unique minimum exists as long as u is not degenerate.

$$Z_t = \sum_{i=1}^{n} \sum_{k=1}^{K} D_t(i,k) \exp(-\alpha_t u_t(i,k))$$ (5.21)

6) Update the importance distribution as shown in Equation (5.22). The denominator normalizes D to sum to one.

$$D_{t+1}(i,k) = \frac{D_t(i,k) e^{-\alpha_t u_t(i,k)}}{Z_t}$$ (5.22)

7) If $t<T$, let $t=t+1$ and go to step 1. Otherwise, we are finished.

When the AdaBoost classifier is used on a test case, all KT component models are evaluated. Each model is given a weight equal to its alpha as computed in step 5. The weighted sum of the models' predictions is computed, each class (submodel output) separately. The class having maximum sum defines the AdaBoost classification decision. This is expressed in Equation (5.23).

$$h_{AdaBoost}(x) = \arg\max_k \left[\sum_{t=1}^{T} \alpha_t h_t[k](x) \right] \qquad (5.23)$$

The code for this algorithm can be found in the file ARCING_M.CPP on my web site. Here is the class declaration for the class that implements AdaBoost.MH:

```
class AdaBoostMH {

public:
  AdaBoostMH (int n, int nin, int nout, double *ts et, int nmods);
  ~AdaBoostMH ();
  int class_predict (double *input);

private:
  int nmodels;        // Number of models (nmods in constructor call)
  int nout;           // Number of outputs (classes)
  double *work;       // Work area for holding output and training case
  double *alpha;      // Nmods long alpha constant for each model
  double *dist;       // N by nout long probability distribution
  double *u;          // N by nout long work area for saving model's error products
};
```

The constructor trains all of the component models and prepares for classification. Here is the constructor code. An explanation follows.

```
AdaBoostMH::AdaBoostMH (
  int n,              // Number of training cases
  int nin,            // Number of inputs
  int nouts,          // Number of outputs (classes)
  double *tset,       // Training cases, ntrain by (nin+nout)
  int nmods           // Number of models (T)
  )
{
  int i, iout, imodel, ngood, nbad, n_nout;
  double *tptr, *dptr, temp, sum, h, uu;
  double x1, y1, x2, y2, x3, y3;
```

```
nout = nouts;
nmodels = nmods;
n_nout = n * nout;                      // KT

work = (double *) malloc ((nin + nout) * sizeof(double));
alpha = (double *) malloc (nmodels * sizeof(double));
dist = (double *) malloc (n_nout * sizeof(double));
u = (double *) malloc (n_nout * sizeof(double));

/*
  Initialize distribution to be uniform
*/

temp = 1.0 / n_nout;
for (i=0; i<n_nout; i++)
  dist[i] = temp;

/*
  Main training loop trains sequence of models
*/

for (imodel=0; imodel<nmodels; imodel++) {

  for (i=0; i<nout; i++)               // K submodels
    models[imodel*nout+i]->reset();   // Prepares the reusable model

  for (i=0; i<n; i++) {                // Build this training set
    tptr = tset + i * (nin + nout);    // Point to this case
    dptr = dist + i * nout;            // Its output probability weights
    memcpy (work, tptr, nin * sizeof(double)); // This case's input

    for (iout=0; iout<nout; iout++) {   // Each output handled by separate model
      work[nin] = tptr[nin+iout];       // This output, +/-1
      models[imodel*nout+iout]->add_case (work, dptr[iout]);
    }
  }
```

```
    for (iout=0; iout<nout; iout++)
      models[imodel*nout+iout]->train();   // Train the model set
/*
   Compute and save u, the error array.
*/

      ngood = nbad = 0;                      // Degenerate if all cases good or bad

    for (i=0; i<n; i++) {
      tptr = tset + i * (nin + nout);        // Point to this case

      for (iout=0; iout<nout; iout++) {
        models[imodel*nout+iout]->predict (tptr, &h);

        if (h > 1.0)                         // This hard limiting aids stability
          h = 1.0;                           // Design your model so that [-1,1] is its
        if (h < -1.0)                        // natural range, so this has minimal impact
          h = -1.0;

        uu = h * tptr[nin+iout];             // Error is predicted times true
        u[i*nout+iout] = uu;                 // Save it for optimization of alpha

        if (uu > 0.0)                        // Class prediction is correct
          ++ngood;                           // This lets us detect degenerate situation
        if (uu < 0.0)                        // Class prediction is incorrect
          ++nbad;                            // If degenerate, optimal alpha does not exist

        }
      }

    if (nbad == 0) {                         // Unusual situation of model never failing
      nmodels = imodel + 1;                  // No sense going any further
      alpha[imodel] = 0.5 * log ((double) n); // Heuris tic big value
      break;                                 // So exit training loop early
      }
```

```
      if (ngood == 0) {              // Unusual situation of model being worthless
         nmodels = imodel;          // No sense going any further
         break;                      // So exit training loop early
         }

/*
   Compute the optimal alpha.
*/

      local_n = n;
      local_nout = nout;
      local_dist = dist;
      local_u = u;

      glob_min (-1.0, 1.0, 3, 0, 0.0, alpha_crit, &x1, &y1, &x2, &y2, &x3, &y3);
      brentmin (20, 0.0, 1.e-6, 1.e-4, alpha_c rit, &x1, &x2, &x3, y2);
      alpha[imodel] = x2;

/*
   Adjust the relative weighting, then rescale to make it a distribution
*/

      sum = 0.0;
      for (i=0; i<n_nout; i++) {
         dist[i] *= exp (-alpha[imodel] * u[i]);
         sum += dist[i];    // Cumulate total probability for rescale
         }

      for (i=0; i<n_nout; i++)      // A probability distribution must sum to 1.0
         dist[i] /= sum;            // Make it so
         } // For all models

   free (dist);
   free (u);
   }
```

The constructor begins by allocating work areas, some of which could be automatics instead of private class members if they are not needed for more general use. The importance distribution is initialized to uniformity, not just across cases but across classes as well. Then the main training loop begins.

This loop is similar to what has been seen before. However, this time we need to reset nout models rather than just one. All of the models share the same input vector, so this is copied to the work vector. Each model handles one output, so their training sets are built accordingly. After the training set is completely built in each model, they are trained.

The next step is to compute the performance matrix, u. In the binary AdaBoost algorithm shown in the previous section, u was a vector. Here it is a matrix because each class is considered along with each case. As before, we count the number of positive and negative entries so that we can detect the rare but troublesome degeneracy of a model being perfect or worthless. If degeneracy is not encountered, the optimal alpha is computed in exactly the same way as in the binary case.

The final step is to update the importance distribution. The update is identical to the binary case, except that once again we deal with a matrix rather than a vector.

When it comes time to classify a test case, we invoke all of the trained models and sum their alpha-weighted predictions. The class (output) having maximum sum defines the final decision, as was shown in Equation (5.23) on page 237. Here is the classification code:

```
int AdaBoostMH::class_predict (double *input)
{
   int iout, imodel, ibest;
   double sum, best, h;

   if (nmodels == 0)          // Abnormal condition of no decent models
      return -1;              // Return an error flag

   for (iout=0; iout<nout; iout++) {
      sum = 0.0;

      for (imodel=0; imodel<nmodels; imodel++) {
         models[imodel*nout+iout]->predict (input, &h);
         if (h > 1.0)              // This hard limiting aids stability
            h = 1.0;               // Design your model so that [-1,1] is its
         if (h < -1.0)             // natural range, so this has minimal impact
```

```
      h = -1.0;
    sum += alpha[imodel] * h;  // Weighted sum of models' predictions for iout
      }
  if ((iout == 0) || (sum > best)) {
      best = sum;
      ibest = iout;
      }
    }

  return ibest;
}
```

You might observe that in one sense this implementation of AdaBoostMH fails to take full advantage of the importance distribution when training the component models. By separating the columns of the distribution matrix, sending each column alone to a single model for training, differences between the columns are ignored. At first glance, this may seem to impose a handicap on the models. However, in practice, it usually does not. Cases that are particularly easy or difficult in regard to certain classes will tend to correctly differentially weight in each column as iterations continue. Also, suppose we regard the total error of a model (all submodels for model t) as the correctly weighted error across all classes, taking the importance distribution fully into account. If each submodel can be trained independently, as is done here, then the minimum total error is obtained by separately minimizing the error of each column (output). Thus, we are implicitly using all of the information in the distribution. It is only if we link the submodels, making a case's error in one submodel dependent on the predictions of the other submodels for that case, that we fail to utilize the entire importance distribution. This is a more complex scenario for the component models, and it adds dramatically to training difficulty. Also, much practical experience indicates that training quality can be seriously compromised by such linkage. For these reasons, it is avoided in this implementation. However, if the user has such a fully linked model, it is easy to modify the AdaBoost code to accommodate the model. Simply refrain from splitting the trained output across submodels. Include the entire output vector in each training case, and pass the entire distribution array to the training algorithm. Everything else remains exactly the same.

In the same vein, observe that the importance vectors passed to the component models indicate relative importance but do not sum to one. The models must be able to handle this.

AdaBoost.OC for More Than Two Classes

The previous section described a method for generalizing binary AdaBoost to be able to handle more than two classes. This section describes a completely different way of achieving the same goal. Unfortunately, the AdaBoost.OC algorithm presented here is unable to take advantage of confidence information provided by the underlying model. On the other hand, this version seems to be more rugged than AdaBoost.MH so that in practice the two versions often perform roughly equivalently even when confidence is available from the model. If the model cannot provide confidence, this version is usually superior to the previous version. But as always for these resampling algorithms, it pays to test both versions if possible and select the best.

This algorithm first appeared in [Schapire, 1997]. It is a hybrid of the original AdaBoost algorithm with an earlier technique for reducing a multiple-class problem to a series of binary problems. The underlying model needs to be only a binary classifier, which not only simplifies programming but almost always speeds training. At first glance, AdaBoost.OC appears to be much more complicated than AdaBoost.MH. In fact, the code for this algorithm is considerably longer than that in the previous section. On the other hand, this is deceptive. The algorithm is really quite simple. It's just that a lot of picky details are involved.

To help understand AdaBoost.OC, suppose for a moment that we need to design a classifier that can handle four classes. A nearest-neighbor method, for example, would take the direct approach of finding the closest member of the training set and assigning its class. This often works well, but it also often fails because it ignores a huge amount of potentially useful information in the training set. The AdaBoost.MH algorithm presented in the previous section would take the approach of constructing four models, each of which attempts to decide the plausibility of a test case's membership in the corresponding class. The most plausible class is the winner.

Here is a third approach to separating four classes: Suppose we design two binary classifiers. One classifier decides if a test case is a member of class 1 or class 2, as opposed to being a member of class 3 or class 4. The other classifier decides if the case is a member of class 1 or 3, as opposed to being a member of class 2 or 4. It should be apparent that if we can answer these two questions, we can decide which of the four classes supplied the test case. This is the principal philosophy underlying AdaBoostOC.

Using just two binary decisions to separate four classes is minimal, but it is probably not optimal. We could almost certainly train a set of models that partition the classes in a variety of ways. By doing so, we lose the nice property that the combined decision is unambiguous. With so many partitions, conflicting decisions are possible. But we can always handle conflict by using intelligently weighted voting. What we gain in return is a set of effective specialists and a larger group from which to form a consensus opinion. The trade-off is worthwhile.

AdaBoost.OC doesn't just choose its partition scheme (called *coloring*) randomly. At each step, it partitions the classes in such a way as to provide maximum information gain relative to the information already in its possession from prior partitions. In addition, it assigns importance probabilities to individual cases exactly as in previous AdaBoost algorithms. This importance distribution is updated continuously to focus the model's effort on those cases whose separation is most valuable given the current coloring (partition scheme). It's elegant and beautiful.

A key to understanding AdaBoost.OC is understanding how it computes its error for a particular round (round t of a total of T rounds). We start by examining the error that a single binary classifier component model will be asked to minimize. As before, define the notation that a logical expression enclosed in double brackets $[\![\ ]\!]$ evaluates to 1.0 if the enclosed expression is true, and 0.0 if false. Assume that we have a coloring vector c_t consisting of K elements, where K is the number of classes. Each element of this vector is either -1 or $+1$, and at least one element is different from the others. This defines a binary partition of the set of classes. When a case x is presented to a binary model h_t, this model's output is designated $h_t(x)$ and is equal to $+1$ or -1. Let y be the true class of the case whose predictor is x, where y is an integer from 1 through K. This model makes a correct decision if and only if it places this case in the correct partition: $h_t(x)=c_t(y)$.

It should now be clear how the binary model is to be trained. Let the original training set be $\{(x_1, y_1), ..., (x_n, y_n)\}$, where each y_i is an integer from 1 through K designating the correct class. Each x_i is a predictor, a scalar, or a vector. Create a new training set in which each dependent variable takes on the value $+1$ or -1 according to the current coloring: $\{(x_1, c_t(y_1)), ..., (x_n, c_t(y_n))\}$ As with previous AdaBoost algorithms, we employe a vector D_t containing the importance probability of each training case. The binary classifier h_t is trained to minimize the error shown in Equation (5.24).

$$err_t = \sum_{i=1}^{n} D_t(i)[\![h_t(x_i) \neq c_t(y_i)]\!] \qquad (5.24)$$

This equation defines the error of a single binary model. It is important because it tells us how the component model is to be trained. However, we also need to be able to define a more general error for iteration t, an error that encompasses as much class information as possible while still living within the restriction that the best we can do on any given iteration is based on a strictly binary model. This general error measure is what will be used to adjust the importance distribution now as well as weight the models for classification of a test case later.

To define a general error measure, we need to first define an error weighting or loss matrix that we will call E_t. This matrix has n rows (one for each training case) and K columns (one for each class). Element (i, k) of this matrix contains the error contribution associated with case i and class k. By definition, $E(i, y_i)$ is zero because there is no error associated with a case being correctly classified. Initially, this matrix is made uniform, as shown in Equation (5.25). As rounds pass, it will be adjusted.

$$E_1(i,k) = \frac{[\![y_i \neq k]\!]}{n(K-1)} \tag{5.25}$$

Let us put this error weight matrix aside briefly and consider the types of decisions that a binary component model can make with respect to the individual classes. Suppose we apply the model to training case i and thereby produce a decision $h_t(x_i)$ that is either $+1$ or -1. The correct decision would be $c_t(y_i)$. Specify some class k against which we want to discriminate, and define success (versus failure) as whether $h_t(x_i)$ successfully discriminated against that class. The situation with regard to this case and the specified class is one of three mutually exclusive and exhaustive categories.

- Total success in that class k is in one partition, the case is in the other partition, and the model predicted that partition.

$$c_t(y_i) = h_t(x_i) \neq c_t(k)$$

We would not want to penalize this category in any way.

- Total failure in that class k is in the partition predicted by the model, but the case is actually in the other partition.

$$c_t(y_i) \neq h_t(x_i) = c_t(k)$$

We would apply a full penalty, say 1.0, to this category.

245

- Indeterminate in that class k is in the same partition as the case. The model is not capable of discriminating this way.

$$c_t(y_i) = c_t(k)$$

We choose to apply a half penalty, say 0.5, to this category.

It is easy to write this penalty as an arithmetic expression involving the various equalities and inequalities. This is done in Equation (5.26).

$$penalty = \frac{1}{2}\left(\left[\!\left[h_t(x_i) \neq c_t(y_i)\right]\!\right] + \left[\!\left[h_t(x_i) = c_t(k)\right]\!\right]\right) \tag{5.26}$$

Looked at this way, we can see another interpretation for the penalty involving a particular training case and specified class to be discriminated against. The first logical expression says that the partition chosen by the model is not the partition to which this case truly belongs, clearly a situation deserving a penalty. The second logical expression says that this class (k) is in the partition chosen by the model. This is an error deserving penalty for all classes except the correct one, y_i, even though it is unavoidable because of the nature of the partition. What about this one exception? We will see that it is innocuous.

It is time to return to the error weight matrix and pull everything together. Recall that $E_t(i,k)$ is an Importance figure given to an error involving case i and class k. This importance will be zero for $k=y_i$, the correct class of this case. You can look back at Equation (5.25) to see how it is initialized. Equation (5.26) specified a reasonable way to apply a penalty based on the decision of the model. We can combine these quantities to define the grand error measure shown in Equation (5.27). It is now clear why the exception mentioned in the previous paragraph is of no consequence: the single inappropriate penalty is multiplied by a weight of zero.

$$\varepsilon_t = \frac{1}{2}\sum_{i=1}^{n}\sum_{k=1}^{K} E_t(i,k)\left(\left[\!\left[h_t(x_i) \neq c_t(y_i)\right]\!\right] + \left[\!\left[h_t(x_i) = c_t(k)\right]\!\right]\right) \tag{5.27}$$

Similar to what was done in previous AdaBoost algorithms, we use the error to compute and save an alpha weight for this model. Smaller errors lead to larger alpha weights, as shown in Equation (5.28). We will use this weight to adjust the error weight matrix for the next round, and we will save the weight for use later when classifying a test case.

$$\alpha_t = \frac{1}{2}\log\left(\frac{1-\varepsilon_t}{\varepsilon_t}\right) \tag{5.28}$$

The adjustment of the error weight matrix for the next round is roughly analogous to the adjustment of the importance distribution in the AdaBoost.MH algorithm. Case/class components that have a large penalty factor, as defined by Equation (5.26), need to have their error contribution increased, while those with no penalty should have their error contribution kept the same or decreased. However, this adjustment must be tempered by the performance of the model. If the model is an excellent performer (large alpha), its successes and failures should be taken very seriously, while poor performers (alpha near zero) do not merit much attention in the adjustment process. With these intuitive guidelines, along with rigorous theory that can be found in [Schapire, 1997], Equation (5.29) is an appropriate update method. The denominator Z_t is simply the sum of the adjusted terms, acting as a normalizer so the error weights sum to one. We already saw this in Equation (5.21) on page 236.

$$E_t(i,k) = \frac{E_t(i,k)e^{\alpha_t\left(\llbracket h_t(x_i)\neq c_t(y_i)\rrbracket + \llbracket h_t(x_i)=c_t(k)\rrbracket\right)}}{Z_t} \tag{5.29}$$

The error weight matrix E_t defines a degree of importance for each training case and class combination, telling AdaBoost where to focus the majority of its discrimination effort. But there is no way to use it directly in training a model because each component model is responsible for a strictly binary decision, and E_t encompasses all K classes. How can we extract from E_t the importance information needed to optimally train a binary model? Recall that the purpose of the model is to discriminate between the two partitions defined by the current coloring scheme, c_t. The model has no means of discriminating between classes that are in the same partition. This tells us how to weight a given case. We do not care about the elements of E_t that are in the same partition as the true class of this case. The model will not be able to do anything about these classes, so they should not have any effect on the importance given to the case. To compute the importance of this case, sum the error weights of the classes that are in the other partition. This is shown in Equation (5.30). If this vector is normalized to unit sum, it makes an excellent importance probability vector for training the binary model of this round. Equation (5.31) sums these weight sums, and Equation (5.32) defines the binary model's importance probability vector.

$$U_t(i) = \sum_{k=1}^{K} E_t(i,k) \left[\!\left[c_t(y_i) \neq c_t(k) \right]\!\right]$$

(5.30)

$$U_t(\cdot) = \sum_{i=1}^{n} U_t(i)$$

(5.31)

$$D_t(i) = \frac{U_t(i)}{U_t(\cdot)}$$

(5.32)

One last issue remains. So far, we have been saying things like "Given the current coloring c_t..." without worrying about how the coloring came about. Does the coloring matter? The answer is that it does matter, though not perhaps as much as might be thought. Suppose that on each round t the coloring for that round were chosen at random. Because it is likely that certain classes are inherently more difficult to separate than others, it is also likely that certain colorings are more useful than others. If the coloring is chosen randomly, sometimes a useful coloring will be found, and sometimes it will not. Despite this lack of continuing optimality, everything will work out in the end. This is because after many rounds, the evolution of the E_t matrix and the computation of the alpha weight for each model will conspire to produce a good AdaBoost ensemble. Still, it should be apparent that the use of random colorings is not the best approach. Much computer time will be wasted training binary models that contribute little to the total solution. It is in our best interest to make at least a modest attempt at intelligent partitioning on each round.

If we are to choose an optimal (or at least decent) coloring, we need a means of evaluating the quality of a trial coloring. An appropriate criterion is already in hand. Recall that Equation (5.30) reveals the importance of a given case in regard to contributions to error, and Equation (5.31) sums these figures across the entire training set. This sum is dependent on the current coloring. If some coloring produces a small value of $U_t(\cdot)$, this means that the coloring focuses on case/class combinations that do not represent a large source of problematic error. This is not good. We want the binary model for this round to focus on the worst remaining sources of error. We want $U_t(\cdot)$ to be as large as possible. So we are left with a numerical optimization problem.

There are three approaches to finding a coloring that maximizes $U_t(\cdot)$. The brute-force method is to try every possible coloring and choose the best. This is the approach taken in the sample code provided with this text. The brute-force approach is the best in the sense that it optimizes ultimate AdaBoost performance relative to the amount of time spent training component models. Unfortunately, it is not an effective general approach because if the number of classes is large, the time taken to compute $U_t(\cdot)$ for every possible coloring becomes intractable.

A simpler, faster, and totally general approach is to randomly generate a reasonable (whatever that is deemed to be) number of colorings, evaluate $U_t(\cdot)$ for each, and go with the best. Because this is so simple, it is left to you to modify the supplied code suitably if needed for your application. The third and probably best but certainly most complex approach is to use an efficient constrained optimization algorithm. This is more thorough than random search, yet it does not suffer speed problems when given a large number of classes. On the other hand, constrained optimization adds a layer of complexity that may not be warranted here, given that even totally random colorings perform fairly well. The choice is yours.

Computation of $U_t(\cdot)$ with Equation (5.31) is not terribly efficient, which is a consideration when the optimization process requires that many colorings be evaluated. Some artful algebraic manipulation provides an alternative representation of $U_t(\cdot)$ that is much faster to compute. This is given in Equation (5.33).

$$U_t(\cdot) = \sum_{k,\,k' \in \{1,\,\ldots,\,K\}} [\![c_t(k) \neq c_t(k')]\!] w_t(k,k') \tag{5.33}$$

$$w_t(k,k') = \sum_{i=1}^{n} E_t(i,k) [\![k' = y_i]\!] \tag{5.34}$$

This presentation of AdaBoost.OC has followed a logical development, but it does not show the algorithmic flow. Training of the AdaBoost.OC classifier is as follows:

Let $t=1$.

Initialize the error weight matrix E_t to be uniform across all errors and zero for all correct classes, as shown in Equation (5.25).

1) Compute a good or optimal coloring c_t by (at least approximately) maximizing $U_t(\cdot)$ as defined in Equation (5.31). For maximum efficiency, use Equation (5.33) to compute $U_t(\cdot)$.

2) Compute the casewise importance probability vector D_t using Equation (5.32).

3) Train the binary model h_t to predict the colorings, with the training cases weighted according to D_t.

4) Compute ε_t using Equation (5.27).

5) Compute α_t using Equation (5.28).

6) Update the error weight matrix E_t according to Equation (5.29).

7) If $t<T$, increment t and go to step 1.

After AdaBoost.OC has been fully trained, it is ready to classify a test case. To do this, submit the test case to all T models and note the prediction (+1 or −1) of each. For each class, note the coloring of this class and then sum α_t for all models whose prediction is equal to the coloring of this class. Whichever class has the greatest sum is the winner. This is expressed in Equation (5.35).

$$h_{AdaBoost}(x) = \arg \max_k \left[\sum_{t=1}^{T} \alpha_t \left[\!\left[h_t(x) = c_t(k) \right]\!\right] \right] \tag{5.35}$$

The source code for the AdaBoost.OC algorithm can be found in the file ARCING_M. CPP on my web site. The essential parts are now listed and discussed. Instead of presenting an entire subroutine and reviewing it at the end, the code will be given in short sections. This is because the complete subroutine is too long to digest in one shot. Here is the declaration for the AdaBoostOC class. Note that as has been done previously, all work vectors are declared as private members even though many of them could simply be automatics. This allows the user to easily move the memory freeing to the destructor and write additional member functions to access these vectors if desired. If you are annoyed by this, you should change the declaration.

```
class AdaBoostOC {

public:

  AdaBoostOC (int ncases, int nins, int nouts, double *tset, int nmods);
  ~AdaBoostOC ();
  int class_predict (double *input);

private:
  double colorcrit (int *coloring);  // Compute coloring criterion
  int n;                             // Number of training cases
```

```
int nmodels;              // Number of models (nmods in constructor call)
int nin;                  // Number of inputs
int nout;                 // Number of outputs (classes)
double *work;             // Work area for holding output and training case
double *alpha;            // Nmods long alpha constant for each model
double *dist;             // N long case probability distribution
double *err_dist;         // N by nout long error weight distribution
int *colormap;            // Nout long coloring map
int *bestcolor;           // Nout long best coloring map
int *tclass;              // N long class membership of training cases
int *h;                   // N long predictions for training set
double *w;                // Nout by nout 'w' matrix
};
```

The constructor starts by allocating the needed work areas. It also passes through the training set and computes a vector that holds the class identifier for each case. This information is accessed many times in the remainder of the algorithm. For conformity with other models (and common convention), class information in the training data array is stored by coding an output for each class. Converting this to a single integer identifier is slow, so it is done once at the beginning.

```
AdaBoostOC::AdaBoostOC (
    int ncases,      // Number of training cases
    int nins,        // Number of inputs
    int nouts,       // Number of outputs (classes)
    double *tset,    // Training cases, ncases by (nins+nouts)
    int nmods        // Number of models (T in the text)
    )
{
    int i, k, kk, iout, jout, imodel, *h, icolor, *coloring;
    double *tptr, *dptr, temp, best, sum, out, denom;

    n = ncases;              // Save in private area for later use
    nin = nins;
    nout = nouts;
    nmodels = nmods;
```

```
work = (double *) malloc ((nin + nout) * sizeof(double));
alpha = (double *) malloc (nmodels * sizeof(double));
dist = (double *) malloc (n * sizeof(double));
err_dist = (double *) malloc (n * nout * sizeof(double));

colormap = (int *) malloc (nmodels * nout * sizeof(int));
bestcolor = (int *) malloc (nout * sizeof(int));
tclass = (int *) malloc (n * sizeof(int));

w = (double *) malloc (nout * nout * sizeof(double));
h = (int *) malloc (n * sizeof(int));

/*
   Compute and save class of each training case.
   This speeds repetitive operations later.
*/

   for (i=0; i<n; i++) {                // For entire training set
     tptr = tset + i * (nin + nout); // Point to this case
     for (iout=0; iout<nout; iout++) {
       if (tptr[nin+iout] > 0.0) {            // +1 for this class, -1 for others
         tclass[i] = iout;
         break;
         }
       }
     }
```

The error weight distribution must be initialized to be uniform over all errors, and zero for correct classes. This is expressed in Equation (5.25) on page 245. The main training loop (*t* from 1 to *T* in the text) is now ready to begin.

```
/*
   Initialize error distribution to be uniform over errors
*/

   temp = 1.0 / (n * (nout - 1));
   for (i=0; i<n; i++) {                // For entire training set
     tptr = tset + i * (nin + nout);    // Point to this case
     dptr = err_dist + i * nout;        // Its error weights
```

```
for (iout=0; iout<nout; iout++)
    dptr[iout] = (tptr[nin+iout] > 0.0) ? 0.0 : tem p;
    }
```

```
/*
   Main training loop trains sequence of models
*/
```

```
for (imodel=0; imodel<nmodels; imodel++) {      // t from 1 to T
    coloring = colormap + imodel * nout; // Each model has its own coloring
```

When the optimal color map is sought later, it will be necessary to repeatedly compute $U_t(\cdot)$ Instead of using the inefficient definition given by Equation (5.31) on page 248, we will use Equation (5.33). This requires computation of w via Equation (5.34).

```
for (iout=0; iout<nout; iout++) {
    for (jout=0; jout<nout; jout++) {
        sum = 0.0;
        for (i=0; i<n; i++) {
            if (jout == tclass[i])
                sum += err_dist[i * nout + iout];
        }
        w[iout*nout+jout] = sum;
    }
}
```

It is now time to bite the bullet and perform the most computationally intensive part of the AdaBoost.OC algorithm: finding the optimal coloring. The exhaustive search used in this code is surely the best method when not more than a half dozen or so classes are involved. It may also be the best if significantly more classes are present but the model training is exceptionally slow, because exhaustive search provides the maximum average gain per training session. If you expect more classes, you should substitute an alternative as discussed in the text.

This algorithm simply counts in binary. Every time a new coloring is to be generated for testing, k is initialized so that it will start by pointing to the lowest bit in the coloring. If this bit has already been flipped up, it is returned to its low state, and the loop carries to the next bit. But if it is in its low state, it is flipped to its high state, and the flipping loop is exited. The search for new colorings is terminated the moment the highest bit is flipped. This is because any subsequent colorings would define partitions identical to those already tested, having opposite signs only.

```
for (i=0; i<nout; i++)
   coloring[i] = -1;

best = -1.e80;
for (;;) {
   k = nout;                  // Start flipping with bit nout-1

   while (k--) {              // Count bits
      if (coloring[k] > 0)    // If this bit is 1
         coloring[k] = -1;    // Flip it to -1 and continue to next bit
      else {                  // If this bit is -1
         coloring[k] = 1;     // Flip it to 1
         break;               // And do not go to next bit
      }
   } // While flipping bits for next trial coloring

   if (k <= 0)                // Happens when all combinations tested
      break;                  // (Keeping highest bit at -1)
   temp = colorcrit (coloring); // Test this coloring
   if (temp > best) {         // If it is the best so far
      best = temp;            // Update best and save this coloring
      memcpy (bestcolor, coloring, nout * sizeof(int));
   }
} // For all possible colorings

memcpy (coloring, bestcolor, nout * sizeof(int)); // Get best
```

Now that we have found the coloring that will be used for this round's binary model, the model must be trained. Use Equation (5.32) on page 248 to compute the importance probability weights that are to be assigned to each case. Then copy the training cases to the model and train it.

```
/*
    The optimal coloring has been found.
    Compute the casewise probability distribution.
*/

    denom = 0.0;
    for (i=0; i<n; i++) {                    // For every training case
        dptr = err_dist + i * nout;          // Its error weights
        icolor = coloring [ tclass[i] ];     // Its color
        sum = 0.0;
        for (iout=0; iout<nout; iout++) {    // Equation (5.30)
            if (coloring[iout] != icolor)
                sum += dptr[iout];
            }

        dist[i] = sum;
        denom += sum;                        // Equation (5.31)
        }

    for (i=0; i<n; i++)                      // Equation (5.32)
        dist[i] /= denom;

/*
    Build the training set and train the model.
*/

        models[imodel]->reset()              // Prepares the reusable model

        for (i=0; i<n; i++) {                // Build this training set
            tptr = tset + i * (nin + nout);  // Point to this case
            memcpy (work, tptr, nin * sizeof(double)); // This case's input
            work[nin] = coloring[tclass[i]]; // Target output
            models[imodel]->add_case (work, dist[i]);
            }
        models[imodel]->train();             // Train the model
```

This model, h_p, has been trained. It is now evaluated by classifying every training case. The sign of the binary model's prediction defines its color decision. Equation (5.27) on page 246 is used to compute the error of this round. The alpha for this model is computed by Equation (5.28) and saved for later use.

```
sum = 0.0;
for (i=0; i<n; i++) {              // For every training case
   tptr = tset + i * (nin + nout); // Point to this case
   dptr = err_dist + i * nout;    // Its error weights
   models[imodel]->predict (tptr, &out); // Make its predictions
   if (out > 0.0)                 // Only the sign of the prediction is used
      h[i] = 1;                   // To determine the binary decision
   else
      h[i] = -1;
   kk = (coloring[tclass[i]] != h[i]) ? 1 : 0;    // Equation (5.27) term 1
   for (iout=0; iout<nout; iout++) {
      k = kk;

      if (coloring[iout] == h[i])            // Equation (5.27) term 2
        ++k;
      sum += k * dptr[iout];

      }
   }

sum *= 0.5;        // Equation (5.27)
if (sum < 1.e-12)   // Prevent undefined fpt operations below
   sum = 1.e-12;
if (sum > 1.0 - 1.e-12)
   sum = 1.0 - 1.e-12;

alpha[imodel] = 0.5 * log ((1.0 - sum) / sum);   // Equation (5.28)
```

The alpha of this round is used to update the error weight distribution. Equation (5.29) on page 247 is evaluated in two steps. First, each element is updated by multiplying by the appropriate factor. The sum of the updated elements is cumulated so that it can be used as a divisor to normalize the matrix to unit sum.

```
denom = 0.0;
for (i=0; i<n; i++) {
   dptr = err_dist + i * nout;     // Its error weights
   kk = (coloring[tclass[i]] != h[i]) ? 1 : 0;     // Term 1
   for (iout=0; iout<nout; iout++) {
      k = kk;
      if (coloring[iout] == h[i])               // Term 2
         ++k;
      dptr[iout] *= exp (alpha[imodel] * k);
      denom += dptr[iout];
      }
   }

for (i=0; i<n; i++) {               // Normalize to sum to 1.0
   dptr = err_dist + i * nout;
   for (iout=0; iout<nout; iout++)
      dptr[iout] /= denom;
   }
} // For all models
```

The final step is to free the work areas that were needed only for training. Those areas needed for classifying will be freed by the destructor. Recall that these were all declared as private class members for algorithmic clarity and to facilitate reference by optional member functions. Feel free to make them automatic or free them in the destructor if you want.

```
free (dist);
free (err_dist);
free (w);
free (h);
free (tclass);
free (bestcolor);
}
```

In the process of seeking the best coloring, a function called colorcrit() is called. This routine employs Equation (5.33) on page 249 to compute $U_t(\cdot)$.

```
double AdaBoostOC::colorcrit (int *coloring)
{
   int i, j;
   double sum;

   sum = 0.0;

   for (i=0; i<nout; i++) {                    // Equation (5.33)
      for (j=0; j<nout; j++) {
         if (coloring[i] != coloring[j])
            sum += w[i*nout+j];
      }
   }

   return sum;
}
```

When this class is called upon to classify a test case, Equation (5.35) on page 250 is employed. The vector work is used to sum the alpha weights of the models that choose the corresponding class. Whichever class has the greatest sum of alphas is declared the winner.

```
int AdaBoostOC::class_predict (double *input)
{
   int iout, imodel, ibest, hh, *coloring;
   double best, out;

   for (iout=0; iout<nout; iout++)
      work[iout] = 0.0;

   for (imodel=0; imodel<nmodels; imodel++) {
      coloring = colormap + imodel * nout;   // Each model has its own coloring
      models[imodel]->predict (input, &out);
      if (out > 0.0)                         // Only the sign of the prediction is used
         hh = 1;                             // The magnitude is ignored
      else
         hh = -1;
```

```
  for (iout=0; iout<nout; iout++) {      Equation (5.35)
    if (coloring[iout] == hh)
      work[iout] += alpha[imodel];
    }
  }

  for (iout=0; iout<nout; iout++) {
    if ((iout == 0) || (work[iout] > best)) {
      best = work[iout];
      ibest = iout;
      }
    }

  return ibest;
}
```

Comparing the Boosting Algorithms

A potentially serious problem with boosting algorithms is the fact that they can
sometimes actually *increase* out-of-sample prediction error because of overfitting.
This increase is only rarely dramatic, but users should always attempt to compare
performance of their classifiers with and without boosting. Poor relative performance
of boosting generally appears to be associated with applications in which the classifier
already has enough power to handle the problem entirely on its own. A good rule of
thumb is the following: *it is often better to boost a weak classifier than to use a single
powerful classifier*. The reason for this rule is that boosting seems to enjoy significant
immunity to overfitting. When one employs a complex classifier, the possibility of
overfitting always looms. Boosting a weak classifier provides the power to solve the
problem without (usually!) the danger of overfitting.

A Binary Classification Problem

In this section we explore a binary classification application. The complete C++ code for
this application can be found in the file ARCING.CPP on my web site. You should feel free
to modify the code to conduct your own explorations.

This test program uses random bivariate data. A user-specified number of training cases are generated, and a classification model is trained. Each of the binary boosting algorithms described in this chapter is also trained. Then, a large number of test cases are computed, and the original and boosted classifiers are evaluated. This process is repeated many times, and the mean test-set performance of each competitor is printed.

The test problem is deliberately made difficult in a straightforward way. One of the classes is bimodal. Most of its cases lie on one side of the other class, but a significant portion of its cases lie on the opposite side. This is illustrated in Figure 5-1. Thus, the classifier cannot separate the two classes with a single straight boundary. Two straight lines or a complex curve are required.

The classifier used in this example is deliberately hobbled. It is a multiple-layer feedforward network with a single hidden neuron. Such a classifier is not capable of separating classes that are split as in this example. A single instance of the classifier will be able to separate only the dominant portion of the first class, leaving the majority of the smaller portion misclassified. Additional instances are required to resolve the conflict.

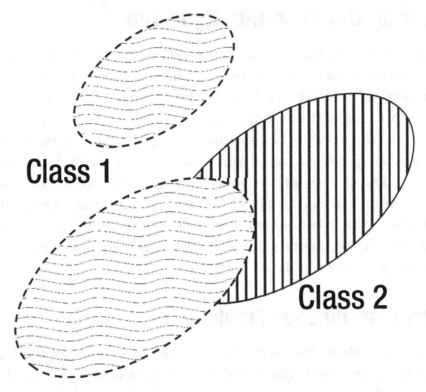

Figure 5-1. *Test problem for comparing binary boosting*

The results for this test clearly demonstrate the power of boosting, as shown in the following tables. Four binary boosting algorithms are compared: bagging, confidence ignored, confidence ignored with sampling, and the full confidence model. These are the *Bagging* (page 206), *AdaBoostBinaryNoConf* (page 215), *AdaBoostBinaryNoConfSampled* (page 223), and *AdaBoostBinary* (page 226) classes, respectively. Classification error rates are shown for no boosting, as well as two, three, five, and ten iterations (component models) of each method. The following table was computed using a moderate degree of separation:

Iterations	None	Bagging	NoConf	Sampled	Confidence
None	0.189				
2		0.216	0.191	0.199	0.112
3		0.198	0.149	0.176	0.101
5		0.197	0.144	0.157	0.098
10		0.196	0.105	0.125	0.100

The following table shows classification error rates when the classes are somewhat better separated. Note that the performance of the single model improves modestly relative to the previous table, but the performance of the boosted versions increases significantly.

Iterations	None	Bagging	NoConf	Sampled	Confidence
None	0.127				
2		0.149	0.136	0.131	0.074
3		0.130	0.086	0.091	0.058
5		0.127	0.032	0.058	0.011
10		0.127	0.016	0.023	0.007

These two tables demonstrate some important aspects of boosting. Some of these are as follows:

- The potential power of boosting is enormous. When the full boosting model (confidence information, no sampling) is employed for the situation of wide but difficult separation, the misclassification rate of 0.127 for the raw model was reduced to 0.007. Wow.

- The importance of multiple iterations depends on the application. For the moderate separation case, minimal boosting (two iterations) provided the majority of error reduction, 0.189 to 0.112. Subsequent iterations contributed almost nothing. For the high separation case, going from three to five iterations resulted in more than a fivefold error reduction, and going to ten iterations cut that error nearly in half again.

- Not only was bagging not beneficial, but it actually increased the error in most cases. This is common but not universal. Bagging is moderately effective in some applications and almost never causes serious problems.

- If confidence information is available from the primitive model, as is the case in this demonstration, it should definitely be used. The difference in performance is substantial.

- Use of sampling to avoid modification of the primitive classifier is fairly costly in terms of performance. Performance almost always deteriorates, often significantly.

A Multiple-Class Problem

We now explore a multiple-class application. The complete C++ code for this application can be found in the file ARCING_M.CPP on my web site. You should feel free to modify the code to conduct your own experiments.

As in the binary example, random bivariate data is used for the predictors. A user-specified number of training cases are generated, and a classification model is trained. Each of the multiple-class boosting algorithms described in this chapter is also trained. Then, a large number of test cases are computed, and the original and boosted classifiers are evaluated. This process is repeated many times. The mean test-set error rate of each competitor is printed.

In this example, there is no need to contrive a difficult class distribution as was done for the binary situation. The existence of multiple classes is sufficiently difficult to demonstrate the effectiveness of boosting. Five classes are employed, each an ellipsoid, with their centers linearly offset from one another by varying amounts. Thus, each class will experience different degrees of confusion with its neighbors, depending on proximity. Fifty training-set cases are employed.

As in the binary example, four levels of boosting are tested, involving two (the minimum possible), three, five, and ten iterations (component models). The simple multiple-class extension of bagging is tested, along with the *MH* and *OC* versions of *AdaBoost*. The results are shown in the following table:

Iterations	None	Bagging	MH	OC
None	0.097			
2		0.162	0.039	0.045
3		0.128	0.029	0.032
5		0.119	0.023	0.021
10		0.111	0.019	0.017

This table highlights some important points of interest. These include the following:

- Once again, bagging is the grand loser, underperforming the raw model.

- Using boosting with even its minimum of two iterations provides dramatic improvement in performance.

- The two algorithms (*MH* and *OC*) have similar performance, with *MH* doing better with few iterations, but *OC* ultimately winning. This changeover should not be taken too seriously, as it depends on the application and is far from universal.

Final Thoughts on Boosting

You may by now be overwhelmed by the multitude of boosting algorithms presented here. However, in most applications, there should be little or no conflict over the most appropriate algorithm. Keep the following ideas in mind:

- The use of boosting with a weak classifier, sometimes even a deliberately weak classifier, is a powerful technique. If the weakness of your classifier is beyond your control, boosting may save the day. If you can control the power of the classifier, remember that a boosted weak classifier will very often outperform an unboosted strong classifier.

- Boosting a strong classifier is often a bad move. Always test the suitability of your choice. If test data is scarce, use cross validation to compare the boosted and unboosted versions of your classifier.

- Avoid bagging. It's one of those things that sounds good in theory, but it rarely works in practice.

- If your application has more than two classes, you are limited to the *MH* and *OC* versions of the algorithm. They usually perform equivalently, so don't sweat the choice too much. If convenient, try both and choose the best. If not, don't worry.

- If your classifier provides numerical values whose magnitude relates to confidence in the decision, use this information. The *AdaBoost* algorithm can get a lot of help from it.

- Try hard to modify your classifier so that it can make direct use of casewise probabilistic weighting. Simulating this weighting by Monte Carlo sampling can be better than nothing, but the cost is substantial.

Permutation Training and Testing

When we train a model to predict a quantity or make a classification decision, we typically find a set of parameter values that result in some performance criterion being optimized. The hope is that we thereby create a model that will do a good job when it encounters future data.

Unfortunately, the training process has no guaranteed way of knowing which patterns in the training data represent authentic patterns that will reappear in the future, and which are random noise that will not likely be seen again. A model that is too weak will not be able to learn the authentic patterns, and one that is too strong will learn too many noise patterns. Such an overly strong model is said to *overfit* the data, and it will perform poorly in the future when the noise patterns that it learned fail to reappear.

In any case, the performance results from training will nearly always overestimate the performance that can be expected in the future, and this is especially true if the model is too powerful relative to the amount of noise in the training set. It is necessary that we evaluate the quality of the model before it is put to actual use. To do so, we must answer two different but equally important questions.

1. What is the performance that can be expected in the future? We answer this by computing an unbiased estimate of our performance measure in the population that generates the data.

2. What is the probability that the performance we observed could have been obtained from a truly worthless model because of pure good luck?

The first question is the one dwelt upon by the vast majority of developers. But what if our great-looking expected future performance was just a lucky streak? An answer to this question is vital, and it's troubling that the question is often ignored. This text has already covered cross validation, walk-forward testing, and some other standard and not-so-standard procedures for answering the first question. Some of these procedures also included, as subprocedures, the ability to compute an approximate answer to the second question.

However, there are at least three situations in which we should strive to answer the second question as thoroughly as possible.

- As discussed earlier in this text, the final step in most model-building endeavors is to trust the quality of the model factory and then use the entire available dataset to build the final model. It is an act of faith that if the model construction was successful in the past (as indicated by out-of-sample testing), it will continue to be successful this one last time. That's usually a reasonable assumption, but this would be a great time to answer question 2 by means of permutation training.

- It may be that our data is limited, in which case extensive out-of-sample evaluation is impractical, and even the quality of our testing procedure for question 1 is suspect. In this situation, permutation training to answer question 2 would provide much needed additional information.

- For out-of-sample procedures that do not provide significance tests, or whose significance tests are suspect, permutation testing is an effective way to answer question 2 in regard to expected future performance.

The bottom line is that unless training time is impracticably long, at a minimum the procedure described here should *always* be performed as the final step before an apparently satisfactory model is put into use. Optionally, this procedure can also be used early in the model development as a screening tool. If, in either case, there is anything but a tiny probability that your good results could have arisen from good luck, you should sit up and take notice. Seriously.

Before we delve into details, one important caveat must be mentioned. Permutation training/testing absolutely requires either that *all* predictor vectors be independent and identically distributed or that the predicted variable be independent and identically distributed. If dependence (typically serial correlation) occurs in the predicted variable *and* one or more predictors, the computed probability will be anti-conservative (too small), meaning that it will fool you into thinking that things are better than they are. This is very dangerous, and it happens easily, becoming serious with only small dependencies.

The Permutation Training Algorithm

This algorithm is a standard Monte Carlo permutation test. The idea is that if you randomly permute the predicted variable relative to the predictors, all predictable patterns are destroyed. Thus, any model trained on permuted data is, by definition, worthless. Suppose that the model is worthless on the original (unpermuted) data. If you train (and perhaps test) the model once on the original (worthless) data and then train/test it k-1 more times on permuted data (also worthless), the order in which the performances fall is random. There is a $1/k$ probability that the original performance happens to be the best, $2/k$ that it is at least second best, and so forth. So we simply count the number of times the permuted performance equals or exceeds that of the original and divide by the total number of times the model was trained. If the model is worthless, this fraction will be less than or equal to any number p with probability p. If we've done a large number of permutations, this fraction is the approximate probability that results as good as those obtained might have occurred by random good luck if the model were truly worthless. The basic algorithm outline looks like this:

```
for permutation from 0 through n_permutes-1

        if permutation > 0
            shuffle predicted variable

        train (and optionally test) the model and compute its performance

        if permutation = 0
            original performance = performance
            count = 1

    else
            if performance >= original performance
                count = count + 1

probability = count / n_permutes
```

The model is trained (and optionally also tested) n_permutes times, with trial zero being the original, unpermuted data. We need to count the number of times a performance (including the original) equals or exceeds that of the original data. Thus, if permutation=0, we save the performance in original_performance. We also initialize the count to 1. This is because the original performance is obviously greater than or equal to itself. Another way to think of this is to realize that even if the original performance is the best, greater than all of the permutations, there is still a 1/n_permutes chance that this could happen by pure good luck. So, the count must always begin at one, not zero.

Partitioning the Training Performance

There are a few more bits of information that we can often extract from permutation training (though not testing). Be warned that the material in this section is extremely heuristic. It is applicable in only a limited number of situations, and the exact conditions under which it is valid are not well known. Still, this information is fast and easy to compute along with the probability just discussed, and it is usually worth examining, if for no other reason than to open ourselves to discovering some possibly startling results.

Suppose the performance figure we are optimizing is a measure of gain, as opposed to a measure of error such as mean squared error. Also suppose it is at least moderately linear in the sense that doubling the observed gain corresponds to a real-life doubling of approximate value. For example, dollar profits made in market trading, cost saved in an industrial process, and yield in a chemical reactor vessel all may satisfy these requirements. R-square probably does not because it is not linear, at least not near its upper limit of one.

The gain we observe after training the model can be roughly (very roughly!) partitioned into three components, as shown in Equation (5.36). Their definitions follow.

$$TrainingGain = Ability + InherentBias + TrainingBias \qquad (5.36)$$

TrainingGain is the gain achieved by the model after being optimized on a dataset.

Ability is the gain attributable to the model's taking advantage of authentic patterns in the data.

InherentBias is the gain (which may be negative) that would be obtained on average by making random decisions. For example, if we are trading a market with a strong upward trend and our model-driven trading system takes only long positions, even a coin-toss model would make money on average.

TrainingBias is the gain attributable to the model's learning of random noise as if it contained authentic patterns.

The key to teasing some additional information from the permutation training algorithm is to note that because permutation destroys any authentic patterns in the data, the gain after training a permuted dataset lacks the *Ability* component of gain, but it still has the *InherentBias* and the *TrainingBias*. In other words, the gain from a model trained on a permuted dataset is expressed in Equation (5.37).

$$PermutedGain = InherentBias + TrainingBias \qquad (5.37)$$

At a minimum, we can subtract Equation (5.37) from Equation (5.36) in order to obtain the useful Equation (5.38).

$$Ability = TrainingGain - PermutedGain \qquad (5.38)$$

However, we can break things down even further if we know or have a way of computing the *InherentBias*. This is often the case. In the next section we'll see a contrived example of credit card fraud detection in which the *InherentBias* is easy to compute. As another example, in automated market trading the *InherentBias* is the fraction of the total bars long minus the fraction short times the bar return. This is expressed in Equation (5.39).

$$InherentBias = \frac{BarsLong - BarsShort}{TotalNumberOfBars} * BarReturn \qquad (5.39)$$

Suppose we know or can compute the *InherentBias*. Then we can shuffle Equation (5.37) to get Equation (5.40).

$$TrainingBias = PermutedGain - InherentBias \qquad (5.40)$$

This is nice, for it tells us how much the process of training optimistically inflates the gain. If the *TrainingBias* is large, this constitutes evidence that the model may be too powerful relative to the amount of noise in the data. This information can be particularly handy early in model development, when it is used to compare some competing modeling methodologies to judge whether additional data is required.

We can also use the *TrainingBias* in conjunction with the basic breakdown of Equation (5.36) to give the definition in Equation (5.41) and the computational formula in Equation (5.42).

$$UnbiasedGain = Ability + InherentBias \qquad (5.41)$$

$$UnbiasedGain = TrainingGain - TrainingBias \qquad (5.42)$$

It's useful to compare *Ability*, which is computed with Equation (5.38), to the *UnbiasedGain* of Equation (5.42). They both remove the seriously harmful effects of the *TrainingBias*, thus giving us an approximate estimate of future performance. But the *UnbiasedGain*, which includes the effect of the *InherentBias*, is a reflection of what will be seen in the future, while the *Ability* is a more pure measure of performance.

The relative utility of these two unbiased performance measures is the cause of passionate debate in some circles. The most famous debate is perhaps in the field of automated market trading. Suppose we are developing a system for trading blue-chip equities, which have a strong upward bias. One school of thought says that in evaluating performance we should use the *UnbiasedGain*, which includes the effect of the upward market bias. Models that do well on this performance measure will likely favor long positions over short, and these proponents will argue that this is good: if the market trends upward over time, we might as well take advantage of this trend and favor long positions.

The opposing school of thought says that trends are fickle and may vanish for years, so we must not rely on them as a major contributing force. Instead, we should favor the more pure *Ability* in judging model quality, which will usually result in trading systems that are more balanced in long/short positions. Which opinion is correct? Sorry, I won't go there.

A Demonstration of Permutation Training

This section presents an artificial example of the use of permutation training in detecting fraudulent credit card transactions. Because it is designed to demonstrate training bias elimination as well as probability calculation, this example permutes only the training phase of development. This would most often be used when the final model is trained for deployment, or perhaps in early explorations. Remember that permutation can also be used with cross validation, walk-forward testing, and most other out-of-sample evaluation methods in order to compute the probability that good observed out-of-sample results could have arisen from a worthless model. The principle is exactly the same, although of course it makes no sense to eliminate training bias from out-of-sample results!

The complete code for this demonstration program can be found in the file MC_TRAIN.CPP on my web site. If you're interested in the details, you are advised to study this code along with the material in this text. Here we will omit aspects of the program that do not relate to the task at hand: training the model with permuted data, computing the probability, and estimating statistics related to elimination of training bias.

In this example, three predictor variables are used: THIS_CHARGE (the amount of the charge in question), AVG_CHARGE (the average charge made in the past by the card holder), and FOREIGN (one if the charge is made in a foreign country, zero otherwise). To simulate predictive power, the FOREIGN predictor can have a user-specified relationship with the predicted variable.

The predicted variable, FRAUD, is one if the charge is fraudulent and zero if it is legitimate. In such a binary situation, a logistic model is more appropriate than linear regression. However, linear regression is so fast and simple that we use it in this example. We use singular value decomposition (SVDCMP.CPP on my web site) to fit the linear model. So, we allocate this object as shown here. There are four columns because there are three predictors plus the constant term.

```
svdptr = new SingularValueDecomp (ncases, 4, 0);
```

To clarify the layout of the synthetic data in this program, here is how it is generated and stored:

```
for (i=0; i<ncases; i++) {
   is_fraud = (unifrand() < 0.01) ? 1 : 0;    // True situation
   data[4*i+0] = 1000 + 1900 * (unifrand() - 0.5); // THIS_CHARG E (never predictive)
   data[4*i+1] = 1000 + 500 * (unifrand() - 0.5);  // AVG _CHARGE (never predictive)
   data[4*i+2] = (unifrand() < 0.01) ? 1 : 0; // FOREIGN (modified for predictive power)
   data[4*i+3] = is_fraud;                     // FRAUD (true situation)
   data[4*i+2] = power * is_fraud + (1.0 - power) * data[4*i+2];
   }
```

The card company has historical experience that lets it assign a gain to each of the four possible outcomes (truly fraud or legitimate; classified as fraud or legitimate). They want to optimize the decision threshold for the linear model's predictions in such a way as to maximize the gain. Here are these gain figures. They have been chosen for this example to produce informative results, not necessarily for reality.

```
   gain_ll = 10;         // Legitimate and predicted legitimate
   gain_lf = -10;        // Legitimate and predicted fraud
   gain_fl = -1000;      // Fraud and predicted legitimate
   gain_ff = 500;        // Fraud and predicted fraud
```

We now do some one-time preparation. The three predictors and constant term must be stored in the singular value decomposition object. The predicted variable (FRAUD) must be stored in a work vector for later shuffling. And we must count the number of fraudulent transactions.

```
   k = 0;
   dptr = svdptr->a;
   for (i=0; i<ncases; i++) {
      *dptr++ = data[4*i+0];     // THIS_CHARGE
      *dptr++ = data[4*i+1];     // AVG_CHARGE
      *dptr++ = data[4*i+2];     // FOREIGN
      *dptr++ = 1.0;             // Constant term
      work[i] = data[4*i+3];     // FRAUD (predicted variable)
      if (work[i] > 0.5)         // If this case is fraud
         ++k;                    // Count it for probability
      }
```

Later, to compute the *InherentBias*, we will need the probability that any transaction is fraudulent. The occurrence of fraud in the training set is used to estimate this probability. Also, we need to perform the singular value decomposition of the predictor matrix to prepare for repeated linear regression later.

```
p_fraud = (double) k / ncases; // Probability that a case is fraud
p_legit = 1.0 - p_fraud;        // And legitimate

svdptr->svdcmp ();
```

The replication loop begins now. The first pass through the loop (irep=0) trains with the original, unpermuted data. All subsequent passes permute the data. For later elimination of training bias, we must cumulate the *InherentBias* and the *PermutedGain*, so they are initialized to zero. The shuffling algorithm shown here is a standard algorithm that is known to perform correctly in the sense that every possible permutation is equally likely.

```
mean_inherent_bias = 0.0;      // Computed from permuted only
mean_permuted_gain = 0.0;      // Ditto

for (irep=0; irep<nreps; irep++) {

  // Shuffle dependent variable if in permutation run (irep>0)

  if (irep) {          // If doing permuted runs, shuffle
    i = ncases;        // Number remaining to be shuffled
    while (i > 1) {    // While at least 2 left to shuffle
      j = (int) (unifrand () * i);
      if (j >= i)      // Never happens if unifrand() guaranteed less than one
        j = i - 1;
      dtemp = work[--i];
      work[i] = work[j];
      work[j] = dtemp;
    }
  }
```

The linear regression model is now fitted to the data and predictions made. To do so, the (possibly shuffled) predicted values are copied to the SVD object, and backsub() is called.

```
memcpy (svdptr->b, work, ncases * sizeof(double));
svdptr->backsub (1.e-8, coefs);

for (i=0; i<ncases; i++) {    // Find prediction for each case
   sum = coefs[3];            // Constant term
   for (j=0; j<3; j++)        // Three predictors
      sum += coefs[j] * data[4*i+j];
   pred[i] = sum;
   }
```

We now compute the optimal threshold, that which maximizes the gain. This algorithm is useful in many applications and should be studied. It is a standard method for computing an optimal threshold when the four possible components of gain are known.

The basic algorithm begins by assuming the threshold is so low that all cases are classified as fraud. To do this, we pass through the entire dataset. For each case, if it is legitimate, we cumulate the gain (probably negative) resulting from calling it fraud. If it is fraudulent, we cumulate that gain.

```
gain = 0.0;
for (i=0; i<ncases; i++) {
   if (work[i] < 0.5)     // If this is a legitimate transaction
      gain += gain_lf;    // Legitimate called fraud
   else                   // This is fraud
      gain += gain_ff;    // Fraud called fraud
   }
```

After the previous code is executed, gain will contain the gain resulting from classifying all cases as fraudulent. We will then slowly increase the threshold, keeping track of the threshold that produces maximum gain. The thresholds where change can occur are the unique values of the predictions. Thus, we sort the predictions ascending and simultaneously move the predicted values in work to keep them paired.

```
qsortds (0, ncases-1, pred, work); // Sort predictions ascending, moving work
```

```
best_gain = gain; // Currently, this is the gain from calling all transactions fraud
```

273

The following loop passes through the dataset, raising the threshold as it goes. Each time, the new threshold is pred[i] (though it will not change if it is in a block of tied predictions), and the prior case at i-1 goes from being classified as fraudulent to classified as legitimate. For the last pass through the loop, in which i=ncases, we set the threshold to the largest prediction plus any positive number (1.0 is used here). This handles the situation of all cases being classified as legitimate.

The only potentially confusing part of this simple algorithm is the transition of a case from being classified as fraud to being classified as legitimate. When this happens, the gain that had been due to its being called fraud must be subtracted, and the gain now due its being called legitimate must be added. Finally, we must allow for the fact that ties in the predictions may be present. Therefore, we must never update the best gain when we are in the middle of a block of ties. We must update only
when a new threshold is encountered.

```
for (i=1; i<=ncases; i++) {      // Try all possible thresholds, including all legitimate

   if (i < ncases)               // Usual situation
      thresh = pred[i];
   else                          // Must include possibility of all called legitimate
      thresh = pred[i-1] + 1.0;  // Actual value added makes no difference

   prior_thresh = pred[i-1];     // Case changes from predicted fraud to predicted legit

   if (work[i-1] < 0.5)          // If this transaction is legitimate
      gain += gain_ll - gain_lf; // Went from called fraud to called legitimate
   else                          // This transaction is fraud
      gain += gain_fl - gain_ff; // Went from called fraud to called legitimate

   if (thresh > prior_thresh) {  // Only update when threshold actually changes
      if (gain > best_gain) {    // (Must not break in the middle of a block of ties)
         best_gain = gain;       // Keep track of best
         ibest = i;              // Lets us later compute fraction classified as fraud
      }
   }
} // For all cases, finding optimal threshold
```

The next step in the replication loop is to handle the probability computations. If this is the first, unpermuted pass, we save the gain and initialize the probability counter to one, as per the earlier discussion. But if we are in permutation passes, we see if the gain from the permutation equals or exceeds that of the original. If so, the probability counter is incremented.

```
if (irep == 0) {        // If doing original (unpermuted), save gain
  original_gain = best_gain;
  mcpt_count = 1;    // Original gain equals or exceeds itself, so count it
  }

else {
  if (best_gain >= original_gain) // Count for p-value
    ++mcpt_count;
  }
```

We also handle the training bias computation. The fraction of cases classified as fraud is determined by the location of the best threshold in the sorted array of predictions. Since the data is permuted, destroying any truly predictable patterns, we can assume that the probability of classifying a case as fraud is similar in these permutations to what it would be with a real-life worthless system. Moreover, under the definitional assumption of random classification decisions, the probability of a fraud classification and the probability that a transaction truly is fraudulent are independent. Therefore, the probability of any of the four possible outcomes is just the product of the individual probabilities. Multiplying each of these four probabilities by the associated gain gives the bias that we can expect from a model that makes random classification choices. If we are doing the first, unpermuted replication, we save the inherent bias, but only for display so the user can compare the original to the permuted. If we are in a permutation run, cumulate the gain and the *InherentBias*.

```
c_fraud = (double) (ncases - ibest) / ncases;      // Fraction classified as fraud
c_legit = 1.0 - c_fraud;                           // Ditto legitimate

inherent_bias =     p_legit * c_legit * gain_ll +
                    p_legit * c_fraud * gain_lf +
                    p_fraud * c_legit * gain_fl +
                    p_fraud * c_fraud * gain_ff;
```

```
if (irep == 0)
    original_inherent_bias = inherent_bias;
else {
    mean_inherent_bias += inherent_bias;
    mean_permuted_gain += best_gain;
    }

} // For all reps
```

Finally, when the permutation replications are complete, we divide as needed to convert sums to per-case values for uniformity. The following lines of code are based on the simple equations already presented:

```
original_gain /= ncases;                        // Make it per case, not total
mean_inherent_bias /= nreps-1;
mean_permuted_gain /= ncases * (nreps-1);       // Ditto
training_bias = mean_permuted_gain - mean_inherent_bias;
unbiased_actual_gain = original_gain - training_bias;
unbiased_gain_above_inherent_bias = unbiased_actual_gain - mean_inherent_bias;
```

We end this presentation of a permutation training example with two sample outputs from the program. This first output is produced when the user specifies that the model has no predictive power whatsoever, and a very large number of cases (100,000) are used.

```
Called fraud 141 of 100000 (0.14 percent)
Actual fraud 0.94 percent

p = 0.57600

Original gain = 0.55370 with original inherent bias = 0.52818
Mean permuted gain = 0.56517
Mean permuted inherent bias = 0.50323
Training bias = 0.06194 (0.56517 minus 0.50323)
Unbiased actual gain = 0.49176 (0.55370 minus 0.06194)
Unbiased gain above inherent bias = -0.01147 (0.49176 minus 0.50323)
```

In this no-power example, the number of cases called fraud (0.14 percent) is quite different from the actual number (0.94 percent). Such variation is common when there is little predictive power. The most important number in this table is the p-value, the probability that a worthless model could have performed as well or better just by luck. This p-value is 0.576, a strong indication that the model could be worthless. You really want to see a tiny p-value, ideally 0.01 or less, in order to have solid confidence in the model.

The per-case gain of the unpermuted model is 0.55370, which may seem decent until you notice that 0.52818 of it is inherent bias. So even after optimization of performance, it barely beats what a coin-toss model would achieve. Things get even worse when you see that the mean gain for permuted data is 0.56517, better than the original model! One could stop right there and go back to the drawing board, but we will continue.

The inherent bias computed as the mean of permutations is almost equal to that of the original model. This is common when the model has little or no power, and it is rare when the model has significant power. Their equality (or lack thereof) is much less important than the other figures in the table.

The training bias is quite small compared to the other numbers. This is because the training set is huge. With so many cases, and hence so much variety in the data, the training algorithm has little wiggle room to capitalize on specific noise patterns.

If we remove this small training bias, the estimated actual return shrinks a little, but this pulls it even lower than the inherent bias. So, we see that the actual return of the model with both sources of bias eliminated, -0.01147, is actually negative!

The next example uses fewer training cases (10,000) and small but real power (0.01). Here is its table of results:

Called fraud 654 of 10000 (6.54 percent)
Actual fraud 0.98 percent

p = 0.09800

Original gain = 0.92200 with original inherent bias = -0.23180
Mean permuted gain = 0.51390
Mean permuted inherent bias = -0.30854
Training bias = 0.82244 (0.51390 minus -0.30854)
Unbiased actual gain = 0.09956 (0.92200 minus 0.82244)
Unbiased gain above inherent bias = 0.40810 (0.09956 minus -0.30854)

Again, there is a wide discrepancy between the number of cases called fraud and the number actually fraud. But the key point here is that the p-value is now somewhat significant at 0.098. In other words, if the model were truly worthless, there is slightly less than a 10 percent chance that results as good as those observed would have been obtained. This is not cause for unrestrained celebration, but it is cause for hope.

The original gain is larger than when there was no power, which is not surprising. This is even more impressive given that now the inherent bias is negative, showing that the model overcame a significant counter-current. By the way, in this demonstration program there is a large negative correlation between the actual number of fraud cases randomly generated and the inherent bias. This is why the disparity in inherent gain between this and the prior example is so large. Examination of the four gain constants reveals why this happens and is left as an exercise for you.

The training bias in this example is very large, much larger than in the prior example. This is because this example has one-tenth the number of training cases as the prior example. This inverse relationship between training set size and training bias is nearly universal in all aspects of model development. This is because in a large training set, the noise is extremely varied and no single noise pattern can dominate and fool the procedure. But a small training set will have fewer instances of authentic patterns. Thus, the training algorithm is not so clear on which is which. Even if there are no authentic patterns, in a small dataset it is easier for some noise pattern to be randomly over-represented and masquerade as an authentic pattern.

Once we remove the training bias, the unbiased actual gain is fairly small. This is a warning that in real life we should not expect great things from this model. But if we then subtract the inherent bias, the gain jumps up considerably. So, the reason the expected future performance of 0.09956 is small is that the model is fighting against a strong counter-current. The model is fairly good, gaining about 0.40810 against the current. It's just being beaten down by the prevalence of fraud and the high price paid for it.

If this program is run with a perfect model, the computed p-value will be the reciprocal of the number of replications, the smallest possible. The gain will be so large that it swamps out all other numbers. There is nothing educational in that demonstration, so it is avoided here.

CHAPTER 6

Combining Numeric Predictions

- Simple Average

- Unconstrained Linear Combinations

- Constrained Linear Combinations

- Variance-Weighted Interpolation

- Combination by Kernel Regression Smoothing

- Comparing the Combination Methods

In many applications, several competing models are developed. The instinctive plan is to compare the performance of these models and choose the best for the final use. But there is usually a better approach: keep and use many or all of the models. In all likelihood, some models will have weaknesses that can be alleviated by the strengths of others. By intelligently combining the predictions made by multiple models, a consensus prediction can be made that is nearly always superior to that made by the single best model. This chapter discusses a variety of methods for combining numeric predictions.

Simple Average

The most straightforward approach to combining numeric predictions is to average them. Surprisingly, this method is often as good as or better than far more sophisticated methods. The beauty of using a simple arithmetic average is that overfitting is not a

© Timothy Masters 2018
T. Masters, *Assessing and Improving Prediction and Classification*,
https://doi.org/10.1007/978-1-4842-3336-8_6

possibility. We all know that whenever one or more parameters are estimated as part of a modeling process, there is the possibility that the model will excessively accommodate unique characteristics of the training set, to the detriment of performance in the general population. This overfitting is the bane of modeling. Because computing an average does not involve any parameter estimation, this algorithm is immune to the problem. The other algorithms that appear in this chapter involve varying degrees of parameter estimation, leaving them exposed to potentially serious performance degradation. Never scorn the simple average because of its lack of sophistication. There is much to recommend it.

Although it should be clear what is meant by the average of a set of predictors, later algorithms will demand rigorous nomenclature. We might as well start now. The training set consists of n cases, $\{(x_1, y_1), ..., (x_n, y_n)\}$. Each x is a predictor, often a vector. Each y is the corresponding scalar variable that we want to predict. It is called the *dependent* variable. We also have K trained models. When one of these models, say model k, is presented with a predictor, it makes a prediction $f_k(x)$. We want to find a function g that combines the K predictions into a superior consensus. In other words, we hope that the *consensus function* $f(x)$ as defined by Equation (6.1) is a better predictor than any of its component models.

$$f(x) = g\left[f_1(x), ..., f_K(x) \right]$$

(6.1)

This chapter will explore various methods of finding effective g functions. At the moment we are interested in g being the mean of its arguments. This is expressed in Equation (6.2).

$$f(x) = g\left[f_1(x), ..., f_K(x) \right] = \frac{1}{K} \sum_{k=1}^{K} f_k(x)$$

(6.2)

There is an important theoretical justification for choosing g as defined in this equation. A famous mathematical result (a corollary of the *Cauchy inequality*) states that the square of the sum of K numbers cannot exceed K times the sum of their squares. This is expressed in Equation (6.3). Note that equality is obtained only when all of the a_k values are identical.

$$\left[\sum_{k=1}^{K} a_k \right]^2 \leq K \sum_{k=1}^{K} a_k^2$$

(6.3)

Suppose that for some value x of the predictor variable the corresponding (correct) value of the dependent variable is y. Let a_k in the inequality shown previously be the error made by model k when it attempts to predict y given x. In other words, let $a_k = f_k(x) - y$. This leads to Equation (6.4). As shown in Equation (6.5), the left side of the inequality can be rewritten by splitting the summands, letting $f(x)$ be the average that we are currently exploring, and factoring out the constant K. Substituting the rightmost term of this equation into the left side of the inequality of Equation (6.4) and then dividing both sides of this inequality by K^2 gives Equation (6.6).

$$\left[\sum_{k=1}^{K}\left(f_k(x) - y\right)\right]^2 \le K\sum_{k=1}^{K}\left[f_k(x) - y\right]^2 \tag{6.4}$$

$$\left[\sum_{k=1}^{K}\left(f_k(x) - y\right)\right]^2 = \left[Kf(x) - Ky\right]^2 = K^2\left[f(x) - y\right]^2 \tag{6.5}$$

$$\left[f(x) - y\right]^2 \le \frac{1}{K}\sum_{k=1}^{K}\left[f_k(x) - y\right]^2 \tag{6.6}$$

Each summand on the right side of Equation (6.6) is the squared error of a single component model. Adding the squared errors and dividing by K gives the mean of the squared errors. The left side of this inequality is the squared error of the consensus model of this section, the mean of the K predictions.

This is a very powerful result. It tells us that for any case (x, y), the squared error of the consensus model cannot exceed the average squared error of the component models, with actual equality being obtained only when all of the component models have the same error. To the extent that the component models differ, the squared error of the consensus model will be less than the average squared error of the components. Using the average as a consensus model gives us a method that cannot be worse than the components (on average) and that in practice will virtually always be better. This holds true for *any* point (x, y). Do not take it lightly.

Code for Averaging Predictions

This chapter will build a library of C++ classes for combining multiple numeric predictions. Averaging is obviously a trivial operation, but you should study this code carefully, as it lays the structural foundation for the more complex algorithms that will appear later. This code can be found in the file MULTPRED.CPP on my web site.

```cpp
class Average {
public:
   Average (int n, int nin, double *tset, int nmods);
   ~Average ();
   void numeric_predict (double *input, double *output);

private:
   int nmodels;        // Number of models (nmods in constructor call)
};

Average::Average (
   int n,              // Number of training cases
   int nin,            // Number of inputs
   double *tset,       // Training cases, ntrain by (nin+nout) where nout=1 here
   int nmods           // Number of models in 'models' array
   )
{
   nmodels = nmods;          // Nothing else to do here
}

void Average::numeric_predict (double *input, double *output)
{
   int imodel;
   double outwork;

   *output = 0.0;

   for (imodel=0; imodel<nmodels; imodel++) {
      models[imodel]->predict (input, &outwork);
      *output += outwork;       // Average output is mean across all models
      }

   *output /= nmodels;
}
```

This class, like all others in this chapter, expects the user to have prepared an array of nmods models that are externally referenced. Note that all the constructor has to do is internally preserve the number of models. The other parameters in its call list are ignored. They are provided here because they will be used in the other more complex classes, and consistency is always desirable. Note also that all of the routines in this chapter implicitly assume that the model predicts one output. The generic model, used elsewhere in examples throughout this text, is able to predict any number of outputs.

Unconstrained Linear Combinations

The problem with using the average of the component models as the consensus function is that the quality of the individual models may vary widely. If all of the models are roughly equivalent in prediction power, then simply averaging their predictions is tempting, even desirable. But if their power varies widely or if some of the models consistently tend to predict with overly narrow or wide swings, we can do better.

The most obvious choice for combining models having varying prediction characteristics is simple linear regression. The technique is to compute the final prediction as a weighted sum of the models' predictions, plus a constant term to compensate for any inherent bias in the component models. This is shown in Equation (6.7).

$$f(x) = \sum_{k=1}^{K} w_k f_k(x) + w_0 \qquad (6.7)$$

The usual method for computing the weights is least squares regression. This is discussed in most statistics texts and will not be repeated here. The file LINREG.CPP on my web site contains a class that does the job. Here is a C++ class for combining multiple predictions using unconstrained weights. This code is located in the file MULTPRED.CPP on my web site.

```
class Unconstrained {

public:

   Unconstrained (int n, int nin, double *ts et, int nmods);
   ~Unconstrained ();
   void numeric_predict (double *input, double *output);
```

```
private:
   LinReg *linreg;      // The linear regression object
   int nmodels;         // Number of models (nmods in constructor call)
   double *coefs;       // Computed coefficients here, constant last
};
```

The constructor, destructor, and member function have the same parameter lists as the simple averaging class shown in the previous section. In addition, the coefs member variable saves the computed weight vector. The LinReg object, defined in LINREG.CPP, is used to compute the weights. The constructor is shown here, and an explanation of its operation follows:

```
Unconstrained::Unconstrained (
   int n,              // Number of training cases
   int nin,            // Number of inputs
   double *tset,       // Training cases, ntrain by (nin+nout) where nout=1 here
   int nmods           // Number of models in 'models' array
   )
{
   int i, imodel;
   double *casevec, *outs, *tptr;

   nmodels = nmods;

   casevec = (double *) malloc ((nmodels+1) * sizeof(double));
   outs = (double *) malloc (n * sizeof(double));
   coefs = (double *) malloc ((nmodels+1) * sizeof(double));

   linreg = new LinReg (n, nmodels + 1);

   casevec[nmodels] = 1.0; // This is the regression constant term
   for (i=0; i<n; i++) { // Build the design matrix
      tptr = tset + i * (nin+1); // This case is here
      for (imodel=0; imodel<nmodels; imodel++)
         models[imodel]->predict (tptr, &casevec[imodel]);
      linreg->add_case (casevec);
      outs[i] = tptr[nin]; // Corresponding true value
      }
```

```
linreg->solve (1.e-6, outs, coefs);

free (casevec);
free (outs);
delete linreg;
}
```

The constructor starts by allocating three vectors. The first, casevec, is a work vector that will hold each case so that it can be appended to the LinReg training set via add_case(). The second, outs, holds the n true values of the dependent variable. Finally, coefs is the computed vector of weights for use in Equation (6.7).

Although Equation (6.7) uses a subscript of zero for the constant, the algorithm is most easily coded by placing the constant at the end of the weight vector. The statement casevec[nmodels] = 1.0 initializes the corresponding element of the work vector used for passing the cases to the LinReg training set. We then pass through the entire training set. For each case, each of the nmodels models is invoked to make a prediction. These predictions are placed in the case vector, and the vector is appended to the regression training set. Simultaneously, the outs vector is built.

When the entire training set has been processed, the solve() member function of the LinReg object is called to compute the optimal weights, saving them in coefs. At this time, the two work vectors and the LinReg object can be deleted. If you do not want to access linreg outside the constructor, you can make it an automatic variable.

The almost trivial prediction member function is now shown. It simply computes Equation (6.7). Recall that w_0 in this equation is stored at the end of the coefs vector.

```
void Unconstrained::numeric_predict (double *input, double *output)
{
  int imodel;
  double out;

  *output = coefs[nmodels]; // Regression constant

  for (imodel=0; imodel<nmodels; imodel++) {
    models[imodel]->predict (input, &out);
    *output += coefs[imodel] * out;
  }
}
```

Constrained Linear Combinations

The combination method shown in the previous section has strong intuitive justification, and it does in fact work well in some applications. However, it has so many potential problems that it is almost never used.

The most obvious drawback to combining predictions by means of an unconstrained linear combination is that there are $K+1$ parameters, the weights, that are free to be optimized. This means that the training data can usually be fit quite well. If K is anything beyond tiny, overfitting will likely occur, with the result that data outside the training set will not be handled well.

There is another problem that is less obvious but in practice often far more serious: colinearity. It is not at all unusual for two or more models in an ensemble to have nearly identical predictions across the training set. This inevitably leads to instability in the weight estimates. The *singular value decomposition* algorithm employed by the LinReg class helps control this problem, but instability can never be avoided entirely because it is inherent in linear regression. Simply stated, if two or more models make nearly the same predictions for most cases in the training set, then the two models are largely interchangeable. Any combination of their weights that sums to a constant will perform equivalently.

Why is this a problem? It might not be serious if the similarity of the component models obtained in the training set were guaranteed to also be reflected across all possible test cases. But this is rarely the situation. Sooner or later, a test case will come along for which the formerly similar models react differently. Now suppose that numerical instability has conspired to produce a huge positive weight estimate for one model and a huge negative weight of practically the same magnitude for the other model. The disparate weights would cancel for every case in the training set. But as soon as a test case is encountered for which the models generate significantly different predictions, the wild weights would no longer cancel, and Equation (6.7) would generate an extreme prediction. This is not acceptable behavior.

Luckily, this problem is easily solved by imposing the constraint that all weights except the constant term be nonnegative. Why does this work? In practical terms, it means that extreme weights become impossible. Recall that unconstrained regression normally produces extreme weights only of opposite sign because the weights of correlated models must sum to reasonable values in order for the training data to be fit well. The weight for one member of a correlated pair of models cannot be driven to a large positive value unless the weight for its mate is simultaneously driven to a large negative value. If extreme negative weights are explicitly prevented, extreme positive weights are automatically prevented as well.

A minor benefit of imposing this constraint is that the number of degrees of freedom in weight optimization is reduced, thereby making overfitting less likely. But the major benefit comes not from the actual reduction of degrees of freedom but from the means by which it is obtained. The stability of the combined predictor is greatly enhanced. Consider what happens when the models' weights are all small and positive. For any test point, it will be nearly impossible for the weighted sum to be much less than the minimum prediction of the component models. Similarly, it will be difficult for the prediction to be much greater than the maximum component prediction. This is a valuable property. In fact, many would say that it is a mandatory property.

We now present a class for computing linear regression weights subject to non-negativity constraints. Note that complex explicit formulae exist for computing such weights. See, for example, [Hashem, 1997]. Here, we take the simpler and far more stable approach of iteratively optimizing an appropriate criterion function. The class header is as follows:

```
class Biased {

public:
    Biased (int n, int nin, double *tset, int nmods);
    ~Biased ();
    void numeric_predict (double *input, double *output);

private:
    int nmodels;        // Number of models (nmods in constructor call)
    double *coefs;      // Computed coefficients here
};
```

This class is named Biased because it is appropriate for combining prediction models that may be biased. In the next section, we will see a better class for the unbiased case.

The training process will iteratively minimize an error criterion function. Here is code for this function:

```
static int biased_ncases;        // Number of training cases
static int biased_nvars;         // Number of component models
static double *biased_x;         // Ncases by nvars array of training cases
static double *biased_y;         // Vector of associated true values
```

```
double biased_crit (double *wts)
{
   int i, j;
   double err, pred, *xptr, diff, penalty;

   err = 0.0;
   for (i=0; i<biased_ncases; i++) {          // For all training cases
      xptr = biased_x + i * biased_nvars;      // Point to this case
      pred = wts[biased_nvars];                // Constant term
      for (j=0; j<biased_nvars; j++)           // For all model outputs
         pred += xptr[j] * wts[j];             // Weight them per call (Eq (6.7))
      diff = pred - biased_y[i];               // Predicted minus true
      err += diff * diff;                      // Cumulate squared error
      }

   penalty = 0.0;                              // Impose nonnegativity constraint
   for (j=0; j<biased_nvars; j++) {            // For all models' weights
      if (wts[j] < 0.0)                        // If it is negative
         penalty -= 1.e30 * wts[j];            // Impose a huge penalty
      }

   return err + penalty;                       // Squared error plus constraint penalty
}
```

This criterion function employs static variables to access required values from the calling function. This simplifies the use of canned optimization routines to minimize the function.

The algorithm passes through the entire training set, cumulating the total squared error that results from the supplied weight vector. For each case, the vector of component model predictions is weighted according to the supplied weights, with the constant term w_0 appearing at the end of the weight vector. The loop over j sums Equation (6.7), this prediction is compared to the true value of the predicted variable, and the squared difference is cumulated.

The final step is to see whether any of the trial weights are negative. If so, apply a large penalty. The criterion function returns the total squared error plus any penalty arising from negative weights. Imposing the constraint this way allows use of a simple unconstrained optimizer. Note that constrained optimization algorithms exist and would, in fact, be somewhat faster in execution. However, in nearly all applications, the speed difference is negligible, while the complexity difference is significant.

The constructor is responsible for training by computing the optimal weights. Here it is:

```
Biased::Biased (
   int n,                       // Number of training cases
   int nin,                     // Number of inputs
   double *tset,                // Training cases, ntrain by (nin+nout) where nout=1 here
   int nmods                    // Number of models in 'models' array
   )
{
   int i, imodel;
   double *cases, *outs, *tptr, *base, *p0, *direc, ystart, yend;

   nmodels = nmods;

   cases = (double *) malloc (n * nmodels * sizeof(double));
   outs = (double *) malloc (n * sizeof(double));
   coefs = (double *) malloc ((nmodels+1) * sizeof(double));
   base = (double *) malloc ((nmodels+1) * sizeof(double));
   p0 = (double *) malloc ((nmodels+1) * sizeof(double));
   direc = (double *) malloc ((nmodels+1) * (nmodels+1) * sizeof(double));

/*
   Find and save each model's prediction for each case
*/

   for (i=0; i<n; i++) {
     tptr = tset + i * (nin+1); // This case is here
     for (imodel=0; imodel<nmodels; imodel++)
       models[imodel]->predict (tptr, &cases[i*nmodels+imodel]);
     outs[i] = tptr[nin]; // Corresponding true value
     }

/*
   Optimize
*/

   biased_ncases = n;
   biased_nvars = nmodels;          // Don't include constant term
```

289

```
biased_x = cases;
biased_y = outs;

for (i=0; i<nmodels; i++)
   coefs[i] = 1.0 / nmodels;       // The simple mean is a good starting guess
coefs[nmodels] = 0.0;              // Constant term

ystart = biased_crit (coefs);
yend = powell (20, 0.0, 1.e-6, biased_crit, nmodels+1, coefs, ystart, base, p0, direc);

free (cases);
free (outs);
free (base);
free (p0);
free (direc);
}
```

The first step is to allocate some work areas, as well as the coefs vector that will hold the computed weights throughout the life of the class object. The first two allocations, cases and outs, hold the component models' outputs and the corresponding true values of the predicted variable, respectively. The last three allocations, base, p0, and direc, are work areas for the optimization routine powell().

The algorithm passes through the n training cases. For each case, a row in tptr, all nmodels component models are invoked. Their predictions are stored as a row in cases, and the corresponding true value is stored in outs.

In preparation for optimization, the static variables that will be used by the criterion function are initialized. The starting values for the weight vector are set to reasonable values: the model weights are set so that the prediction would be the mean across the models, and the constant is set to zero. Perturbation for optimization will start with this vector.

The criterion function is called using this weight set because powell() requires this to be done before it is called. Finally, Powell's algorithm is used to optimize the weights. The code for this function can be found in the file MINIMIZE.CPP on my web site. After optimization is complete, all work vectors are freed. Note that the weight vector must be retained until it is freed in the destructor.

The majority of the work for this class is done in the constructor. The function for computing a combined prediction is trivial. All it does is invoke the component models and sum their weighted predictions, plus the constant term. Here is that code:

```
void Biased::numeric_predict (double *input, double *output)
{
   int imodel;
   double out;

   *output = coefs[nmodels]; // Regression constant

   for (imodel=0; imodel<nmodels; imodel++) {      // Equation (6.7)
      models[imodel]->predict (input, &out);
      *output += coefs[imodel] * out;
   }
}
```

Constrained Combination of Unbiased Models

The algorithm just shown is appropriate when it is reasonable to believe that one or more of the component models may be biased. However, if it is known in advance that none of the models has a significant tendency to make predictions that are consistently above or below the true value, there is no need to include a constant term. In fact, the presence of a constant term in such a situation can be problematic, as the optimizer has one more degree of freedom with which to overfit the data. Whenever possible, the constant term should be avoided.

There is one additional constraint that is usually imposed when we are willing to assume that the component models are unbiased or nearly so. Not only are negative weights forbidden, but the weights must sum to one. This weighted average has obvious intuitive beauty, but there are two useful theoretical reasons for it as well. First, as long as the component predictions are unbiased, a weighted sum in which the weights sum to one is unbiased as well. Second, imposition of a unit sum guarantees us a property that is only approximated by the previous combination method: the grand prediction is an interpolation between the component predictions. This is expressed in Equation (6.8).

$$\min_k f_k(x) \le \sum_{k=1}^{K} w_k f_k(x) \le \max_k f_k(x) \qquad (6.8)$$

291

It should be mentioned that practical experience indicates that explicit imposition of the unit-sum constraint is not usually required. In the vast majority of applications, weights computed by minimizing the squared error when there is no constant offset automatically sum to nearly one. However, imposition of this constraint is so easy that it might as well be done. Explicit optimization of one fewer parameters, with the last being determined by the unit sum constraint, is elegant but unnecessary overkill.

The code for computing these optimal weights is practically identical to that given in the previous section, so it will not be fully shown here. See the file MULTPRED.CPP for the complete class listing. However, the error criterion subroutine is notable because it demonstrates a simple yet effective way of imposing the unit-sum constraint. Here is that subroutine:

```
double unbiased_crit (double *wts)
{
   int i, j;
   double sum, err, pred, *xptr, diff, penalty;

   // Normalize weights to sum to one
   sum = 0.0;
   for (j=0; j<unbiased_nvars; j++)
      sum += wts[j];

   if (sum < 1.e-60)          // Should never happen
      sum = 1.e-60;           // But be prepared to avoid division by zero

   for (j=0; j<unbiased_nvars; j++)
      unbiased_work[j] = wts[j] / sum;

   // Compute criterion
   err = 0.0;

   for (i=0; i<unbiased_ncases; i++) {
      xptr = unbiased_x + i * unbiased_nvars;    // Point to this case
      pred = 0.0;                                // Will cumulate prediction

      for (j=0; j<unbiased_nvars; j++)           // For all model outputs
         pred += xptr[j] * unbiased_work[j];     // Weight them per call
      diff = pred - unbiased_y[i];               // Predicted minus true
      err += diff * diff;                        // Cumulate squared error
      }
```

```
penalty = 0.0;
for (j=0; j<unbiased_nvars; j++) {
   if (wts[j] < 0.0)
      penalty -= 1.e30 * wts[j];
   }

return err + penalty;
}
```

Variance-Weighted Interpolation

We have just seen that combining predictions using weights that sum to one is useful in that the result is an interpolation between the component models. The method of the previous section computed the optimal weights by minimizing the squared error within the training set. There is an alternative method that is sometimes superior (and sometimes not). This is to compute the optimal weights according to the overall accuracy of the component models. Models having large error are assigned weights that are less than those for models having small error.

This method tends to be superior to minimizing squared error when there is great disparity between the quality of the component models and the component models are largely uncorrelated. When the models all have roughly similar quality, the method of the previous section tends to be superior. This is especially true if the component models are correlated to any significant degree.

Weighting according to model quality makes intuitive sense. As usual, there is also a strong theoretical basis for this method. Let the relative weighting of each model be computed in inverse proportion to its error, as shown in Equation (6.9). Then scale the weights to unit sum, as shown in Equation (6.10). Some fairly straightforward calculations show that if the f_k are unbiased and uncorrelated, these weights minimize expected squared error. See, for example, [Taniguchi and Tresp, 1997].

$$w_k^* = \frac{1}{\sum_{i=1}^{n} \left[f_k(x_i) - y_i \right]^2} \tag{6.9}$$

$$w_k = \frac{w_k^*}{\sum_{k=1}^{K} w_k^*} \tag{6.10}$$

Note that this optimality property deteriorates as bias or correlation increases, with the methods of the previous sections becoming more appropriate in such situations. In case of doubt, it would be wise to try both methods and compare the results.

The code for this algorithm is much simpler than the code for the previous algorithms. For this reason, only the class constructor, which is responsible for computing the weights, is shown here. See the file MULTPRED.CPP on my web site for a complete listing.

```cpp
Weighted::Weighted (
    int n,          // Number of training cases
    int nin,        // Number of inputs
    double *tset,   // Training cases, n by (nin+nout) where nout=1 here
    int nmods       // Number of models in 'models' array
    )
{
    int i, imodel;
    double *tptr, out, diff, sum;

    nmodels = nmods;

    coefs = (double *) malloc (nmodels * sizeof(double));

    for (i=0; i<nmodels; i++)
        coefs[i] = 1.e-60;              // Theoretically 0, but prevent division by 0

    for (i=0; i<n; i++) {               // Cumulate error variance of each model
        tptr = tset + i * (nin+1);      // This case is here
        for (imodel=0; imodel<nmodels; imodel++) {
            models[imodel]->predict (tptr, &out);
            diff = out - tptr[nin];         // Predicted minus true
            coefs[imodel] += diff * diff;   // Equation (6.9)
        }
    }

    sum = 0.0;                          // Will sum normalizing factor
    for (imodel=0; imodel<nmodels; imodel++) {
        coefs[imodel] = 1.0 / coefs[imodel]; // Weight is recip of variance
        sum += coefs[imodel];                // Cumulate normalzier
    }
```

```
for (imodel=0; imodel<nmodels; imodel++)  // Normalize to unit sum
    coefs[imodel] /= sum;                  // Equation (6.10)
}
```

Operation of the constructor is simple. The coefficient vector, which will be freed by the destructor, is allocated. The squared errors will be cumulated in this vector as an intermediate step. Thus, it should be initialized to zero. However, as cheap insurance against division by zero later, it is actually initialized to a tiny positive number that will vanish in subsequent floating-point arithmetic.

For each case, each model's prediction is made and compared to the correct value. The squared error is cumulated. When done, the sum of all unnormalized weights is found, and the weights are divided by the sum to produce a weight set that sums to one.

Combination by Kernel Regression Smoothing

Constrained regression, with or without a constant as appropriate, is almost certainly the most general and powerful method for combining multiple predictions. However, it does suffer from a potential problem when used with data that is very noisy. Because it is so powerful in that it embodies many degrees of freedom, it is susceptible to overfitting. If the training set contains a significant number of nonsense cases or if the supposedly true values are in actuality true values plus considerable noise, combination by regression will tend to model the noise along with the truth. Generalization will suffer.

There is a technique originally called *kernel regression*, more recently called the *general regression neural network* (GRNN), that has much of the power of traditional regression (and more in one way) but that is far less prone to overfitting. Although a typical GRNN has about the same number of optimizable parameters as ordinary regression, they are weak parameters in the sense that varying their values does not impact the model nearly as much as in the case of traditional regression. A small amount of fitting power is lost, but the compensation in lowered likelihood of overfitting is often more than adequate compensation. As a bonus, the GRNN is inherently nonlinear, so it does not suffer the restrictions of a linear model.

The idea behind the GRNN is strongly intuitive. When an unknown case (in the current context, a set of component model predictions) is presented to a GRNN, its prediction is an interpolation among all true values in the training set. The interpolation weight for each training case is determined by the similarity between that case and the unknown case. Training cases that are very similar to the unknown case receive high relative weights, while dissimilar cases receive low weights.

This scheme is easily visualized in one dimension. Naturally, in the current context, one dimension is not terribly relevant because this corresponds to using just one component model. However, once the principle is understood, generalization to multiple dimensions is easy.

Suppose Figure 6-1 depicts a function that we want to model. Unfortunately, our training data is heavily contaminated with noise. Figure 6-2 is a scatterplot of the observed cases, each of which is the true function plus random error.

Figure 6-3 shows the result of using a GRNN to smooth the data. For each horizontal location (predictor value), the corresponding vertical location (predicted value) is computed as a weighted average of the observed true values of training cases that are nearby horizontally.

The theoretical derivation of the GRNN is best accomplished from a statistical point of view. If you are uncomfortable with this development, you may safely skim over the details, confident that nothing vital is being missed.

Some notation is temporarily modified to make the equations easier to understand. A vector **x** of p independent variables, the component model's predictions here, is being used to predict a dependent scalar variable y. If we happen to know the joint density of these quantities, the prediction having minimum expected squared error is given by the conditional expectation shown in Equation (6.11).

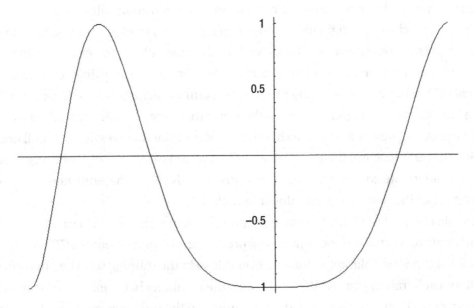

Figure 6-1. *True underlying function*

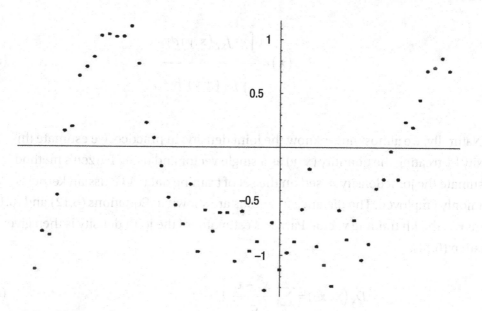

Figure 6-2. *Observed cases*

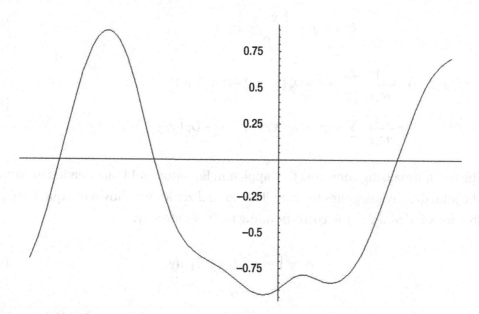

Figure 6-3. *Prediction via GRNN smoothing*

$$E_{Y|X}(\mathbf{x}) = \frac{\displaystyle\int_{-\infty}^{\infty} y \cdot f_{XY}(\mathbf{x}, y)\, dy}{\displaystyle\int_{-\infty}^{\infty} f_{XY}(\mathbf{x}, y)\, dy} \tag{6.11}$$

Naturally, we almost never know the joint density. In practice, we estimate this density by treating the quantity $(\mathbf{x}, y)$ as a single vector and using Parzen's method to estimate the joint density based on the set of training data. A Gaussian kernel is commonly employed. The distance functions are shown in Equations (6.12) and (6.13), where $\mathbf{x}_i$ is the ith training vector. Parzen's estimator of the joint density is then given by Equation (6.14).

$$D_{\mathbf{X}}(\mathbf{x}, \mathbf{x}_i) = \sum_{j=1}^{p} \left(\frac{x_j - x_{ij}}{\sigma_j} \right)^2 \tag{6.12}$$

$$D_Y(y, y_i) = \left(\frac{y - y_i}{\sigma_y} \right)^2 \tag{6.13}$$

$$g(\mathbf{x}, y) = \frac{1}{nc_{\mathbf{x}}c_Y} \sum_{i=1}^{n} \exp\left[-D_{\mathbf{X}}(\mathbf{x}, \mathbf{x}_i) - D_Y(y, y_i) \right]$$

$$= \frac{1}{nc_{\mathbf{x}}c_Y} \sum_{i=1}^{n} \exp\left[-D_{\mathbf{X}}(\mathbf{x}, \mathbf{x}_i) \right] \exp\left[-D_Y(y, y_i) \right] \tag{6.14}$$

The two normalizing constants that appear in Equation (6.14) are needed to ensure that the joint density integrates to unity. The normalizer for y is shown in Equation (6.15), and that for $\mathbf{x}$ is defined in the corresponding multivariate way.

$$c_Y = \int_{-\infty}^{\infty} \exp\left[-D_y(y, 0) \right] dy \tag{6.15}$$

When the theoretically optimal prediction method given by Equation (6.11) is modified by replacing the exact densities with the Parzen approximations just shown, we arrive at Equation (6.16). The numerator and denominator of this expression are shown in Equations (6.17) and (6.18), respectively.

$$\hat{y}(\mathbf{x}) = \frac{N(\mathbf{x})}{D(\mathbf{x})} \tag{6.16}$$

$$N(\mathbf{x}) = \int_{-\infty}^{\infty} \frac{y}{nc_{\mathbf{x}}c_Y} \sum_{i=1}^{n} \exp\left[-D_{\mathbf{x}}(\mathbf{x}, \mathbf{x}_i)\right] \exp\left[-D_Y(y, y_i)\right] dy$$

$$= \frac{1}{nc_{\mathbf{x}}c_Y} \sum_{i=1}^{n} \exp\left[-D_{\mathbf{x}}(\mathbf{x}, \mathbf{x}_i)\right] \int_{-\infty}^{\infty} y \cdot \exp\left[-D_y(y, y_i)\right] dy \tag{6.17}$$

$$= \frac{1}{nc_{\mathbf{x}}} \sum_{i=1}^{n} y_i \cdot \exp\left[-D_{\mathbf{x}}(\mathbf{x}, \mathbf{x}_i)\right]$$

$$D(\mathbf{x}) = \int_{-\infty}^{\infty} \frac{y}{nc_{\mathbf{x}}c_Y} \sum_{i=1}^{n} \exp\left[-D_{\mathbf{x}}(\mathbf{x}, \mathbf{x}_i)\right] \exp\left[-D_Y(y, y_i)\right] dy$$

$$= \frac{1}{nc_{\mathbf{x}}c_Y} \sum_{i=1}^{n} \exp\left[-D_{\mathbf{x}}(\mathbf{x}, \mathbf{x}_i)\right] \int_{-\infty}^{\infty} \exp\left[-D_y(y, y_i)\right] dy \tag{6.18}$$

$$= \frac{1}{nc_{\mathbf{x}}} \sum_{i=1}^{n} \exp\left[-D_{\mathbf{x}}(\mathbf{x}, \mathbf{x}_i)\right]$$

When the expressions for the numerator and denominator are inserted in Equation (6.16), the constants cancel, and we are left with the final GRNN model shown in Equation (6.19).

$$\hat{y}(\mathbf{x}) = \frac{\sum_{i=1}^{n} y_i \exp\left[-D(\mathbf{x}, \mathbf{x}_i)\right]}{\sum_{i=1}^{n} \exp\left[-D(\mathbf{x}, \mathbf{x}_i)\right]} \tag{6.19}$$

We have now come full circle. At the beginning of this section, the action of the GRNN was described as smoothing by interpolating among the true values of the training cases. We then immediately departed from that intuitive justification, presenting a theoretical development based on statistical expected values. But now that the theory is complete, we can see how the GRNN indeed acts as a data smoother.

Look closely at Equation (6.19). The numerator is a weighted sum of the true values in the training set. The denominator is the sum of the weights, making this model a true interpolator. Now look back at Equation (6.12). The weight given to each training case is based on the similarity of the test case to the training case. The rigorous statistical derivation of the GRNN completely agrees with the intuitive idea of a weighted average.

Because the GRNN method of combining multiple prediction models is considerably different from the methods shown earlier in this chapter, it is good to review on a more global scale what this method is doing. Recall that we have a collection of training cases, each of which is a predictor vector (usually) and a predicted scalar. This training set is used to train each component model. At this point, we are no longer interested in the original predictors. We are now concerned with the predictions made by the models. These vectors comprise the predictors for the GRNN. When an unknown test case arrives, each component model is invoked. The resulting vector of predictions is compared to the prediction vector for each training case. If the test case's prediction vector is unlike that of a training case, that training case is largely ignored. But if it is similar, then the true value of the predicted variable for that training case is heavily weighted in the computation of the grand prediction. Thus, the GRNN is acting like a pattern matcher, seeking training cases whose component model predictions resemble those of the test case.

Code for the GRNN

The files MULTPRED.CPP and GRNN.CPP on my web site contain the complete source code for a GRNN combiner. This code is too long and complex to present here in its entirety. However, it is instructive to study selected portions of the code. The core of the algorithm is the subroutine that implements Equation (6.19). Note that this implementation is more general than needed here, as it allows for vector outputs. When used for combining models making a scalar prediction, noutputs is one.

```
void GRNN::predict (
   double *input,                        // Input vector
   double *output                        // Returned output
   )
{
   int icase, iout, ivar;
   double *dptr, diff, dist, psum;

   for (iout=0; iout<noutputs; iout++)   // For each output
      output[iout] = 0.0;                 // Will sum kernels here, Equation (6.17)
   psum = 0.0;                            // Denominator sum, Equation (6.18)
```

```
for (icase=0; icase<ncases; icase++) {      // Do all training cases
  dptr = tset + (ninputs + noutputs) * icase; // Point to this case
  dist = 0.0;                                 // Will sum distance here, Equation (6.12)
  for (ivar=0; ivar<ninputs; ivar++) {        // All vars in this case (model predictions)
    diff = input[ivar] - dptr[ivar];          // Input minus case
    diff /= sigma[ivar];                       // Scale per sigma
    dist += diff * diff;                        // Cumulate Euclidean distance, Eq (6.12)
  }
  dist = exp (-dist);                          // Apply the Gaussian kernel, Eq (6.14)

  dptr += ninputs;                             // Outputs stored after inputs
  for (ivar=0; ivar<noutputs; ivar++)          // For every output variable
    output[ivar] += dist * dptr[ivar];         // Cumulate numerator, Equation (6.17)
  psum += dist;                                // Cumulate denominator, Equation (6.18)
  } // For all training cases

for (ivar=0; ivar<noutputs; ivar++)            // Final, Equation (6.19)
  output[ivar] /= psum;

}
```

This code implements Equation (6.19) and its predecessors. The numerator is cumulated in output[0] and the denominator in psum. The scalar output is returned.

One very large piece of the GRNN puzzle is still missing. How do we determine effective sigma weights? These scale factors, which appear in Equation (6.12) on page 298, play a vital role in the performance of the GRNN. If they are excessively small, little averaging will take place. The grand prediction for an unknown test case will be practically equal to the true value of whichever training case's component model predictions most closely resemble those of the test case. Under the operative assumption that these supposedly true values are badly contaminated with noise, this is not a good response.

Conversely, if the sigma weights are too large, little discrimination among training cases will take place. For any unknown test case, the grand prediction will approximately equal the average of the true values across all training cases. This bland response is clearly worthless.

An algorithm that has been found to be generally effective is to identify the sigma weights that produce the minimum cross validation error. To compute this error for a trial sigma vector, temporarily remove the first case from the training set. Use Equation (6.19) across the remaining training cases to make a prediction for the held-out case. Compare this prediction to the true value of the case that was held out. Square this difference. Now replace the first case and hold out the second case, repeating the prediction and cumulating the squared difference. Do this for the entire training set, holding out one case at a time. The sum of these squared errors is an indicator of the quality of the sigma vector. This algorithm can be found in the subroutine GRNN::execute() in the file GRNN.CPP.

How can we find the sigma vector that minimizes the cross validation error? The error surface usually has a moderate number of false local minima, so a naive descent algorithm is not effective. Probably the fastest and most accurate method is the combination of gradient descent and differential evolution described in [Masters and Land, 1997]. However, if speed is not important, the primitive version of simulated annealing given here works acceptably well. This little workhorse algorithm is versatile for a wide variety of recalcitrant problems, so it is worth listing and briefly discussing here. Again, this code is in the file GRNN.CPP.

```
void GRNN::anneal_train (
    int n_outer,              // Number of outer loop iterations, perhaps 10-20
    int n_inner,              // Number of inner loop iterations, perhaps 100-10000
    double start_std // Starting standard deviation of log weights, about 3.0
    )
{

    int i, inner, outer;
    double error, best_error, std, *best_wts, *test_wts, *center;

/*
    Best_wts keeps track of the best (log) sigma weights.
    Center is the center around which perturbation is done.
    It starts at zero. After completion of each pass through the inner loop
    it is changed to best_wts.
*/

    best_wts = (double *) malloc (ninputs * sizeof(double));
    test_wts = (double *) malloc (ninputs * sizeof(double));
    center = (double *) malloc (ninputs * sizeof(double));
```

```
   for (i=0; i<ninputs; i++)
     center[i] = 0.0;

   best_error = -1.0;                    // Will be min error; -1 flags start
   std = start_std;                      // Standard deviation of perturbation

   for (outer=0; outer<n_outer; outer++) {
     for (inner=0; inner<n_inner; inner++) {

       for (i=0; i<ninputs; i++) {                 // Compute a trial sigma by perturbing
         test_wts[i] = center[i] + std * normal(); // randomly around the current center
         sigma[i] = exp (test_wts[i]);             // We actually work with log(sigma)
         }

       error = execute ();                // Compute cross validation error for this trial sigma

       if ((best_error < 0.0) || (error < best_error)) { // If first trial or found new minimum
         best_error = error;       // Keep track of min error and associated sigma vector
         memcpy (best_wts, test_wts, ninputs * sizeof(double));
         }
       } // For inner loop iterations

     memcpy (center, best_wts, ninputs * sizeof(double)); // New center is best so far
     std *= 0.7;                  // Each outer pass shrinks search area
     } // For outer loop iterations

   for (i=0; i<ninputs; i++)              // We are done. Return actual sigma weights
     sigma[i] = exp (best_wts[i]);   // Must exp() because we work with logs here

   free (best_wts);
   free (test_wts);
   free (center);
}
```

Even if you do not want to use this algorithm to optimize the sigma weights, you will find it worth studying. Despite its simplicity, it is a marvelously versatile method for finding the approximate minimum of a function that is plagued with multiple false minima. It is considerably slower than need be, and it is not guaranteed to find the exact minimum, but it is very good at finding a minimum that is reasonably close to the global value, and its simplicity is a great virtue.

Rather than work with sigma values, we work with the log of sigma, which is appropriate because of the multiplicative nature of the parameter. At any given time, trial values will be computed by means of random offsets around a central value stored in the vector center. The standard deviation of this random perturbation is initialized to a relatively large value so that a wide area is searched.

An inner loop tries many perturbations. For each, the actual sigma values are computed by exponentiation, and the cross validation error for this trial sigma vector is computed. Every time a new minimum is found, the log sigma vector that gave this new minimum is preserved in best_wts. Thus, at the end of a complete pass through all inner loop iterations, we have a sigma vector that produces a relatively small error.

At this time, in preparation for the next pass through the outer loop, two actions are taken. First, the best (log) sigma vector is copied to center so that the next round of inner loop perturbations will be centered around this better vector. Second, the standard deviation of the perturbations is reduced somewhat, resulting in a narrower search.

By the time many passes through the outer loop are complete, the search distance will have been reduced to a fairly small value, so the best weight vector should be fairly close to the true minimum. If a sufficient number of inner loop iterations were done each time and if the initial standard deviation was large enough, it is very likely that at some point the iterations will have fallen within the domain of attraction of the global minimum and stayed there until the end. Naturally, this brute-force method could be improved in many ways if speed is important. However, as quick-and-dirty algorithms go, this one is quite respectable.

Note that the starting standard deviation is dependent on the scaling of the model outputs. Be sure to start at a value that is commensurate with the natural variation of the outputs so that Equation (6.13) is sensible. Otherwise, some variables may not be adequately searched.

Comparing the Combination Methods

A wide variety of algorithms for combining multiple predictions has been given. The strength and weakness of each were discussed. However, it is interesting to see how these methods compare when put to work. To compare and contrast the methods in a methodical way, 12 synthetic datasets were constructed by varying the design considerations, as shown here:

Model Quality:

- Three moderately good models

- These, plus a fourth completely worthless model

- These, plus a fifth good but strongly biased model

Number of training cases:

- A small dataset consisting of 20 cases

- A large dataset consisting of 200 cases

Noise contamination:

- A clean dataset, in which the true values are totally uncontaminated

- A noisy dataset, in which the true values are heavily contaminated with random noise

Seven different combination methods were tested, representing all of the techniques discussed in this chapter. One thousand replications were used. A single model was also included as a basis of comparison. The columns in the summary table are labeled as follows:

Raw: Mean squared error of a single model

Avg: The predictions are simply averaged

Uncons: Unconstrained linear regression

Unbias: Constrained linear regression with no bias offset term

Bias: Constrained linear regression including a bias offset term

VarWt: Variance weighting

Bag: Bagging, as discussed in Chapter 5

GRNN: General regression neural network smoothing

For each row in the table, the first label is the number of classes (with class 4 being worthless and class 5 being good but biased). The second label is the number of training cases, and the third label indicates whether the true values are clean or contaminated.

	Raw	Avg	Uncons	Unbias	Bias	VarWt	Bag	GRNN
3, 20, clean	1.68	0.87	1.64	0.63	0.64	0.70	2.21	0.79
4, 20, clean	5.24	1.69	0.92	0.67	0.70	0.60	4.19	0.81
5, 20, clean	9.35	2.48	2.52	1.45	0.57	1.45	1.11	0.83
3, 20, noisy	30.92	15.04	70.75	13.86	23.27	15.06	112.60	4.49
4, 20, noisy	16.33	7.23	35.91	9.76	11.21	7.89	418.76	4.50
5, 20, noisy	23.61	9.40	242.36	17.66	17.42	10.56	178.87	4.60
3, 200, clean	0.19	0.15	0.13	0.13	0.13	0.14	0.17	0.26
4, 200, clean	1.02	0.38	0.12	0.13	0.13	0.14	0.16	0.26
5, 200, clean	1.06	0.42	0.11	0.13	0.12	0.15	0.16	0.28
3, 200, noisy	2.31	2.27	2.25	2.25	2.25	2.27	2.33	2.36
4, 200, noisy	3.12	2.46	2.27	2.25	2.26	2.31	2.32	2.38
5, 200, noisy	3.16	2.51	2.27	2.25	2.25	2.35	2.30	2.38

This table is very revealing. Each combining method has its own favorite conditions. The following observations may be made:

- In the vast majority of cases, any combination method is better than none. The only exception is that for small noisy datasets, unconstrained regression fares poorly, and bagging is a disaster. This should not be surprising.

- As was seen in Chapter 5, bagging is never the best option, although it may be better than nothing.

- No one method is universally best. Each performs well in some situations and poorly in others. Beware. Test.

- The most dangerously volatile method, unconstrained regression, is the best performer only when the dataset is large and clean. Even then, it does not outperform its safer competitors by enough to make the risk worthwhile. This method is almost never recommended, even though it can occasionally shine.

- GRNN smoothing is by far the best method when the dataset is small and noisy. The linear regression methods, even constrained, are severely disrupted because they overfit the noise. If the GRNN is not available, then simple averaging or possibly variance weighting is the best choice.

- Even when the GRNN is not the best method, it is usually a good performer, with one notable exception. When the dataset is both large and clean, it gives up too much fitting power.

- The only situation in which the choice of combining method is not particularly crucial is when the data is noisy but the training set is large. In this case, the problem lies primarily with noise, and just about any algorithm that allows noise to cancel will work. The large training set discourages overfitting, which is the demon that haunts some methods under some conditions.

The complete source code for the program that produced this table of experiments can be found in the file MULTPRED.CPP on my web site. You are encouraged to modify this code to run your own experiments.

CHAPTER 7

Combining Classification Models

- The Majority Rule

- The Borda Count

- The Average and Product Rules

- The MaxMax and MaxMin Rules

- The Intersection and Union Rules

- Logistic Regression

- Model Selection by Local Accuracy

- Maximizing the Fuzzy Integral

- Pairwise Coupling

Chapter 6 discussed methods for combining several models that are designed to make numeric predictions. For classification models that base their decisions on numeric predictions, the methods of that chapter are often a good choice. However, some models are inherently strict classifiers in that they produce a class decision and nothing more. Also, many number-based classifiers produce numeric predictions that are unstable in some way. In such situations, we must use more specialized techniques. This chapter discusses model-combination algorithms for applications in which the ultimate goal is classification. Component models in which the prediction is and is not numeric are both considered. Also, we see how to take advantage of component classifiers in which class ranks of ordinal scale are produced.

© Timothy Masters 2018
T. Masters, *Assessing and Improving Prediction and Classification*,
https://doi.org/10.1007/978-1-4842-3336-8_7

Introduction and Notation

Assume that we have K mutually exclusive and exhaustive classes, which in this chapter will be indexed with the letter $k=1, ..., K$. By mutually exclusive, we mean that no case can belong to more than one class, and by exhaustive we mean that each case belongs to some class. Thus, a *reject* category is not allowed. If the possibility of "none of the above" is required for an application, then either a numerical combination method from Chapter 7 must be employed so that a membership threshold can be specified or the reject category must be defined as a class in its own right. One or both of these options is nearly always available.

There are M trained classification models, indexed by $m=1, ..., M$. Each of these models, when presented with an input vector x, produces K scalar outputs. (Some authorities prefer to view the output of a model as a K-vector, which of course is equivalent to K scalars. The scalar approach is used here because it simplifies notation.)

There are an infinite number of possibilities for the nature of the K scalars produced by a component model. Some of the most common possibilities are the following:

- Each is a real number representing a probability or confidence assessment regarding membership in each of the K classes.

- Exactly one of the K outputs is 1.0 (or *true*), and all of the others are 0.0 (or *false*). This is the situation for a classifier that by nature must choose a single class as its decision.

- The K outputs are the integers 1 through K ordered such that the class having the output 1 is considered the least likely, 2 the second least likely, and so forth. (For some algorithms, clarity is improved by reversing the rank ordering, with 1 being the best. It should be obvious from the context which ordering standard is employed.)

One of the most important points to note is that the third possibility, component models that rank likelihood of class membership, is in many or most applications the best choice. Certainly, models that produce good estimates of class membership probability are the most desirable of all. However, such models are far rarer than is commonly thought (or admitted!). There is considerable danger in treating model outputs as probabilities when they really are not. On the other hand, the information lost by converting real-valued outputs to ranks is relatively small. The trade-off is definitely worthwhile if there is any doubt at all about the quality of the information in the raw outputs.

Models that provide rank information are clearly more useful than models that produce a single class decision. What may not be so obvious is that rank information is often easily obtained if only a little effort is expended, and this rank information may be far more useful than believed at first glance. Granted, if there are only two classes, ranks are meaningless. For three classes, ranks have only modest value. But if there are dozens of classes, which is the case for applications like handwriting or speech recognition, a single winning class may be nearly worthless if there is significant possibility of error. The list of runner-ups in each model's decision contains a tremendous amount of useful information. So, for example, it is nearly always worth modifying a nearest-neighbor classifier to also provide the second-nearest neighbor and the third and so forth.

There is yet another problem to consider and for which ranks are an effective solution. When combining real-valued outputs of classification models, it is crucial to verify that the outputs of the models have the same scale of importance. A classic example comes from handwriting classification. Suppose one component model is particularly skilled at discriminating between the letter C and the letter O, a task for which small differences are crucial. Another component model may be concerned with gross differences, such as choosing between X and Q. This model may be operating on such a large scale that its information concerning likely membership in the O and C classes is noise compared to the subtlety of the first model. Careless combination of the outputs of these two models will be disastrous. Conversion of the class decision confidences to ranks will go a long way toward rescuing the situation.

Regardless of the nature of the model's outputs, we will refer to the outputs produced by input x as $f_{k,m}(x)$ where $k=1, ..., K$ indexes the class and $m=1, ..., M$ indexes the component model.

Reduction vs. Ordering

In a great many applications, the ultimate goal of the combined classifier is to produce a single class decision. However, there is a subtle issue that should be considered as part of the design. At a minimum, consideration of this issue will clarify thought processes, and in some cases the issue will exert significant control over the choice of combination algorithm.

It is useful to go beyond the most naive measure of classification accuracy: the probability of a correct decision. One should also consider a more general measure of success that includes situations in which the correct decision is not made. With this in mind, it is possible to divide the general task of classification into two related but different objectives. Either of these measures may be used to evaluate the performance of any class combination scheme, and several of the schemes to follow directly address the issue.

Class set reduction is the goal of finding a subset of the original K classes that is as small as possible but that has high probability of containing the true class. We don't care about ranking within this subset, as long as the subset is manageably small.

Class ordering is the goal of ranking the class membership likelihoods in such a way that the true class is as near the top as possible. No explicit rank threshold is specified, so performance is measured by the average distance of the true class from the top.

Why should we consider these goals? First, it may be that the approach taken in the application can be beneficially modified to take advantage of one or another of these schemes. But even if not, deciding which of these goals is more appropriate to the application, and using the corresponding error measure, will probably provide a better performance measure than simply using naive classification accuracy.

Finally, it may be that both of these goals can be employed simultaneously. We may first use a combination method aimed at class set reduction. This provides a small subset that is highly likely to contain the true class. Then we can use another combination method to order this subset, with the top-ranked class being the grand choice. Many applications can benefit from a classification scheme that produces a small ordered set of runner-ups to the winner.

The Majority Rule

The most general and intuitive combination scheme is democracy: choose whichever class is chosen by the majority of component models. This is expressed in Equation (7.1), which will be explained. Do not fear.

$$Choice = \underset{k}{\arg\max} \left[\sum_{m=1}^{M} \Delta \left(k = \underset{i}{\arg\max} f_{i,m} \right) \right] \tag{7.1}$$

As sometimes happens, a simple concept can be rigorously expressed only by a fierce-looking equation. However, this equation introduces notation that will be used throughout the chapter, so it should be studied.

The *argmax* function returns whichever value of its subscript variable produces the maximum value of the expression that follows argmax. The Δ function returns the value 1.0 if its argument is true and 0.0 if false. In Equation (7.1), the argument of Δ is true if and only if model m chooses class k. The summation counts how many models made this choice. Thus, the combined choice is whichever class was the winner the most times. Please do not go on until you understand this equation, as this notation will be used regularly.

The obvious advantage of the majority rule is that the component models do not need to produce numeric predictions. All that is needed is for the component models to choose a class. In the unfortunate situation that better models are not available, the majority rule is almost certainly the best we can do.

There are two disadvantages to the majority rule. First, the lack of information beyond the nominal class choice can be very costly, especially if there are many classes. Second, ties in the winner count present a theoretical, if not necessarily practical, difficulty. Note that we do not worry about ties in the argument of the Δ function. It is assumed that either the model returns an explicit class choice or it makes numeric predictions that are sufficiently fine that ties do not occur. This is a reasonable assumption in nearly all applications. We only need to worry about ties in the count of winners, though this can occasionally be a serious problem.

Code for the Majority Rule

Although this combination scheme is almost trivial, we will examine source code to implement the rule. The complete code is in the file MULTCLAS.CPP on my web site. This code provides a simple foundation from which more complex combination rules will be derived. It also illustrates a clever way of handling ties. Here is the class header:

```
class Majority {

public:

  Majority (int n, int nin, int nclasses, double *tset, int nmods);
  ~Majority ();
  int classify (double *input, double *output);

private:
  int nout;                 // Number of outputs (nclasses in constructor call)
```

```
int nmodels;             // Number of models (nmods in constructor call)
double *outwork;         // Work vector nout long
};
```

The constructor is given the number of training cases, number of classes, number of component models, and training data. The object keeps a private copy of the number of classes, which is also the number of outputs produced by each component model. In addition, it preserves the number of models, and a work vector is allocated.

The classify() function is given an input case. It returns an integer in the range zero through nclasses−1, which signifies the chosen class. A vector of scaled output confidences is also returned in case you are interested. The constructor and classification function now follow. As in the numeric prediction examples of Chapter 7, the array of models is an external reference.

```
Majority::Majority (
    int n,                   // Number of training cases
    int nin,                 // Number of inputs
    int nclasses,            // Number of outputs (classes)
    double *tset,            // Training cases, n by (nin+nout)
    int nmods                // Number of models in 'models' array
    )
{
    nout = nclasses;
    nmodels = nmods;
    outwork = (double *) malloc (nout * sizeof(double));
}
/*
    This returns the integer 0 through nout-1 for its class decision.
    It also returns class responses, normalized to sum to one.
*/

int Majority::classify (double *input, double *output)
{
    int i, imodel, ibest;
    double best, sum, temp;

    for (i=0; i<nout; i++)
        output[i] = 0.0;
```

314

```
for (imodel=0; imodel<nmodels; imodel++) {
  models[imodel]->predict (input, outwork);

  for (i=0; i<nout; i++) {    // Find which class wins for this model
    if ((i == 0) || (outwork[i] > best)) {
      best = outwork[i];
      ibest = i;
    }
  }

  output[ibest] += 1.0; // Tally this winning class
}

/*
  Output[i] now contains the number of models that chose class i.
  Find the winner, and normalize to unit sum.
  Note that we employ an interesting tie-breaking scheme. By adding a
  uniform random number less than 1.0 to each count, we ensure that
  no true winner is ignored, but ties are randomly broken.
*/

sum = 0.0;
for (i=0; i<nout; i++) {
  temp = output[i] + 0.999 * unifrand ();
  if ((i == 0) || (temp > best)) {
    best = temp;
    ibest = i;
  }
  sum += output[i];
}

if (sum > 0.0) {
  for (i=0; i<nout; i++)
    output[i] /= sum;
}

return ibest;
}
```

The only especially interesting part of this code is the tie-breaking scheme near the end. In the loop that determines which class received the most votes, a random number in the range 0.0 to 0.999 is added to each class vote count before the comparison is made. If two classes have different counts, the addition of this random number will not affect the comparison. But for comparisons that are tied, the random number will virtually always break the tie in a way that gives equal probability to either decision.

The Borda Count

A significant problem with the majority rule is that it discards a tremendous amount of potentially useful information when it considers only the top-ranked class in each model. Relative rankings below the top almost always contribute valuable information that may aid in making the combined decision. In the next section, we will see how a simple arithmetic average of the class outputs captures such information. However, that method is extremely susceptible to noise and other problems discussed later. The *Borda count* is an effective compromise between the two extremes of totally discarding information below the top rank and excessively valuing this information.

The Borda count for a class is defined as the sum across all models of the number of classes in that model ranked lower (worse) than the class in question. This is expressed as the argument in Equation (7.2).

$$Choice = \underset{k}{\text{argmax}} \left[\sum_{m=1}^{M} \sum_{i=1}^{K} \Delta\left(f_{k,m} > f_{i,m} \right) \right] \tag{7.2}$$

Notice that no provision is made for ties within a model. This should almost never be a problem in practice, as most classification models either produce continuous real outputs or make a single class choice that could not be used to compute a Borda count anyway. However, ties for the final best model can occur and should be taken into account. For example, one might employ one of the other methods of this chapter to break the tie.

If there are only two classes, the Borda count is equivalent to the majority rule. Thus, we are concerned only with applications in which there are three or more classes. A class that scores at the bottom for every model will have a Borda count of zero. A class that scores at the top for every model will have a Borda count of $m(k-1)$. Counts outside these limits are not possible.

Code for computing the Borda count is remarkably similar to the majority-rule code just shown. Again, no training is needed. All of the work is done at classification time. The biggest difference is that the algorithm is most efficient if we sort the class outputs of each model, simultaneously switching their indices. For this, we need an additional work vector iwork whose length is equal to the number of classes.

The complete source code for a Borda-count algorithm can be found in the file MULTCLAS.CPP on my web site. For now, you need only review the majority-rule code shown in the previous section and then study the following code fragment that replaces most of the majority classification code:

```
for (i=0; i<nout; i++)
   output[i] = 0.0;

for (imodel=0; imodel<nmodels; imodel++) {
   models[imodel]->predict (input, outwork);
   for (i=0; i<nout; i++)
      iwork[i] = i;    // Initialize for sorted indices
   qsortdsi (0, nout-1, outwork, iwork);   // Sort ascending
   for (i=0; i<nout; i++)
      output[iwork[i]] += i;
}
```

We start by initializing the output vector, which will sum the Borda counts, to zero. Each component model is invoked, and an index vector is initialized. Then the outputs are sorted from smallest to largest. The sorting routine, found in the file QSORTD.CPP, simultaneously moves the indices. Finally, the Borda counts are summed. After sorting, iwork[0] contains the index of the class having smallest (worst ranked) output. The Borda count for this class, in the corresponding element of output, is incremented by zero. This loop advances until the best-ranked class, identified by iwork[nout-1], is incremented by nout-1. Note, by the way, that this loop could start at one, not zero, because there is not much point in incrementing a quantity by zero! However, the loop was written this way for clarity. Feel free to change it if you want.

The Average Rule

If the component models produce outputs whose relative values are meaningful and commensurate across models, it is almost certainly beneficial to use the numeric values. The majority rule, Equation (7.1), discards much useful information, and the Borda count discards quite a lot also. The most straightforward way to take advantage of relative information is to average the output for each class across the component models. Since the number of models is fixed, this is equivalent to summing the outputs as expressed in Equation (7.3).

$$Choice = \underset{k}{\arg\max} \left[\sum_{m=1}^{M} f_{k,m} \right] \qquad (7.3)$$

This method treats each classification model as a numeric predictor and then uses the simple average discussed on page 279 to combine them. The class choice is determined by the highest output of the combined predictor.

There is a special problem to think about when this method is employed. When the ultimate goal is numeric prediction, each of the component models will surely have been trained with the same goal in mind. But when the models are intended for classification, implying that only ordering of outputs is important to each, it is easy to become careless and have component models whose outputs are not commensurate. At best, this results in implicit computation of a weighted average rather than a straight average. At worst, one or a few models may dominate the sum, weakening the ensemble. In case of doubt, it never hurts to study histograms of all outputs for all component models to confirm that they are commensurate.

Code for the Average Rule

There is no need to list the class header or constructor for this class, as they are nearly identical to those of the majority rule. But for the sake of completeness, here is the classification member function for the average rule:

```
int Average::classify (double *input, double *output)
{
   int i, imodel, ibest;
   double best, sum;
```

```
   for (i=0; i<nout; i++)
     output[i] = 0.0;                // Will cumulate sums here

   for (imodel=0; imodel<nmodels; imodel++) {
     models[imodel]->predict (input, outwork);
     for (i=0; i<nout; i++)
       output[i] += outwork[i];   // Sum output across all models
     }

/*
   Output[i] now contains the sum of all models' predictions for class i.
   Find the largest, and normalize to unit sum.
   This normalization only makes sense if the model's predictions are 0-1
   probabilities or some such thing.
*/

   sum = 0.0;                      // For normalization
   for (i=0; i<nout; i++) {
     if ((i == 0) || (output[i] > best)) {
       best = output[i];
       ibest = i;                  // For returning class decision
       }
     sum += output[i];
     }

   if (sum > 0.0) {                // Cheap insurance against division by zero
     for (i=0; i<nout; i++)
       output[i] /= sum;
     }

   return ibest;                   // 0 through nout-1 signifies class decision
}
```

The Median Alternative

It is well known that the mean of a set of numbers is sensitive to outliers and that the median is a frequently useful alternative. A small amount of information is lost, but the median is nevertheless an excellent indicator of central tendency without the burden of being adversely affected by outliers. If there is reason to believe that any of the component models may produce outlying values, the median should be considered. But remember that conversion to ranks also eliminates outlier problems. Computing the mean of each class's rank, with the ranking done separately for each model, is another possible solution.

The Product Rule

It is intuitively obvious that a reasonable way to compute the combined likelihood of a class is to average the likelihoods of the class across all models. The method of the previous section makes a lot of sense, and it often works well. However, if we can honestly regard the outputs of the component models as probabilities, then there is some theoretical justification for *multiplying* them instead of adding them. In fact, [Kittler, Hatef, Duin, and Matas, 1998] provides a detailed examination of the conditions under which this product derives directly from Bayes' method.

Unfortunately, in practice multiplication proves to be a harsh overseer if the probability assumption is not strictly correct. If even one model seriously underestimates the probability for a class, this class receives a blow from which it cannot possibly recover. Multiplication by a number near zero, when all other multipliers are bounded above by one, invariably produces a product near zero. This sensitivity proves to be an unacceptable handicap in nearly all applications, so the product rule is almost never used. It is only mentioned here because occasionally someone is tempted to use it based on its theoretical appeal. Be warned.

The MaxMax and MaxMin Rules

Many times we will have a collection of models that respond primarily to different subsets of the class set. Cases from a class that is outside the area of specialization of a particular model may produce output responses from that model that are small or moderate and whose actual value carries little meaning. But when this model encounters a case within its area of specialization, it produces a high output value for the correct class. Note, by the way, that we may or may not have deliberately planned for the component model (or models) to exhibit this behavior. Sometimes it happens automatically, perhaps even invisibly.

In this situation, we are interested only in high scores. It is counterproductive to muddy the waters by considering moderate scores that may be meaningless. A good choice in this case is to judge each class by its maximum output across all models. This is shown in Equation (7.4).

$$Choice = \underset{k}{\arg\max} \left[\underset{m}{\max} \left(f_{k,m} \right) \right] \tag{7.4}$$

Note that this method, called the *MaxMax* rule, explicitly discards a lot of information. When the models specialize as described, this is precisely what we want to do because the information being discarded is largely noise. But if outputs other than the maximum are important, this rule should not be used.

Less often, we may have exactly the opposite situation. The models may be unusually accurate in eliminating one or more classes, with different models specializing in black-balling certain subsets of the class set. In other words, suppose a case is from some particular class. We can (in an ideal situation) be sure that for every other class, there is at least one model that will produce a very low output for that incorrect class. When we are in this position, the best way to judge class membership is to examine the *minimum* output across all models. This is expressed in Equation (7.5).

$$Choice = \underset{k}{\arg\max} \left[\underset{m}{\min} \left(f_{k,m} \right) \right] \tag{7.5}$$

The Intersection Method

The methods described in this and the next sections are designed specifically to perform *class set reduction* as described on page 311. The intersection method, shown here, is not often an effective method for combining classifiers. However, it is sometimes useful. More importantly, it provides an intuitive foundation for the union method, which performs quite well in some applications.

The intersection and union methods require that each component model be able to respond to a case by ranking the entire class set in order from most to least likely. Very few classifiers are unable to do this. In fact, it is often true that classifiers that produce real-valued outputs can be improved by ranking the outputs, a process that may well discard more noise than useful information.

The process of training the intersection method involves finding, for each component model, the fewest number of best-ranked outputs that must be retained to guarantee that the true class will be included for every case in the training set. Then, when an unknown case appears, the combined decision for a minimal subset containing the class of the unknown is defined as the intersection of the minimal subsets for all of the component models.

For an illustration of the training process, suppose we have many classes (say about 100) and three models to combine. The following chart shows the rank (1 for the best here) of the true class for each model. Only the worst (most poorly classified) nine of the presumably hundreds of training cases appear. So, for example, the true class of the second case was ranked eighth best by the first and third models and was ranked fourth best by the second model.

Case	Mod 1	Mod 2	Mod 3
1	3	22	4
2	8	4	8
3	1	17	12
4	2	16	6
5	1	8	6
6	6	1	3
7	5	12	2
8	3	2	1
9	7	4	9
...			
Max	8	22	2

The maximum entry in each column is determined. These are 8, 22, and 12, respectively. When an unknown case arrives, all three models are evaluated. The 8 classes ranked best by the first model, the 22 best for the second, and the 12 best by the third are intersected. Those classes that are in common define the class subset returned by the intersection rule.

You may well have noticed some weaknesses of this method. In fact, the intersection method is only rarely a good means of combining classifiers. The following items of interest should be noted:

- As a direct consequence of the definition of the final subset, the training set will be classified "correctly" in the sense that for each training case, the final subset will always contain that case's true class. This is good but not sufficient reward for the weaknesses of the system.

- For cases outside the training set, it is always possible that the final class set will be empty. This happens when the top-ranked subsets in the component models are disjoint (have no classes in common). The more component models employed, the more likely this disturbing outcome will be.

- Because so much depends on *worst-case* behavior, this method can often produce unnecessarily large class subsets.

The last point in the previous list is, in practice, the killer. In a great many applications, we will have a set of specialist models. For any class, there will be at least one model that is particularly good at identifying this class by giving it a very good rank. Such specialist models typically perform poorly on other classes, but this should be acceptable as long as other models step in to fill the void. But this can't happen here because each model's subset size is determined by its behavior across the entire training set. Cases outside a model's area of specialization will ruin performance by distorting the subset size.

In defense of the method, it should be noted that the intersection rule can be an acceptable means of class set reduction when none of the component models ever performs really badly. As long as each model can be expected to give a fairly good rank to the true class of every training case, this method may be worth trying.

Because of the complexity of the algorithm, it is worth studying the code. A complete listing can be found in the file MULTCLAS.CPP on my web site. Here is the class header:

```
class Intersection {

public:

   Intersection (int n, int nin, int nclasses, double *tset, int nmods);
   ~Intersection ();
   int classify (double *input, double *output);

private:
   int nout;                // Number of outputs (nclasses in constructor call)
   int nmodels;             // Number of models (nmods in constructor call)
   double *outwork;         // Work vector nout long
   int *iwork;              // Work vector nout long
   int *rank_cuts;          // Work vector nmodels long
};
```

The constructor is responsible for allocating the work areas and computing rank_cuts, the worst class ranking achieved by each model. Here is the constructor:

```
Intersection::Intersection (
   int n,                   // Number of training cases
   int nin,                 // Number of inputs
   int nclasses,            // Number of outputs (classes)
   double *tset,            // Training cases, n by (nin+nout)
   int nmods                // Number of models in 'models' array
   )
{
   int i, j, k, iclass, imodel, icase, nbad;
   double *case_ptr, best;

   nout = nclasses;         // Save in private member
   nmodels = nmods;         // Ditto
   outwork = (double *) malloc (nout * sizeof(double));
```

```
rank_cuts = (int *) malloc (nmodels * sizeof(int));
iwork = (int *) malloc (nout * sizeof(int));

/*
   Pass through the training set, invoking all models for each case.
   For each model, keep track of the worst rank across the training set.
   In other words, for each model, find the minimum number of best-ranked
   classes we would need to keep in order for that subset to contain the
   correct class for every case in the training set. Naturally, the
   holy grail is one, meaning that for this outstanding model its highest
   output is always the correct class. Dream on. The worst situation
   is when this threshold is nout, meaning that we would need to keep the
   entire set of all classes in order to get the entire training set right.
   This disastrous situation means the model is worthless as far as this
   algorithm is concerned.
*/

for (imodel=0; imodel<nmodels; imodel++)
   rank_cuts[imodel] = 0;               // Will keep track of minimal size needed

for (icase=0; icase<n; icase++) {
   case_ptr = tset + icase * (nin + nout);   // Point to this case
   // Find the true class of this case
   for (j=0; j<nout; j++) {                   // Scan case's output vector
      if ((j == 0) || (case_ptr[nin+j] > best)) { // Outs after nin inputs
         best = case_ptr[nin+j];
         k = j;                           // Keep track of index of best so far
      }
   }

   // At this time, k is the true class of this case. Invoke each model.
   for (imodel=0; imodel<nmodels; imodel++) {
      models[imodel]->predict (case_ptr, outwork);
```

```
    // Count how many of this model's outputs equal or exceed the output
    // corresponding to the true class.
    best = outwork[k];                    // Output of true class
    nbad = 1;                  // Counts i=k, skipped below so we don't have to trust fpt

    for (i=0; i<nout; i++) {
      if (i == k)
        continue;
      if (outwork[i] >= best)             // This should be true for i=k
        ++nbad;                           // Count number needed in subset
    }

  if (nbad > rank_cuts[imodel])           // Keep track of the worst performance
    rank_cuts[imodel] = nbad;             // of this model across training set
  } // For all models
 } // For all cases
}
```

The constructor starts by allocating three vectors. Rank_cuts will contain the computed rank cutoffs for each model. The other two vectors are for scratch use during training and classification.

We then pass through the training set. The true class of each case is determined by finding the largest output value. Then, each model is invoked so that the rank of the true class can be found for each model. To be conservative, we count the number of outputs that exceed *or equal* the output corresponding to the true class. By counting equality, we score ties in the most damaging manner, a reasonable safety precaution that in practice will almost never come into play.

There is one subtle point in the counting loop. In theory, the test of an output being greater than or equal to that of the true class should always be true for the output of the true class (i=k). However, a few carelessly written compilers can fail to guarantee this result. To be safe, this known event is counted in advance (nbad=1) and then skipped in the counting loop.

The classification routine is responsible for invoking all of the models and computing the intersection of the classes that fall within the highest-rank subset of each model. Here is a listing of that routine:

```
int Intersection::classify (double *input, double *output)
{
   int i, imodel, n;

/*
   Compute the class output vectors for each model, then sort them.
*/

   for (i=0; i<nout; i++)
      output[i] = 1.0;                      // Will kill these off below

   for (imodel=0; imodel<nmodels; imodel++) {
      models[imodel]->predict (input, outwork);
      for (i=0; i<nout; i++)
         iwork[i] = i;                      // Initialize for sorted indices
      qsortdsi (0, nout-1, outwork, iwork); // Sort ascending
      n = nout - rank_cuts[imodel];         // This many classes in models bad part
      for (i=0; i<n; i++)                    // Pass through this bad part
         output[iwork[i]] = 0.0;            // And nix every class there
   }

   n = 0;                                    // Will count survivors
   for (i=0; i<nout; i++) {                  // Check each class
      if (output[i] > 0.5)                   // 1.0 if in intersection, else 0.0
         ++n;
   }

   return n;
}
```

The easiest way to find the intersection of several sets is to set a flag vector to indicate that all classes are included and then reset these flags as it becomes known that certain classes are excluded. Whatever remains is the intersection. Thus, the output vector is initialized to all ones.

All models are invoked for the test case. A good way to find the indices of the highest-ranked classes is to build an identity vector (iwork) and then sort the outputs with a subroutine that simultaneously moves the elements in the index vector. The source code for qsortdsi() is in the file QSORTD.CPP on my web site.

The constructor already computed rank_cuts[imodel] as the number of highest-ranked classes needed to guarantee inclusion of the true class throughout the training set. Subtracting this quantity from the total number of classes gives the number of classes not in the good subset. We zero each element in the output flag vector that corresponds to a class that didn't make the cut.

The final act of the classification routine is the busywork of counting the number of classes remaining in the intersected subset. This is a useful quantity to return to the caller.

The Union Rule

We saw how the intersection rule, while intuitively reasonable, suffers from the weakness of being overly dependent on worst-case performances. In particular, component models that specialize can almost never be effectively combined by intersection. This is almost always an overwhelming problem.

A fairly small modification of the intersection rule enables us to significantly reduce the impact of models that perform poorly for some cases. The *union rule* largely (though not totally) shifts the focus away from those worst cases and onto the best cases. This is ideal when we have a collection of specialists of varying abilities.

The union rule begins the same way as the intersection rule: for each case in the training set, the rank of its true class in each component model is computed. But instead of tallying worst ranks for each model, the best model for each case is found. Then, the worst of these best cases is tallied across the training set, separately for each model. When an unknown case is to be classified, the combined class subset is defined as the union of the best subsets from each model.

To make this clearer, consider again the small dataset used to illustrate the intersection rule. It is reproduced here, with three additional columns:

Case	Mod 1	Mod 2	Mod 3	Mod 1	Mod 2	Mod 3
1	3	22	4	3	0	0
2	8	4	8	0	4	0
3	1	17	12	1	0	0
4	2	16	6	2	0	0
5	1	8	6	1	0	0
6	6	1	3	0	1	0
7	5	12	2	0	0	2
8	3	2	1	0	0	1
9	7	4	9	0	4	0
. . .						
Max				3	4	2

For the first case, model 1 is clearly the best, ranking its true class at third. This value is entered in the corresponding column in the right half of the table. This process is repeated for each case. The maximum of these minimum values appears at the bottom of each model's column. Several properties of this algorithm should be apparent.

- Unlike the intersection method, this combination scheme can never produce an empty subset. At least one model will be best for some case.

- It is possible for one or more models to never be best, resulting in a column of zeros. This is an indicator that the offending model is worthless as far as the union algorithm is concerned.

- Although computing a subset by union produces a larger subset than intersection when all other things are equal, all other things certainly are not equal here. Because the right half of the table comes from minimums, the column maximums will almost surely be much smaller than those in the left half. Thus, we will be finding the union of subsets that are dramatically smaller than those used in the intersection method.

- The effect of specialists performing badly for cases outside their specialization has been completely eliminated at training time, because their large ranks will be ignored when the correct specialist takes over. Unfortunately, this may not carry through to classification, when such problematic models may contribute incorrect classes to the union. Luckily, it should not be many classes.

- There is still one type of worst-case performance not handled well by the union rule. This is outlier cases, those whose ranking is large for all component models. In a well-designed application, this should not occur.

There is significant duplication of code for the union and intersection methods, so a complete listing is not needed now. See the file MULTCLASS.CPP for details. Here is a listing of the most important part of the union-rule constructor, which may be compared with the corresponding part for the intersection rule on page 324.

```
for (imodel=0; imodel<nmodels; imodel++)
    rank_cuts[imodel] = 0;                      // Will keep track of minimal size needed

for (icase=0; icase<n; icase++) {
    case_ptr = tset + icase * (nin + nout);    // Point to this case

    // Find the true class of this case
    for (j=0; j<nout; j++) {                    // Scan case's output vector
        if ((j == 0) || (case_ptr[nin+j] > best)) { // Outs after nin inputs
            best = case_ptr[nin+j];
            k = j;                              // Keep track of index of best so far
        }
    }

    // At this time, k is the true class of this case. Invoke each model.
    for (imodel=0; imodel<nmodels; imodel++) {
        models[imodel]->predict (case_ptr, outwork);
        // Count how many of this model's outputs equal or exceed the output
        // corresponding to the true class.
```

```
best = outwork[k];                   // Output of true class
nbad = 1;          // Counts i=k, skipped below so we don't have to trust fpt
for (i=0; i<nout; i++) {
   if (i == k)
      continue;
   if (outwork[i] >= best)           // This should be true for i=k
      ++nbad;                        // Count number needed in subset
   }
// Nbad is now the rank of this case's true class according to imodel.
if ((imodel == 0) || (nbad < bestrank)) {
   bestrank = nbad;                  // Keep track of the best model's rank
   ibestrank = imodel;              // And know which model was best for this case
   }
} // For all models
// The best rank was 'bestrank' and it was achieved by model 'ibestrank'
if (bestrank > rank_cuts[ibestrank])     // Keep track of worst best performance
   rank_cuts[ibestrank] = bestrank;      // of this model across training set
} // For all cases
```

Understanding of this constructor may be aided by studying the example on page 329. We begin by initialize the rank_cuts array to zero. This array will keep track of the maxima of the columns on the right side of the table.

Much of the remainder of the code is identical to that for the intersection rule. We pass through the entire training set. For each case, its true class is found. The case is given to each component model, and the rank of the true class for each model is computed. This is all done exactly as before.

The difference involves what is done with nbad, the model's ranking of the true class. We keep track of the minimum of this rank across all models for this case. After all models have been checked and we know which one performed best, the maximum in rank_cuts is updated if needed.

Much of the classifier code is similar to that shown before. For the intersection rule, we initialized the output flags to indicate inclusion of all classes, and then we reset the flags of excluded classes. In the union algorithm, we initialize the flags to indicate exclusion and then set the flags of included classes. The code segment that does this is shown here:

```
for (i=0; i<nout; i++)
   output[i] = 0.0;                 // Will activate these below

for (imodel=0; imodel<nmodels; imodel++) {
   models[imodel]->predict (input, outwork);
   for (i=0; i<nout; i++)
      iwork[i] = i;                 // Initialize for sorted indices
   qsortdsi (0, nout-1, outwork, iwork);
   for (i=nout-rank_cuts[imodel]; i<nout; i++) // Pass through good part
      output[iwork[i]] = 1.0;       // And select every class there
   }
```

Logistic Regression

If one had to choose the single best *universally* applicable method for combining classifiers of equal power, it would probably be the Borda count algorithm described on page 316. This is because in most applications it captures the majority of the important information by using all outputs from all models. At the same time, it tends to deemphasize noise by converting raw outputs to ranks. This is the best of both worlds.

However, the Borda count method does have one notable drawback: it assumes that all models possess about the same predictive power. When some models are significantly better classifiers than others, it is in our interest to differentially weight the models according to their power. *Logistic regression* is a popular and effective means for doing so.

Before proceeding, it should be stated that this section ventures into a realm that might best be avoided in this text. Entire books have been written on the single topic of logistic regression, and none of these books does a thorough job of addressing the subject. For an excellent introduction, see [Ryan, 1997]. Under no circumstances should the discussion of logistic regression in this section be taken as complete or generally applicable. Here, we focus on the narrow topic of using logistic regression to combine classifiers, using Borda counts as predictors. Even within this narrow constraint, the algorithm has many subtle dangers and should be used only after carefully checking its operation. The algorithms shown here are not suitable for more widespread use. We'll have more to say about this later.

A good way to introduce logistic regression is to start by considering its close and more familiar relative, ordinary linear regression. This is described in the context of numeric prediction on page 283. Ordinary linear regression attempts to predict a variable by means of a weighted sum of predictors as shown in Equation (7.6).

$$f(x) = \sum_{m=1}^{M} w_m f_m(x) + w_0 \qquad (7.6)$$

For classification, we would let the predicted variable be binary, with 1.0 indicating membership in a particular class and 0.0 indicating nonmembership. If each $f_m(x)$ were the rank (or equivalent) of the class for model m in response to case x, we could use the training set to find optimal weights. Equation (7.6) would then be a reasonable way to combine decisions to obtain a consensus.

There are two distinctly different ways by which we could compute and employ the weights. The obvious and most commonly used method is to find separate weight sets for each class. When an unknown case arrives, each class' weight set is used to combine the models' class ranks, and the combined decision is defined as whichever class attains the maximum prediction using its ranks and weight set.

The problem with this approach is that we are estimating a lot of weights, which invites overfitting unless the training set is very large. A frequently better approach is to find a single weight set that is applicable to all classes. Each original training case is used to create K new training cases, where K is the number of classes. Exactly one of these new training cases will have the value 1.0 for its dependent variable, and its predictors will be the rank of the true class for each component model. Each of the remaining $K-1$ cases will have 0.0 as the predicted variable, and the predictors will be the ranks of the corresponding incorrect classes. When an unknown case arrives, the previously computed weights will be used to compute a combined Borda count for each class, with each component model's weight being used to weight this model's contribution to the combined Borda count.

You might complain that computing a single set of weights precludes use of interactions between classes and weights. If some models are better performers for some classes, which does indeed occur in many applications, then using one weight set

333

discards this potentially valuable information. This is definitely a valid consideration. However, if separate weight sets are to be estimated, the training set must be large enough to avoid overfitting. As will be seen, logistic regression has even more problems than ordinary linear regression when the training set is not large relative to the number of weights being estimated. For this reason, this section will focus primarily on computing a combined weight set. The more powerful but risky alternative of using separate weight sets will be briefly discussed at the end of the section.

To clarify the computation and use of the combined weights, consider an example of how to generate the regression training set from the original training data. Suppose our application has three classes, and we employ four models. We apply a training case to the models and obtain the following three outputs for each of the four models:

	Model 1	Model 2	Model 3	Model 4
Output 1	0.7	0.1	0.8	0.4
Output 2	0.8	0.3	0.9	0.3
Output 3	0.2	0.2	0.7	0.2

Since there are three classes (outputs), we will generate three regression training cases for each original training case. Suppose this case truly belongs to class 2. Observe that the first three models do a good job with the case, giving the correct class the highest output. Model 4 does not fare so well, placing the case in the middle.

Rather than using the ranks as regression predictors, we could use any quantity that conveys the same information. A variable that tends to encourage numerical stability in the regression is to use the number of outputs that score below the given output, divided by the total number of outputs. Thus, the minimum possible value, zero, would be obtained for the smallest output. The maximum possible value, $(K-1)/K$, would be obtained by the highest output. Because the hypothetical case just shown belongs to the second class, the regression case corresponding to that class would have a true value of 1.0 for the predicted variable, while this value would be 0.0 for the other two cases. This leads us to generate the following three regression cases for the hypothetical original case, where the last column is the true value of the predicted variable:

0.33	0.00	0.33	0.67	0.0
0.67	0.67	0.67	0.33	1.0
0.00	0.33	0.00	0.00	0.0

The first predictor for the first case is 1/3=0.33 because only one of this model's outputs, 0.2, is less than the output for this class, 0.7. The true value of the predicted variable for this case is 0.0 because this case corresponds to an incorrect decision: it is class 1 when the original case truly belongs to class 2.

It should be obvious that if most of the training cases look like this example, the optimal weights will be relatively large for the first three models, in which large values of the predictor (0.67) correspond to large values of the predicted variable (1.0) and small values of the predictor (0.0 and 0.33) correspond to small values of the predicted variable (0.0). However, the weight for the fourth model will probably be relatively small because the relationship is not so clear.

Once the weights are computed based on the regression training set, how are they best used? The technique suggested here is to classify using Borda counts, as was discussed on page 316. However, when summing each model's contribution to the Borda count, weight the contribution according to the computed regression weight. This is expressed in Equation (7.7), which should be compared to Equation (7.2).

$$choice = \operatorname*{argmax}_{k} \left[\sum_{m=1}^{M} w_m \left(\sum_{i=1}^{k} \Delta\left(f_{k,m} > f_{i,m} \right) \right) \right] \tag{7.7}$$

Code for the Combined Weight Method

Before venturing further, let us look at sample code for building the regression training set and using the combined weights to classify an unknown case. Note that we have not yet discussed how to use logistic regression (or any other technique) to compute the weights. This complex subject will be presented in the next section. First, we need to be clear on how to build a suitable training set and use the computed weights.

This code is extracted from the file MULTCLAS.CPP on my web site. Here is the class header for the Logit class that employs a combined weight set:

```
class Logit {

public:

    Logit (int n, int nin, int nclasses, double *tset, int nmods);
    ~Logit ();
    int classify (double *input, double *output);
```

```
private:
   int nout;                // Number of outputs (nclasses in constructor call)
   int nmodels;             // Number of models (nmods in constructor call)
   double *inwork;          // Work vector nmodels+1 long
   double *outwork;         // Work vector nout long
   int *iwork;              // Work vector nout long
   double *rankwork;        // Work vector nmodels * nout long
   Logistic *logit;         // Logistic regression object
};
```

The regression training set is built by the constructor, which also calls the appropriate algorithm to compute the optimal weights. Here is the constructor:

```
Logit::Logit (
   int n,                   // Number of training cases
   int nin,                 // Number of inputs
   int nclasses,            // Number of outputs (classes)
   double *tset,            // Training cases, n by (nin+nout)
   int nmods                // Number of models in 'models' array
   )
{
   int i, j, k, nbelow, iclass, imodel, icase, nbad;
   double *case_ptr, best;

   nout = nclasses;
   nmodels = nmods;
   inwork = (double *) malloc ((nmodels+1) * sizeof(double));
   outwork = (double *) malloc (nout * sizeof(double));
   iwork = (int *) malloc (nout * sizeof(int));
   rankwork = (double *) malloc (nmodels * nout * sizeof(double));
   logit = new Logistic (n * nclasses, nmodels);

/*
   Pass through the training set, building the logistic regression training
   set. We add nout cases to that set for each original training case.
*/
```

```
for (icase=0; icase<n; icase++) {
    case_ptr = tset + icase * (nin + nout);    // Point to this case
    // Find the true class of this case
    for (j=0; j<nout; j++) {                        // Scan case's output vector
        if ((j == 0) || (case_ptr[nin+j] > best)) { // Outs after nin inputs
            best = case_ptr[nin+j];
            k = j; // Keep track of index of best so far
        }
    }

    // At this time, k is the true class of this case. Invoke each model
    // and save its outputs.
    for (imodel=0; imodel<nmodels; imodel++)
        models[imodel]->predict (case_ptr, rankwork + imodel * nout);
    // The nmodels by nout array contains the outputs for each model
    // Create the nout cases for logistic regression
    for (iclass=0; iclass<nout; iclass++) {         // Add a case for each class
        for (imodel=0; imodel<nmodels; imodel++) { // Each model is an input
            nbelow = 0;
            best = rankwork[imodel*nout+iclass];   // Output for this class
            for (j=0; j<nout; j++) {                        // Compute the number worse
                if (rankwork[imodel*nout+j] < best)
                    ++nbelow;
            }
            inwork[imodel] = (double) nbelow / (double) nout; // Mean # below
        }
        inwork[nmodels] = (iclass == k) ? 1 : 0;
        logit->add_case (inwork);
    } // For all classes
    } // For all cases

    logit->train ();
}
```

Several work areas are allocated. Each regression training case will be built in inwork and appended by a call to add_case(). The two vectors outwork and iwork have their usual use of holding a model's outputs and keeping track of sorted indices, respectively. The matrix rankwork is needed because the regression training cases corresponding to a single original training case cannot be computed until all models' outputs for all classes are known. Finally, we will need a Logistic class object to handle computation of optimal weights. This class will be discussed later.

We pass through the original training set one case at a time. The first step is to determine the true class membership of the case. The method used here, finding the largest output, is not as efficient as simply storing the class index in the training set. However, this method allows us to use a universal training set layout. Note that for each case, the nin inputs appear first, followed by the nout outputs.

This case is then given to each model. The nmodels by nout array holds the outputs of each model for each class.

The core of the algorithm is the creation of nout (the number of classes) regression cases for each original case. We start by computing the nmodels predictor variables for the regression training case corresponding to class iclass. This is done by placing the output corresponding to this class in best and then counting in nbelow how many of the model's outputs are inferior to this one. The count is normalized by dividing by the number of classes. This normalization is not mathematically necessary, but it helps numerical stability when computing the regression weights.

The final step in building the case is to place the true value of the predicted variable at the end. This is 1.0 for the true class and 0.0 for the other classes.

After the optimal weights have been computed, they may be used to aid in combining class decisions for an unknown case. The code for this operation is shown here:

```
int Logit::classify (double *input, double *output)
{
   int i, imodel, ibest;
   double best, sum, temp;

   for (i=0; i<nout; i++)
      output[i] = 0.0;
```

```
/*
    Compute the class output vectors for each model, then sort them.
    The Borda count for each class is the sum across models of the number
    of classes in that model ranked worse.
*/

    for (imodel=0; imodel<nmodels; imodel++) {
      models[imodel]->predict (input, outwork);
      for (i=0; i<nout; i++)
        iwork[i] = i;                    // Initialize for sorted indices
      qsortdsi (0, nout-1, outwork, iwork);
      temp = logit->coefs[imodel];    // Weight for this model
      for (i=0; i<nout; i++)
        output[iwork[i]] += i * temp;    // As before, we could skip i=0
      }

/*
    Output[i] now contains the (weighted) Borda count of class i.
    Find the winner, then normalize to unit sum.
    Note that tie-breaking is not needed because the weights almost guarantee
    no ties.
*/

    sum = 0.0;
    for (i=0; i<nout; i++) {
      temp = output[i];
      if ((i == 0) || (temp > best)) {
        best = temp;
        ibest = i;
        }
      sum += temp;
      }
    if (sum > 0.0) {
      for (i=0; i<nout; i++)
        output[i] /= sum;
      }

    return ibest;
}
```

The weighted Borda counts will be cumulated in the output vector, so it is initialized to zero. The unknown case is applied to each model, and the model's outputs are sorted in increasing order. The previously computed weight for each model is placed in temp, and the weighted Borda count is cumulated. Note that as in the original Borda code, we do not need to make the first pass through the loop (i=0) because we are simply adding zero to the count for the most inferior class. It is only done here for clarity.

The remainder of the code is trivial. The highest Borda count is determined, and the outputs are normalized for the convenience of the user. No tiebreaker is needed here because the weights virtually guarantee that no ties can occur.

The Logit Transform and Maximum Likelihood Estimation

It is now time to think about how best to compute the optimal weights that will be used in Equation (7.7). We could use ordinary linear regression, finding the weights that minimize the mean squared error when Equation (7.6) on page 333 is used to model the regression training set. Actually, this is not a bad method, especially if the weights are constrained to be positive as discussed on page 286. However, substantial theory and considerable experience indicate that in many instances, when the predicted variable is binary, the method of logistic regression provides results that are superior to those obtained by ordinary regression.

Logistic regression derives from an entirely different paradigm than ordinary regression. The latter attempts to predict the actual value of the predicted variable. A simple formula, such as Equation (7.6), is used to estimate the value of the predicted variable. This is fine when this variable is a continuous real number. But when it is binary, limited to the values 0.0 and 1.0 as here, direct prediction seems almost to be an absurd endeavor. This is especially true when the prediction equation is primitive, such as the linear combination needed for combining decisions.

In contrast, logistic regression attempts to compute a quantity related to the *probability* that the predicted variable has the value 1.0 (as opposed to 0.0). For example, some observed value of the predictor variable set may tell us that there is a high, say 0.97, probability that the true value of the predicted variable is 1.0. This relaxed goal has better theoretical properties and often performs better in practice than the strict goal of predicting either the value 1.0 or the value 0.0. The implication is that our computed optimal weights can be more reliable.

If we simply let the regression equation predict the probability just discussed, we would be almost as bad off as before. Continuous values between zero and one would now be allowed, which is an improvement. But the range of sensible values for the predicted variable would still be narrowly restricted. We need a transformation that maps an unlimited domain to the interval [0, 1]. The *logistic* or *logit* transformation shown in Equation (7.8) has been found to work well in a wide variety of applications.

$$\pi(t) = \frac{e^t}{1+e^t} \tag{7.8}$$

When the argument to this transformation becomes extremely negative, the result approaches zero. Conversely, when the argument becomes extremely large, the result approaches one. Interestingly, the result is 0.5 (50-50 chance) when the argument is zero. This is a useful property. It means that when we let t be the predicted variable in the regression, a predicted value of zero means that there is an equal chance that the class membership is true or false, while positive values favor true and negative values favor false.

Instead of speaking of the probability of an event, people often refer to the *odds ratio*. Someone might say that there are, say, two-to-one odds of an event taking place. The odds ratio is the ratio of the probability of an event occurring to the probability of it not occurring, as in $\pi(t) / (1-\pi(t))$. Rearranging Equation (7.8) gives us Equation (7.9).

$$\frac{\pi(t)}{1-\pi(t)} = e^t \tag{7.9}$$

If we take logs on both sides of this equation to remove the exponentiation and explicitly let t be the quantity predicted by the regression equation, the result is expressed in Equation (7.10).

$$\log\left[\frac{\pi}{1-\pi}\right] = \sum_{m=1}^{M} w_m \, f_m + w_0 \tag{7.10}$$

Equation (7.10) says that for each case in the regression training set, a weighted linear combination of the predictors generated by the component models provides the log of the odds ratio for the corresponding class flag.

In case Equation (7.10) is confusing, review the discussion of the regression training set starting on page 334. Each regression training case consists of M predictors, which are the f_m in Equation (7.10). These M predictors are accompanied by a binary flag variable having the value one (true) for the case corresponding to the true class of the original case, and the value zero (false) for the other cases. When the predictors are weighted and summed according to the right side of Equation (7.10), the result is the log of the odds ratio that the binary flag variable for this regression case is one (true).

When we have a regression training set in hand, how can we compute a set of weights that will make Equation (7.10) represent the training set well? The most commonly used method is called *maximum likelihood estimation*. A detailed examination of this approach is beyond the scope of this text. Any good statistics text should provide details. An intuitive justification accompanied by the important formulas will be provided here.

We can judge the quality of the weights by how well the equation describes the training set. If the weights are poor, we will find many cases in which this equation suggests a high probability that the class flag is true, when in fact it is false. The converse will also be true. But if the weights are good, we should find that if the equation tells us there is a high probability that the class flag is true, then the majority of the time it will be true. Conversely, if the equation tells us there is a high probability that the class flag is false, then the majority of the time it will be false. And in those cases in which the equation has a value near zero (50-50 odds), we should find that about half the time the flag is true and half the time it is false. Note that we do not require any guaranties or close fits. All we require is that the probabilities make sense. It can shown that the *likelihood function* given by Equation (7.11) is a good measure of how well the probabilities fit the model.

$$L = \sum_i Y_i \left[\sum_{m=1}^{M} w_m f_{m,i} + w_0 \right] - \sum_i \log\left(1 + \exp\left[\sum_{m=1}^{M} w_m f_{m,i} + w_0 \right] \right) \qquad (7.11)$$

In this equation, the summation is over the cases in the regression training set, indexed by i. The class flag for each case, having the value one or zero, is represented by Y_i. The predictors are indexed by i as well as m because each case has its own set of M predictor values.

Larger values of L in Equation (7.11) indicate better weights. Thus, our goal is to find the weights that maximize L. One common method is to differentiate Equation (7.11), use a crude formula to estimate starting values for the weights, and then iterate with some version of Newton's method. Unfortunately, the iteration equations are often very ill conditioned. It regularly happens that wild weights emerge, and the iteration even fails to converge often enough to be troublesome. In the early days of computing, Newton iteration was the only practical approach because of time constraints. But now that computers are orders of magnitude faster than when logistic regression was devised, we can afford to use less efficient but more stable maximization algorithms.

There are two other aspects of this particular logistic regression application that work in our favor. First, it is reasonable to assume that the weights will never be negative. A negative weight implies that a model is contrary, worse than worthless. It gives an incorrect decision more often than would arise from pure guessing. Only in the most pathological situations would we encounter such a model. No responsible developer would let such a model get past the earliest stages of development.

Second, it is reasonable to assume that all of the weights will be small and roughly commensurate. This is because the predictors we are using (see page 334) have been deliberately normalized to the range zero through $(K-1)/K$, a value just under one. The computed weights are responsible only for importance indication. No dramatic rescaling of predictors is required.

The optimization method used in this text is simulated annealing, performed as follows: a set of M random trial weights is generated. These weights are of the form $\exp(Z)$ where Z follows a standard normal distribution. Thus, the trial weights are centered around $\exp(0)=1$, and they are positive, typically covering the range $\exp(-2)=0.135$ to $\exp(2)=7.39$. Then a fast and efficient linear optimizer is used to find the value of the constant, w_0, that maximizes L with the other weights fixed at their trial values. This is repeated many times, keeping track of the random weight set that provides the maximum value of L. After many tries, the center of the normal distribution shifts to the location of the best weight set, and the variance of the distribution is moderately decreased. The whole process is repeated several times. This algorithm will ultimately converge to a weight set that may not be strictly optimal, but it will almost surely be very close to optimal and good enough. The stability of this algorithm is also excellent because no ill-conditioned iteration is required. It is based on intelligently guided trial and error.

Code for Logistic Regression

The file LOGISTIC.CPP on my web site contains complete source code for a class that is effective at computing model weights using logistic regression. This class is not good for general use because it relies on the fact that all weights are small and positive, an assumption that is valid for this particular application but may be too restrictive in other applications. Here is the class header:

```
class Logistic {

public:

  Logistic (int ncase, int nin);
  ~Logistic ();
  void reset ();
  void add_case (double *newcase);
  void train ();
  void predict (double *input, double *output);

  double execute ();

  int ncases;      // Number of cases
  int ninputs;     // Number of inputs
  int nrows;       // How many times has add_case() been called?
  int trained;     // Has it been trained yet?
  double *tset;    // Ncases by (ninputs+1) matrix of training data
  double *coefs;   // Trained coefficient vector ninputs+1 long
};
```

The constructor is responsible for allocating memory and keeping local copies of important information. The routine add_case() simply adds a new case to the training set. After the training set is built, a call to train() invokes the training algorithm. The training code keeps a copy of the class object pointer to facilitate passing information to the local criterion functions needed by the optimizers. Here is this code:

```
double logit_crit (double *x);       // Local criterion function for optimization
double logit_unicrit (double x);     // Local criterion function for optimization
static Logistic *local_logistic;     // Needed by above
```

```
void Logistic::train ()
{
   int i, inner, outer, first;
   double y, best_y, std, *test_wts, *best_wts, *center;

   local_logistic = this;

/*
   Best_wts keeps track of the (log) best coefs.
   Center is the center around which perturbation is done.
   It starts at zero. After completion of each pass through the inner loop
   it is changed to best_wts.
*/

   test_wts = (double *) malloc (ninputs * sizeof(double));
   best_wts = (double *) malloc (ninputs * sizeof(double));
   center = (double *) malloc (ninputs * sizeof(double));

   std = 1.0;  // Reasonable when predictors are mean ranks

   for (i=0; i<ninputs; i++)
      center[i] = 0.0;          // Good starting point

   first = 1;
   for (outer=0; outer<10; outer++) {
      for (inner=0; inner<10 + 5 * ninputs * ninputs; inner++) {

         for (i=0; i<ninputs; i++)
            test_wts[i] = center[i] + std * normal();

         y = logit_crit (test_wts);
         if (first || (y > best_y)) {
            first = 0;
            best_y = y;
            memcpy (best_wts, test_wts, ninputs * sizeof(double));
            }
         } // For inner loop iterations

      memcpy (center, best_wts, ninputs * sizeof(double));
      std *= 0.7;
```

```
    } // For outer loop iterations

  logit_crit (best_wts); // Needed to set coefs correctly
  trained = 1;           // Training complete

  free (test_wts);
  free (best_wts);
  free (center);
}
```

Operation of the training code should be clear in light of the discussion that preceded it. Of greater interest is the local criterion function logit_crit() that evaluates the log likelihood function, Equation (7.11), for a trial weight set. Here is this code:

```
static double logit_crit (double *x)
{
  int i;
  double x1, y1, x2, y2, x3, y3;

  for (i=0; i<local_logistic->ninputs; i++)
    local_logistic->coefs[i] = safe_exp (x[i]);

  glob_min (-20.0, 20.0, 5, 0, -1.e160, logit_unic rit,
          &x1, &y1, &x2, &y2, &x3, &y3);

  y2 = brentmin (50, -1.e160, 1.e-10, 1.e-10, logit_unic rit,
          &x1, &x2, &x3, y2);

  local_logistic->coefs[local_logistic->ninputs] = x2;
  return -y2;
}

static double logit_unicrit (double t)
{
  local_logistic->coefs[local_logistic->ninputs] = t;
  return -local_logistic->execute ();
}
```

The criterion function exponentiates the trial parameters to define the actual model weights in Equation (7.11). It is then necessary to find the (otherwise unimportant) constant that maximizes the equation. This is done in two steps. First, a call to glob_min()

346

(located in the file MINIMIZE.CPP) finds a trio of points that encloses the maximum. Then, a call to brentmin() utilizes Brent's algorithm to quickly refine this interval. Both of these routines call the local function logit_unicrit() to compute the criterion. This function simply copies the trial constant to the end position in the weight vector and evaluates Equation (7.11) via a call to execute(). Note that because these routines minimize rather than maximize, we must negate the function values.

The execute() function is a straightforward implementation of Equation (7.11). It is as follows:

```
double Logistic::execute ()
{
   int icase, ivar;
   double *tptr, term, sum1, sum2, err;

   sum1 = sum2 = 0.0;

   for (icase=0; icase<ncases; icase++) {
      tptr = tset + (ninputs + 1) * icase;      // This case
      predict (tptr, &term);                     // Log odds ratio
      sum1 += term * tptr[ninputs];              // Output stored after inputs
      sum2 += log (1.0 + exp (term));
      } // For all training cases

   return sum1 - sum2;
}

void Logistic::predict (
   double *input,               // Input vector
   double *output               // Returned output
   )
{
   int i;
   *output = coefs[ninputs];     // Constant term
   for (i=0; i<ninputs; i++)
      *output += input[i] * coefs[i];
}
```

The main loop passes through all training cases, cumulating the sum of the log likelihood. The same term, computed by predict(), occurs in both halves of Equation (7.11). These two halves are cumulated separately in sum1 and sum2, respectively.

Separate Weight Sets

The focus thus far has been on finding a single set of model weights to weight the Borda counts needed for the combined class decision. This method has the advantage of being relatively stable yet powerful. The disadvantage of a single weight set is that it is unable to take advantage of specialist models. It may well be that some models are better than others at detecting certain classes. Whether by design or by accident, such specialization can and does occur. Thus, we may want to consider using a separate weight set for each class decision. This is expressed in Equation (7.12), which should be compared with Equation (7.7) on page 335.

$$Choice = \underset{k}{\arg\max} \left[\sum_{m=1}^{M} w_{k,m} \left(\sum_{i=1}^{K} \Delta \left(f_{k,m} > f_{i,m} \right) \right) \right] \qquad (7.12)$$

The problem with using separate weight sets is that we now have $K*M$ parameters to optimize, a number that is frequently very large. (Note that the K constants are not counted as free parameters here.) This technique should not even be considered unless *each class* is represented by *at least* ten times as many training cases as models. If you can pull together a training set this large (or much larger if possible) and if you have reason to believe that significant specialization is taking place, then separate weight sets may be a viable alternative to a single set. But be warned that this method is potentially dangerous and should be attempted only with caution.

To compute separate weight sets, we need to perform K independent logistic regressions. The training sets for these regressions are computed exactly the same way as for a combined weight set. They are simply distributed differently. Look back at the discussion of training set generation that started on page 334. In that example, three regression training cases were generated from a single original case. There, all three cases went into the same regression training set. But here, we would place each of the three cases in the corresponding specialized training sets. The first case, which is an example of predictors for a case that does not belong to class $k=1$, would go into the first training set. The second, an example of the predictors for a case from class $k=2$ goes into the second set. Finally, the third case, an example of predictors for a case that does not belong to class $k=3$, goes into the third set. In other words, each new regression training case goes into the training set corresponding to the predictors, with the true class of the case determining the one/zero value of the dependent variable.

The code for this separation is practically identical to the code already shown on page 335, so it will not be repeated in full detail here. There are only two small differences. First, instead of allocating one Logistic object, we need to allocate one for each class as follows:

```
Logistic **logit;      // Logistic regression objects (one for each class)
...
logit = (Logistic **) malloc (nout * sizeof(Logistic *));
for (i=0; i<nout; i++)
  logit[i] = new Logistic (n, nmodels);
```

When the case is appended to the training set, the appropriate set must be selected.

```
logit[iclass]->add_case (inwork);
```

When it comes time to classify an unknown, we once again proceed almost exactly as before. The models are all invoked, and the weighted Borda count is computed. The only difference is that now we must use the correct weight set for each class.

```
for (imodel=0; imodel<nmodels; imodel++) {
  models[imodel]->predict (input, outwork);
  for (i=0; i<nout; i++)
    iwork[i] = i;                        // Initialize for sorted indices

  qsortdsi (0, nout-1, outwork, iwork);
  for (i=0; i<nout; i++)
    output[iwork[i]] += i * logit[i]->coefs[imodel];
}
```

By way of final warning, it should be stressed that logistic regression, especially when it involves separate weight sets, should not be used carelessly. At the very minimum, the developer should look at the computed weights to verify that they are not wildly disparate for no apparent reason. Most of the time, logistic regression does a very respectable job of finding effective weight sets. However, it does occasionally happen that one or more weights will be orders of magnitude larger or smaller than the others. This may be a correct indication that the models are truly that disparate in their abilities. On the other hand, it may be a numerical fluke. Trust, but verify.

Model Selection by Local Accuracy

We have already discussed the possibility that certain component models do a better job than others at detecting or rejecting certain classes. Another type of specialization is possible. It is often the case that certain models do a better job of classification (perhaps for certain classes) in different areas of the predictor domain. For example, it may be that when a particular predictor variable has relatively small values, one model is especially effective, while when this predictor has large values a different model is favored. This effect may be by design or by accident. In any case, if such domain specialization is well represented within the training set, the method presented in this section can be useful at capitalizing on it.

A blunt-tool approach is taken here. When an unknown case appears, all component models are called upon to choose a class, and our algorithm selects the single model whose choice is to be most trusted for this particular case. It is likely that even better results could be obtained by weighted voting or some similar technique. However, the algorithm given here already performs excellently, and weighted voting schemes are highly complex in this context. Graduate students in search of a thesis topic should ponder this problem.

Given an unknown case, how does one compare the qualifications of each component model in order to select the one whose choice is most likely to be correct? There are an infinite number of possibilities, but [Woods, Kegelmeyer, and Bowyer, 1997] describe a method that makes a lot of sense and works well in practice.

The algorithm searches through the training set and computes the Euclidean distance between the unknown case and each training case. Those few (typically ten or so) of the training cases that are most similar to the unknown case, as measured by Euclidean distance, are presented to the component models. Each model's local quality is computed by considering only those training cases in the "nearby" subset, which the model classifies into the same class as it chose for the unknown case. The fraction of those cases that it classifies correctly is the model's performance criterion. Whichever model has the highest criterion is the model whose choice is taken.

Here is an example to clarify the algorithm. Assume that we have decided in advance that the ten nearest neighbors will be examined. We locate those ten training cases whose Euclidean distances from the unknown are the smallest. Suppose we present the unknown to a model and it decides that the unknown belongs to, say, class 3. We now present the ten nearby training cases to this model and find that six of these ten cases are assigned to class 3 by the model, and four of these six cases do truly belong to class 3.

The performance criterion for this model is then 4/6=0.67. The same procedure is repeated for the other component models, and the best model has the privilege of deciding the class of the unknown case.

There are many loose ends to be cleaned up. For starters, we are dividing small integers, so the probability of ties among the competing models is high. How do we break ties? The truth is that it probably does not matter very much. Different applications will favor different methods, while the overall variation will likely be small. The method chosen here is to compute a measure of the certainty of the competing models, selecting the model that is most sure of itself. This certainty measure is the ratio of the maximum output to the sum of all outputs. Feel free to try alternatives.

Another loose end is the nature of the criterion itself. We have decided to measure quality as the fraction of correct decisions among those decisions that are the same as that for the unknown case. If we make the often reasonable assumption that the component models may specialize in classes as well as predictor domain areas, this criterion is sensible in that it limits its information to that which is likely to be pertinent. But if we want to assume that no class specialization takes place, the class restraint may discard useful information. We might be better off defining a model's local performance criterion as the fraction of correct decisions within the entire nearby subset. If there is no class specialization, every correct or incorrect decision is vital information, regardless of its class.

Last but certainly not least, we need to decide how many nearby cases to include in the local subset. If we choose just a few, the algorithm will have excellent local sensitivity, but the models' performance criteria may be subject to significant random error. If we choose a large subset of the training set, the criteria will be computed relatively accurately, enabling good model selection, but an accuracy measure that is supposed to be local may be too large to be effective. How can we find a good compromise?

Unless you are dealing with a monstrous training set on a slow computer, the best way to choose the subset size is ordinary cross validation: one training case is removed and treated as an unknown. The remainder of the training set is used by the algorithm just discussed, with various trial subset sizes tested. Then the previously removed case is replaced and another case withdrawn. After repeating this process for every training case, the average performance obtained with each trial subset size is compared, and the best size is selected. In case of a tie, the smallest subset size is usually best when efficiency is considered.

Code for Local Accuracy Selection

The code for model selection by local accuracy is more complex than most of the code in this text, involving numerous small details. It will be presented in three major installments. We will start with the first half of the constructor, which is responsible for preparatory work. Then the classification code will be shown. We conclude with the second half of the constructor, which uses cross validation to compute an optimal local subset size. Here is the class header and first half of the constructor:

```cpp
class LocalAcc {

public:

   LocalAcc (int n, int ninputs, int nclasses, double *tset, int nmods);
   ~LocalAcc ();
   int classify (double *input, double *output);

private:
   int knn;              // This many nearest neighbors will be used
   int ncases;           // Number of cases
   int nin;              // Number of original case inputs
   int nout;             // Number of outputs (nclasses in constructor call)
   int nmodels;          // Number of models (nmods in constructor call)
   double *outwork;      // Work vector nout long
   int *iwork;           // Work vector ncases long
   double *distwork;     // Work vector ncases long
   double *trnx;         // Work vector ncases * nin long holds raw predictors
   int *trncls;          // Work vector ncases * nmodels long holds model decisions
   int *trntrue;         // Work vector ncases long holds true class of each case
   int classprep;        // Very private flag so cross validation can skip prep
};

LocalAcc::LocalAcc (
   int n,                // Number of training cases
   int ninputs,          // Number of inputs
```

```
   int nclasses,              // Number of outputs (classes)
   double *tset,              // Training cases, n by (nin+nout)
   int nmods                  // Number of models in 'models' array
   )
{
   int i, j, k, iclass, imodel, icase, ibest, itemp;
   int *clsptr1, *clsptr2, true_class, *knn_counts;
   int knn_min, knn_max, knn_best;
   double *case_ptr, *last_ptr, *testcase, *clswork, best;

   ncases = n;
   nin = ninputs;
   nout = nclasses;
   nmodels = nmods;

   outwork = (double *) malloc (nout * sizeof(double));
   iwork = (int *) malloc (ncases * sizeof(int));
   distwork = (double *) malloc (ncases * sizeof(double));
   trnx = (double *) malloc (ncases * nin * sizeof(double));
   trncls = (int *) malloc (ncases * nmodels * sizeof(int));
   trntrue = (int *) malloc (ncases * sizeof(int));

/*
   Pass through the training set, saving raw inputs in trnx.
   Invoke all models for each case.
   For each model, save its class decision (highest output) in trncls.
   Also save the true class of each case in trntrue.
*/

   for (icase=0; icase<ncases; icase++) {
      case_ptr = tset + icase * (nin + nout);      // Point to this case
      memcpy (trnx + icase * nin, case_ptr, nin * sizeof(double));
      // Find the true class of this case
      for (j=0; j<nout; j++) {                      // Scan case's output vector
```

```
    if ((j == 0) || (case_ptr[nin+j] > best)) { // Outs after nin inputs
       best = case_ptr[nin+j];
       k = j; // Keep track of index of best so far
       }
    }

// At this time, k is the true class of this case. Save it in trntrue.
// Then invoke each model and save the class decision of each.
trntrue[icase] = k;        // Save true class of this case
for (imodel=0; imodel<nmodels; imodel++) {
   models[imodel]->predict (case_ptr, outwork);
   for (i=0; i<nout; i++) {    // Find which class wins for this model
     if ((i == 0) || (outwork[i] > best)) {
        best = outwork[i];
        ibest = i;
        }
     } // For all outputs of this model
   trncls[icase*nmodels+imodel] = ibest; // Save this model's decision
   } // For all models
} // For all cases
```

Memory areas are allocated first. Then the training set is processed. For each case, the following information is saved for later use:

- The original predictor variables are saved in trnx because they will be needed by the classifier, which must find the training cases that are near the test case.

- The true class of each training case is saved in trntrue because the classifier will need this information in order to count the number of correct decisions made by each model.

- The class decision made by each model for each training case is stored in trncls so that the classifier can locate, for each model, those training cases that had the same class decision as that for the test case.

The classifier code is more complex. It must find the nearest neighbors and compute the performance criterion for each competing model. Note that for now, the clasprep flag should be considered always true so that all of the code is executed. Later we will see when it might be false. Here is the classifier code:

```
int LocalAcc::classify (double *input, double *output)
{
   int i, k, icase, imodel, n, ibest, numer, denom, bestmodel, bestchoice;
   double dist, *cptr, diff, best, crit, bestcrit, conf, bestconf, sum;

/*
   Find the knn nearest neighbors, keeping track of their indices
   If knn is small, there are faster ways than sorting the whole thing.
   The classprep flag should always be true in normal use. The only reason
   it is here is because cross validation in the constructor tries many
   values of knn, and this expensive preparation is only needed the
   first time. The things computed in this block do not change if the
   input case remains the same.
*/

   if (classprep) {
      for (icase=0; icase<ncases; icase++) {
         iwork[icase] = icase;          // Save index for sorting later
         dist = 0.0;                     // Will cumulate Euclidean distance here
         cptr = trnx + icase * nin;      // Point to this case
         for (i=0; i<nin; i++) {         // For all original input variables
            diff = input[i] - cptr[i];   // Input minus training case
            dist += diff * diff;         // Cumulate Euclidean distance
            }
         distwork[icase] = dist;         // Save distance
         }

      qsortdsi (0, ncases-1, distwork, iwork);
      }
```

```
/*
    Invoke all models and find the class decision of each.
    Search the knn nearest neighbors and identify those cases for
    which this model chose this class.
    Calculate the fraction of those cases in which the model was correct.
*/

    for (imodel=0; imodel<nmodels; imodel++) {
    models[imodel]->predict (input, outwork);
    sum = 0.0;
    for (i=0; i<nout; i++) {          // Find which class wins for this model
        sum += outwork[i];           // For computing tie-breaking confidence

        if ((i == 0) || (outwork[i] > best)) {
            best = outwork[i];
            ibest = i;
            }
        } // For all outputs of this model
    conf = best / sum;               // May be needed to break a tie later

    // This model chose class 'ibest' with confidence 'conf'
    denom = 0;          // Counts cases (in knn) in which this model chose ibest
    numer = 0;          // Counts cases (in denom) in which this model was correct

    for (icase=0; icase<knn; icase++) {
        k = iwork[icase];  // Index of this case
        if (trncls[k*nmodels+imodel] == ibest) {
            ++denom;        // This model chose this class
            if (ibest == trntrue[k])
                ++numer;    // And it was a correct choice
            }
        } // For the knn nearest neighbors
```

```
   if (denom > 0)
     crit = (double) numer / (double) denom; // Model's worthiness
   else
     crit = 0.0;

   if ((imodel == 0) || (crit > bestcrit)) {
     bestcrit = crit;
     bestchoice = ibest;
     bestconf = conf;
     memcpy (output, outwork, nout * sizeof(double));
     }

   else if (fabs (crit - bestcrit) < 1.e-10) {   // We must break a tie
     if (conf > bestconf) {                       // Fairly arbitrary
       bestcrit = crit;
       bestchoice = ibest;
       bestconf = conf;
       memcpy (output, outwork, nout * sizeof(double));
       }
     }

   } // For all models

/*
   Normalize the outputs of the best model for user's convenience
*/

   sum = 0.0;
   for (i=0; i<nout; i++)
     sum += output[i];

   if (sum > 0.0) {
     for (i=0; i<nout; i++)
       output[i] /= sum;
     }

   return bestchoice;
}
```

The first step is to pass through the entire training set, computing the Euclidean distance separating the test case from each training case. This distance is saved in the distwork array, and the associated indices are saved in the iwork array. When all distances have been found, the distance array is sorted in ascending order, and the indices in iwork are moved accordingly. Thus, the first knn elements of iwork will contain the indices of the knn training cases nearest the test case.

The next step is to commence an outer loop that considers each competing model. The test case is applied to a model, and then a small loop computes three things: best is the maximum output for the model, ibest is the index of this output and hence the model's class decision, and conf is a rough confidence figure for the decision. This ratio of the best output relative to the total of all outputs will be used only if a tiebreaker is required later.

After these three quantities have been computed for the model, an inner loop passes through the knn nearest neighbors to compute the performance criterion for the model. The number of cases for which the model classified a neighbor into class ibest is counted in denom, and of those cases, the number that were correct decisions is counted in numer. The ratio of these counts is the model's performance criterion.

As each competing model is tried, we keep track of the winning criterion so far, as well as the class choice for the winning model. Because the user may have some interest in the outputs of the winning model, we continually update the output vector.

If there are no ties, the confidence figure in conf is ignored. However, in case of a tie, which is detected by allowing for minor floating-point inaccuracies, the tied model having maximum confidence in its choice is declared the winner.

The final step, as always, is to normalize the outputs for the user's convenience. The index of the winning model's class choice is returned.

To conclude this discussion of the code, we present the second half of the constructor. This code segment uses cross validation to compute an optimal value of knn, the number of nearest neighbors checked.

```
classprep = 1;

if (ncases < 20) {
    knn = 3;
    return;
}
```

```
knn_min = 3;        // Require at least this size
knn_max = 10;       // But at most this size

testcase = (double *) malloc (nin * sizeof(double));
clswork = (double *) malloc (nout * sizeof(double));
knn_counts = (int *) malloc ((knn_max - knn_min + 1) * sizeof(int));

for (knn=knn_min; knn<=knn_max; knn++)
  knn_counts[knn-knn_min] = 0; // Will count correct decisions for each knn

--ncases;               // Cross validation uses reduced training set in classify().
last_ptr = trnx + ncases * nin;   // Last case keeps getting swapped in

for (icase=0; icase<=ncases; icase++) { // Omitted-case loop
  // Copy case icase to testcase, then copy last case to its spot
  // Also swap data in trncls and trntrue
  case_ptr = trnx + icase * nin;          // Point to this case
  memcpy (testcase, case_ptr, nin * sizeof(double));
  true_class = trntrue[icase];            // True class of test case
  if (icase < ncases) {          // Last case is already there, so no need to swap
    memcpy (case_ptr, last_ptr, nin * sizeof(double));
    trntrue[icase] = trntrue[ncases];
    clsptr1 = trncls + icase * nmodels;
    clsptr2 = trncls + ncases * nmodels;

    for (i=0; i<nmodels; i++) {      // Memcpy might be faster
      itemp = clsptr1[i];
      clsptr1[i] = clsptr2[i];
      clsptr2[i] = itemp;
    }
  }

  classprep = 1; // Tell classify() that it must fully prepare
  for (knn=knn_min; knn<=knn_max; knn++) {
    iclass = classify (testcase, clswork);
    if (iclass == true_class)        // Correct decision?
      ++knn_counts[knn-knn_min];     // Score this trial knn
    classprep = 0; // Tell classify() that it does not need to prepare
  }
```

```
      // Done, so move original last case back to its slot and restore icase
      if (icase < ncases) { // Last case does not need repair as nothing changed
         memcpy (last_ptr, case_ptr, nin * sizeof(double));
         memcpy (case_ptr, testcase, nin * sizeof(double));
         trntrue[ncases] = trntrue[icase];
         trntrue[icase] = true_class;
         clsptr1 = trncls + icase * nmodels;
         clsptr2 = trncls + ncases * nmodels;
         for (i=0; i<nmodels; i++) {          // Memcpy might be faster
            itemp = clsptr1[i];
            clsptr1[i] = clsptr2[i];
            clsptr2[i] = itemp;
            }
         }
      }

   ++ncases; // Restore size of training set to its correct value

   // See which trial value of knn had the best score.
   // Do not break ties. Favor the lowest value for classification speed.
   for (knn=knn_min; knn<=knn_max; knn++) {
      if ((knn==knn_min) || (knn_counts[knn-knn_min] > ibest)) {
         ibest = knn_counts[knn-knn_min];
         knn_best = knn;
         }
      }

   knn = knn_best;
   classprep = 1; // Tell classify() that it must fully prepare

   free (testcase);
   free (clswork);
   free (knn_counts);
}
```

First, the classprep flag must be set so that the classifier will do all of its required work. If there are fewer than 20 cases, we make the arbitrary decision to set knn to three. Otherwise, minimum and maximum trial values for knn are established. You should feel free to raise the maximum value if it is felt that it would be useful.

Cross validation will require a test case, so we allocate testcase for this purpose. The work vector clswork will never actually be referenced, but the classifier needs a place to put its output vector. Finally, knn_counts will hold the success score for each trial value of knn so that we can find the best when we are finished.

Because a test case will be withheld each time classification is done, the number of training cases ncases must be temporarily decremented. This private member variable is accessed by classify(). Every time a test case is pulled from the training set array, the last case in the array will be swapped into the vacancy, so we set last_ptr to the address of this frequently used case.

The cross validation loop now commences. The test case index icase must be inclusive of ncases because that count was decremented a moment ago. The location of the current test case is case_ptr. This case, as well as its true class, are copied to known locations. Unless the test case is the last case in the training set, we must fill in the vacancy by copying the last case (along with its true class and models' decisions) to the now empty spot.

We are ready to evaluate all of the trial values of knn. The classprep flag is set so that classify() will compute all of the Euclidean distances and sort them to locate the neighbors nearest this test case. However, this expensive preparatory work needs to be done for only the first trial value of knn. As long as the test case does not change, the quantities computed in the preparatory step will not change. The class decision associated with each value of knn is found by calling classify(). Each time a correct decision is made, the corresponding counter in knn_counts is incremented.

When all trial values of knn have been evaluated, the former last training case that was swapped into the test slot must be swapped back to the end slot, and the test case must be put back where it belongs.

After the entire process is complete, the number of training cases in ncases is incremented back to its correct value. The score for each trial knn is checked, and the winner is found. Note that there is no need to break ties because in the case of a tie, efficiency concerns would incline us to favor the smaller value.

Maximizing the Fuzzy Integral

This is a good news/bad news section. The bad news is that if you do not have a thorough grounding in fuzzy set theory, you will find it nearly impossible to understand exactly why the algorithm described in this section works. Presenting the necessary background is far beyond the scope of this text. The situation is even worse because the fuzzy integral method is not at all intuitive. Even when the algorithm is clearly explained, its operation may seem like magic.

But the good news is that the algorithm works wonderfully. As will be seen at the end of this chapter, it is difficult to find a situation in which the method fails to perform at least reasonably well. And in some cases, the fuzzy integral algorithm outperforms every other algorithm in the chapter. It is well worth plowing through the relatively complex material in order to add a potent weapon to our classifier arsenal.

The fuzzy integral was originally described by [Sugeno, 1977]. The idea was put into practice by (among others) [Cho and Kim, 1995], whose notation we shall mostly follow. For a rigorous theoretical discussion of the fuzzy integral in the broader context of fuzzy sets, see [Pedrycz, 1993].

We start by reviewing the idea of a *fuzzy measure*. Let $\mathbf{X}$ be a universe of discourse, and let $\mathbf{C}$ denote a family of subsets of $\mathbf{X}$ forming a σ-field of $\mathbf{X}$. A mapping g: $\mathbf{C} \rightarrow [0, 1]$ is called a fuzzy measure if it satisfies the following three conditions:

Boundary: $g(\varnothing)=0$ and $g(\mathbf{X})=1$

Monotonicity: For $\mathbf{A}, \mathbf{B} \in \mathbf{C}$, $\mathbf{A} \subset \mathbf{B}$ implies $g(\mathbf{A}) \leq g(\mathbf{B})$

Continuity: For $\mathbf{A}_n \in \mathbf{C}$ an increasing sequence of measurable sets,

$$\lim_{n \to \infty} g(\mathbf{A}_n) = g\left(\lim_{n \to \infty} \mathbf{A}_n\right) \tag{7.13}$$

The first condition means that the measure of the empty set is zero and the measure of the entire universe is one. The second condition means that if one set is a subset of another, then the measure of the former cannot exceed the measure of the latter. The third condition applies only if $\mathbf{X}$ is infinite, and it is a sort of continuity restriction imposed on the fuzzy measure.

[Sugeno, 1977] introduced the idea of a *λ-fuzzy measure* by imposing a fourth condition on a fuzzy measure. Suppose we have $\mathbf{A}, \mathbf{B} \in \mathbf{C}$, with $\mathbf{A} \cap \mathbf{B} = \varnothing$ and $\mathbf{A} \cup \mathbf{B} \in \mathbf{C}$. Loosely, these two subsets are disjoint (no members in common) and their union is what might be termed valid in some sense. Then the λ-fuzzy measure of their union is the sum of the λ-fuzzy measures of each subset, plus a fudge factor, as expressed in Equation (7.14).

$$g_\lambda(A \cup B) = g_\lambda(A) + g_\lambda(B) + \lambda g_\lambda(A) g_\lambda(B) \tag{7.14}$$

If there exists some $\lambda > -1$ such that Equation (7.14) holds true for every possible A and B, then g_λ is said to be a λ-fuzzy measure.

We are now in a position to define a *fuzzy integral*. Let $h: X \to [0, 1]$ be a measurable function, generally what we would call a *membership function*. Then the fuzzy integral (over the universe of discourse) of h with respect to the λ-fuzzy measure g_λ is defined by Equation (7.15),

$$\int h \cdot g_\lambda(\cdot) = \max_{\alpha \in [0,1]} \left[\min\left(\alpha, g_\lambda(F_\alpha)\right) \right] \tag{7.15}$$

where F_α is an α-cut of h as defined by Equation (7.16).

$$F_\alpha = \left[x \in X \,\middle|\, h(x) \ge \alpha \right] \tag{7.16}$$

Assume that X is finite of cardinality n. The brute-force approach to computing Equation (7.15) requires that we enumerate all 2^n possible subsets. We state without proof that there is a vastly more efficient method.

Let the finite set X be defined as $\{x_1, x_2, ..., x_n\}$. We can assume without loss of generality that the membership functions for each individual item in X are arranged in nonincreasing order: $h(x_1) \ge h(x_2) \ge ... h(x_n)$. Such an ordering can always be arranged by simply relabeling the data. Define $A_i = \{x_1, x_2, ..., x_i\}$. In other words, the sequence of subsets A_i is generated by successively appending items from X, starting with the one having the greatest membership function and ending with the one having the minimum membership function. Then Equation (7.15) can be efficiently computed using Equation (7.17).

$$\int h \cdot g_\lambda(\cdot) = \max_{1 \le i \le n} \left(\min\left[h(x_i), g_\lambda(A_i) \right] \right) \tag{7.17}$$

Note that the two terms whose minimum is being found are members of series that are moving in opposite directions. The first, the membership function, is decreasing (by our stipulation). The second, the measure on which the integral is based, is increasing because of the monotonicity property of a fuzzy measure. Thus, one can almost think of the fuzzy integral as the height of an intersection.

The series $g_\lambda(A_i)$ is easily computed using Equation (7.14) recursively, as shown in Equation (7.18).

$$g_\lambda(A_i) = g_\lambda(\{x_i\}) + g_\lambda(A_{i-1}) + \lambda g_\lambda(\{x_i\})g_\lambda(A_{i-1}) \qquad (7.18)$$

How do we go about computing λ? It's actually quite easy. Recall that by the boundary condition included in the definition of a fuzzy measure, the final term in this recursion, $g_\lambda(A_n)$, must equal one. It's not hard to see that when $\lambda=-1$ the final term cannot exceed one as long as the individual terms satisfy the boundary condition. Also, the final term blows up monotonically as λ increases. Thus, all we need to do is find the root of an easily computed and well-behaved equation in one unknown.

What Does This Have to Do with Classifier Combination?

So far, this discussion has been theoretical. It is time to see how fuzzy-integral theory relates to combining classifier decisions. We start by matching terms in equations to practical quantities.

The universe **X** is the set of component models, with $g_\lambda(\{x_i\})$ being what one might call the reliability or quality of model x_i. Now that the theory is over and consistency with other literature is less important, we will change the model index from the traditional i to the more pneumonic m so you can better remember that it refers to a model.

In the context of previous sections, $g_\lambda(\{x_m\})$ would be the weight given to model x_m. In the current context, these values might be known in advance or supplied by human experts. Later, we will see how a simple test performed on the training set can be used to determine reasonable values. For the remainder of this presentation, consider them to be fixed known quantities, with zero meaning that the model is entirely worthless, one meaning that the model is perfect, and intermediate values indicating intermediate model qualities. Note that unlike some previous situations, there is no requirement that the $g_\lambda(\{x_m\})$ values sum to one across all models. It is reasonable to employ many excellent models, whose values are all near one, or to employ nothing but nearly worthless models, whose values are all near zero. Note also that x_m simply designates model m, and it has no numeric value. We care only about its fuzzy measure, $g_\lambda(\{x_m\})$.

To come to a group consensus classification, we will separately compute the fuzzy integral of each class across all models. Unlike some other techniques, the fuzzy integral for each class is computed in isolation, being dependent only on the model outputs for that class. After doing the computation separately for each class, whichever class has the greatest fuzzy integral is designated the winner.

The output of model m for the class in question is $h(x_m)$ in the recent discussion and the especially important Equation (7.17). For economy, we will sometimes take the liberty of abbreviating this as h_m. Note that there is no need to use an index such as k to designate a particular class because the same computation is done separately for each class, one at a time. At any given moment, only one class will be under scrutiny. For this class, we will be computing the fuzzy integral of the model outputs with respect to the measure of model importance.

We now have everything we need. A set of M classifiers can be combined to form a group consensus with the following algorithm:

1) Plug the known fixed model qualities $g_\lambda(\{x_m\})$ (which will be discussed soon) into Equation (7.18) and employ a root finder to compute the value of λ that causes the final fuzzy measure (that of all models combined) to equal one.

2) For each class:

   a) Relabel the models so that $h_1 \geq h_2 \geq \ldots \geq h_M$.

   b) Use Equation (7.18) on the reordered models to recursively compute the series of fuzzy measures. Note that the value of λ that was found in the first step is immune to reordering.

   c) Use Equation (7.17) to compute the fuzzy integral of the class. Note that n in that equation is M here.

3) Choose the class having the maximum fuzzy integral.

The following example may clarify this algorithm. Suppose we have four classification models, and their known reliabilities are 0.9, 0.4, 0.6, and 0.1, respectively. We compute $\lambda=-0.97091$. Now suppose that for some class in question, the four models have the outputs 0.3, 0.2, 0.9, and 0.5, respectively. Equation (7.17) tells us that we compute the fuzzy integral as the maximum of four numbers. The successive computation of these four numbers is illustrated in the following table:

i	h	g	min(g,h)
1	.9	.6	.6
2	.5	.1 + .6 - .97091 * .1 * .6 = .6417	.5
3	.3	.9 + .6417 - .97091 * .9 * .6417 = .9810	.3
4	.2	.4 + .9810 - .97091 * .4 * .9810 = 1.0	.2

We begin the recursion with the model having the greatest output for the class. This output is 0.9, and the reliability of this model is 0.6. The minimum of these two numbers is 0.6. The next greatest output is 0.5, and the corresponding model has reliability 0.1. Applying Equation (7.18) gives a fuzzy measure of 0.6417, and the minimum of the two numbers is 0.5. The table is completed by repeating this two more times. Note that, as expected, the final fuzzy measure is 1.0 because $\lambda=-0.97091$ was computed to guarantee this result. The greatest of the four numbers {0.6, 0.5, 0.3, 0.2} is 0.6, so this is the fuzzy integral for the class.

Code for the Fuzzy Integral

This section contains source code for a class that uses the fuzzy integral method to combine classifiers. This code can be found in the file MULTCLAS.CPP on my web site. We begin with the class header:

```
class FuzzyInt {

public:

  FuzzyInt (int n, int nin, int nclasses, double *tset, int nmods);
  ~FuzzyInt ();
  int classify (double *input, double *output);
```

```
private:
  double recurse (double x); // Recursively compute the final g(A)-1

  int nout;                  // Number of outputs (nclasses in constructor call)
  int nmodels;               // Number of models (nmods in constructor call)
  int *iwork;                // Work vector nmodels long
  double *outwork;           // Work vector nout long
  double *sortwork;          // Work vector nmodels * nout long
  double *g;                 // Model g-values, nmods long
  double lambda;             // Overall lambda
};
```

The following short routine invokes Equation (7.18) repeatedly and then subtracts one from the result. The correct λ will cause this routine to return zero, because the upper boundary condition for a fuzzy measure requires that the measure of the complete set (all models, here) must be one. This is the routine called by the root finder to compute λ:

```
double FuzzyInt::recurse (double x) // x plays the role of λ (lambda)
{
  int i;
  double val;

  val = g[0];   // The equations in the text start index at 1, but C++ starts at 0
  for (i=1; i<nmodels; i++)
    val += g[i] + x * g[i] * val;  // Equation (7.18)

  return val - 1.0;                      // Subtract 1.0 because final value must be 1.0
                                         // Meaning that this routine seeks to return 0.0
}
```

The constructor is relatively long and complex, so we will break it apart and handle each section separately. Here is the beginning, which allocates working memory:

```
FuzzyInt::FuzzyInt (
  int n,                     // Number of training cases
  int nin,                   // Number of inputs
  int nclasses,              // Number of outputs (classes)
```

```
double *tset,         // Training cases, n by (nin+nout)
int nmods             // Number of models in 'models' array
)
{
    int i, j, k, iclass, imodel, icase;
    double *case_ptr, best, xlo, xhi, y, ylo, yhi, step;

    nout = nclasses;
    nmodels = nmods;

    iwork = (int *) malloc (nmodels * sizeof(int));
    outwork = (double *) malloc (nout * sizeof(double));
    sortwork = (double *) malloc (nmodels * nout * sizeof(double));
    g = (double *) malloc (nmodels * sizeof(double));
```

Not much has been said about how to determine suitable values for $g_\lambda(\{x_m\})$, the reliabilities of the M component models. We have stated that each value must range from zero (the model is worthless) to one (the model is perfect). We also know that the values must be known before any of the methods of this section can be applied. The constructor shown here uses a simple but generally effective method for estimating these values. It passes through the training set, testing each model. The correctness rate of each model is computed, the correctness that would be due to random guessing is subtracted to find the model's true contribution, and the result is rescaled to yield a number in the range zero to one. It must be emphasized that other methods, such as using an independent validation set, may be superior. This method, though, is quick and simple. Here is the code:

```
for (imodel=0; imodel<nmodels; imodel++)
    g[imodel] = 0.0;                          // Will sum correct decisions for each model

for (icase=0; icase<n; icase++) {
    case_ptr = tset + icase * (nin + nout);   // Point to this case

    // Find the true class of this case
    for (j=0; j<nout; j++) {                   // Scan case's output vector
        if ((j == 0) || (case_ptr[nin+j] > best)) {  // Outs after nin inputs
            best = case_ptr[nin+j];
```

```
      k = j;                    // Keep track of index of best so far
      }
    }

// At this time, k is the true class of this case. Invoke each model
// and find its decision.

for (imodel=0; imodel<nmodels; imodel++) {
   models[imodel]->predict (case_ptr, outwork);

   for (j=0; j<nout; j++) {              // Scan model's output vector
     if ((j == 0) || (outwork[j] > best)) {
       best = outwork[j];
       iclass = j;                        // Keep track of index of best so far
       }
     }

   // If this model chose correctly, count it
   if (iclass == k)
     g[imodel] += 1.0;

   } // For all models
  } // For all cases
```

```
/*

Divide by the number of cases to get a 0-1 success rate for each model.
Then subtract the rate that would be obtained by simply guessing, and
renormalize to get a 0-1 figure for the contribution of this model.

*/

for (imodel=0; imodel<nmodels; imodel++) {
   g[imodel] /= n;
   g[imodel] = (g[imodel] - 1.0 / nout) / (1.0 - 1.0 / nout);

   if (g[imodel] < 0.0)             // Model may be bad and unlucky
     g[imodel] = 0.0;              // But must respect 0-1 bounds
   }
```

The final step is to compute λ as the unique value that causes the fuzzy measure of the complete set of all models to equal one. We have already seen that recurse (lambda) returns the final recursive value of Equation (7.18), minus one. Thus, our goal is to find the value of lambda that causes the routine to return zero. Here is that code. A brief discussion of its operation follows the listing.

```
xlo = lambda = -1.0;
ylo = recurse (xlo);
if (ylo >= 0.0)                    // Theoretically should never exceed zero
   return;                        // But allow for pathological numerical problems
/*
   Now we must bound the root. Step out until the function becomes positive.
   If all models are worthless, the root is infinite, so we must avoid that!
*/

   step = 1.0;

   for (;;) {

      xhi = xlo + step;
      yhi = recurse (xhi);

      if (yhi >= 0.0)             // If we have just bracketed the root
         break;                  // We can quit the search

      if (xhi > 1.e5) {          // In the unlikely case of extremely poor models
         lambda = xhi;           // Fudge a value
         return;                 // And quit
         }

      step *= 2.0;               // Keep increasing the step size to avoid many tries
      xlo = xhi;                 // Move onward
      ylo = yhi;                 // (Generally, root will be bracketed quickly!)
      }
/*
   We have bracketed the root between (xlo, ylo) and (xhi, yhi).
   Bisection is a primitive way to refine, but it always succeeds.
*/
```

370

```
for (;;) {
    lambda = 0.5 * (xlo + xhi);
    y = recurse (lambda);           // Evaluate the function here
    if (fabs (y) < 1.e-8)           // Primary convergence criterion
        break;
    if (xhi - xlo < 1.e-10 * (lambda + 1.1)) // Backup criterion
        break;
    if (y > 0.0) {
        xhi = lambda;
        yhi = y;
    }
    else {
        xlo = lambda;
        ylo = y;
    }
}
```

It was mentioned earlier that λ will never need to be less than -1. So we begin by testing for this extreme condition. In case pathological floating-point problems cause the final fuzzy measure to (trivially) exceed one even when $\lambda=-1$, we use this extreme value.

The normal situation is that recurse(-1) will be negative. We must bracket the root, so we repeatedly step out until the function becomes positive. When this happens, we know that the correct value of λ lies between the two bracketing values. In the unusual event that all models have extremely tiny values of g, we need to avoid falling into an endless loop. This exit would never be taken in practical applications.

After the root is bracketed, the last step is to refine the interval to close in on the root. Many extremely sophisticated methods exist for doing this, but they are all overkill here. This function is so well behaved in practice, as well as fast to compute, that primitive bisection is sufficient. The bracketing interval is split in half, and the function is evaluated at the midpoint. This midpoint becomes a new endpoint according to its sign. The bisection is considered accurate enough if either of two tests is passed: the fuzzy measure of the complete set is very close to one or the interval surrounding λ is very tiny.

The classification routine reorders the models and evaluates Equation (7.17) for each class. This code is shown now, and a discussion of its operation follows:

```
int FuzzyInt::classify (double *input, double *output)
{
  int i, k, iclass, imodel;
  double sum, gsum, *rptr, minval, maxmin, best;

/*
  Invoke all models and store their (normalized) outputs.
  Note that if we can GUARANTEE that the models' outputs are already
  probabilities, this normalization is counterproductive.
*/

  for (imodel=0; imodel<nmodels; imodel++) {
    models[imodel]->predict (input, outwork);

    sum = 0.0;                          // Will sum this model's outputs

    for (i=0; i<nout; i++)              // So we can normalize them to
      sum += outwork[i];                // probability-like quantities

    for (i=0; i<nout; i++)
      sortwork[i*nmodels+imodel] = outwork[i] / sum;
    }

/*
  This main outer loop computes the fuzzy integral for each class
*/

  for (iclass=0; iclass<nout; iclass++) { // Compute for each class

/*
  Sort the models according to their output for this class
*/

    for (imodel=0; imodel<nmodels; imodel++)
      iwork[imodel] = imodel;           // Initialize index identity vector
    rptr = sortwork + iclass * nmodels;  // Point to this class's outputs
```

```
   qsortdsi (0, nmodels-1, rptr, iwork);    // Sort ascending

/*

Run through the models from highest output (for this class) to lowest.

*/

   maxmin = 0.0;                            // Will keep track of max of mins here
   gsum = 0.0;                              // Will be cumulative g

   for (i=nmodels-1; i>=0; i--) {           // Sorted ascending, so max to min
     k = iwork[i];                          // Index of this model

     gsum += g[k] + lambda * g[k] * gsum;   // Cumulate g(A) with Equation (7.18)

     if (gsum < rptr[k])                    // Compare g so far to model's output
       minval = gsum;                       // We need the smaller of the two
     else
       minval = rptr[k];

     if (minval > maxmin)                   // Keep track of max of this min
       maxmin = minval;                     // (For the current class)

     }

   output[iclass] = maxmin;
   } // For all classes

/*

Return the class having maximum fuzzy integral.
Don't worry about ties, which should be rare.

*/

   for (i=0; i<nout; i++) {
     if ((i == 0) || (output[i] > best)) {
       best = output[i];
       iclass = i;
       }
     }

   return iclass;
   }
```

The first step is to present the test case to all models and save the models' outputs in the matrix sortwork. Note that we do something here that may be counterproductive in some applications. For each model, we sum its outputs and then divide each output by the sum. The result is that the outputs become vaguely (or perhaps remarkably) like probabilities. They all range from zero to one, and they sum to one. In the event that we can guarantee that the models do indeed produce outputs in the range zero to one (though they need not sum to one) and that these outputs approximate probabilities, then the normalization done here should probably be skipped. However, in the absence of such guarantees, it is usually a good idea to normalize this way.

For each class, its fuzzy integral is now computed. The work array iwork is initialized to the identity, rptr is set to point to the models' outputs for this class, and these outputs are sorted, with the indices in iwork moved simultaneously. The sort routine leaves the array in ascending order, so we will access the models from last to first in order to process them starting with the highest output.

The maximum of the g, h minimums will be saved in maxmin. The recursively computed values of the fuzzy measure will cumulate in gsum. As the index k works from the highest output to the lowest, Equation (7.18) will be used repeatedly. Each time, the current recursive values will be compared to the corresponding model output in rptr, and the minimum of these two numbers found as minval. The maximum of minval is retained in maxmin. When all models have been appended, the final value of this maximum is the fuzzy integral for the class, and it is stored in output[iclass].

The final step is trivial: locate the class having the largest fuzzy integral and return its index. In most applications, there is no need to worry about ties because of the large amount of numerical computation being done.

Pairwise Coupling

We conclude this presentation of classifier-combination algorithms with a highly specialized technique. All of the prior algorithms combined classifiers that were individually able to perform the complete task of identifying class membership. Each classifier could be used alone if necessary, because each handled all of the classes.

Pairwise coupling, described in [Hastie and Tibshirani, 1996], performs its task very differently. It combines a collection of $K*(K-1)/2$ binary classifiers to choose the most likely candidate from among the K possible classes. Every possible pair of classes is represented by a single binary model that specializes in discriminating between the members of that particular pair of classes. The obvious advantage of this technique is that each component model can be simple in that it predicts a single output whose value determines which of two competing classes is the more likely. This is the ultimate in specialization.

This algorithm does have one inherent limitation: the binary classification models must each produce a continuous output greater than zero and less than one, and this output must be able to be interpreted as a confidence in the classification decision. If the output is a true probability, this is even better. In other words, if the model is discriminating between Class A versus Class B, an output of 0.7 should imply (at least approximately) that there is a probability of 0.7 that the case presented to the model belongs to Class A and hence a probability of 0.3 that the case belongs to Class B. If we can satisfy this requirement, pairwise coupling may be a useful option for classification.

To be rigorous, let there be K classes and hence $K*(K-1)/2$ pairs of classes. Define r_{ij} as the output produced by the model that specializes in discriminating between Class i versus Class j, and let this output represent the probability that the case presented to the model is a member of Class i. These values can be arranged in a K by K matrix. For example, if $K=3$, the following matrix may be produced by the models in response to a test case:

$$
\begin{matrix}
--- & 0.2 & 0.7 \\
0.8 & --- & 0.4 \\
0.3 & 0.6 & ---
\end{matrix}
$$

In this matrix we see that the model that discriminates between Classes 1 and 2 gives a probability of 0.2 that the case is a member of Class 1 *under the condition that the case is a member of one of these two classes.* Note that the upper-right diagonal elements of this matrix provide complete information on the class decisions because the matrix has a sort of symmetry in that $r_{ji}=1-r_{ij}$.

Given these $K*(K-1)/2$ model outputs, our goal is to compute p_i for $i=1, ..., K$, where p_i is an estimate of the probability that the test case is a member of Class i. It should be obvious that only in special circumstances can we compute these probabilities such that the pairwise probabilities r_{ij} are satisfied exactly. To do so, we would need to find K values that satisfy $K*(K-1)/2$ conditions. Instead, we define u_{ij} as shown in Equation (7.19) and compute p_i for $i=1, ..., K$ such that the set of u_{ij} values matches the models' r_{ij} values as closely as possible.

$$u_{ij} = \frac{p_i}{p_i + p_j} \tag{7.19}$$

One good way to measure the closeness of two probability distributions is their *relative entropy*, which is also called their *Kullback-Liebler distance*. In the general case, this quantity is defined in Equation (7.20).

$$D_{KL}(p,q) = \sum_x p(x)\log\frac{p(x)}{q(x)} \tag{7.20}$$

Even though this is called a distance, it is not a true metric because it is not symmetric and it does not satisfy the triangle inequality. Still, it is never negative and it is zero if and only if p and q are identical. Also, it has some nice theoretical properties that are beyond the scope of this text.

To define the closeness of the models' r_{ij} values to the u_{ij} values implied by $p_1, ..., p_K$, we use the average of the Kullback-Liebler distances for all $K*(K-1)/2$ values. In fact, we can do even better by using a weighted average in which the weights are some measure of the quality of each model. [Hastie and Tibshirani, 1996] suggests that a reasonable weight for each model is the number of training cases that were used for that model. If we designate this quantity as n_{ij}, the distance criterion is expressed in Equation (7.21).

$$D_p = \sum_{i<j} n_{ij} \left[r_{ij} \log\frac{r_{ij}}{u_{ij}} + \left(1-r_{ij}\right) \log\frac{\left(1-r_{ij}\right)}{\left(1-u_{ij}\right)} \right] \tag{7.21}$$

There is no need to directly minimize this fierce quantity using a general-purpose function minimizer, because a simple iterative algorithm exists for doing so indirectly. The first step is to compute starting estimates for $p_1, ..., p_K$ using Equation (7.22).

$$p_i = \frac{2}{K(K-1)} \sum_{j \neq i} r_{ij} \qquad (7.22)$$

If you do not want to differentially weight the individual models according to their reliability and if you only need to find the class having greatest probability, there is no need to go further. The order of the class probabilities will not be changed by iterative improvement. To improve the class probability estimates, use the following algorithm, which is guaranteed to converge to the minimum of Equation (7.21):

1) Compute u_{ij} for $i, j=1, ..., K$ by evaluating Equation (7.19) with the starting estimates of probabilities $p_1, ..., p_K$.

2) Set $i=1$.

3) Update p_i using Equation (7.23).

$$P_i = P_i \cdot \frac{\sum\limits_{j \neq i} n_{ij} r_{ij}}{\sum\limits_{j \neq i} n_{ij} u_{ij}} \qquad (7.23)$$

4) Renormalize $p_1, ..., p_K$ by dividing each by their sum. Then recompute u_{ij} for $i, j=1, ..., K$ as in step 1.

5) If $i<K$ set $i=i+1$ and go to step 3. Else...

6) Find the maximum change of any of the probabilities p_i in the previous loop. If the maximum change is tiny, convergence has been obtained. Otherwise, go to step 2 to perform another iteration.

In practice this algorithm converges quickly. The complete source code for pairwise coupling can be found in the file MULTCLAS.CPP on my web site. Salient points will now be discussed.

Although there are K^2 entries in the r_{ij} and n_{ij} matrices, $K*(K-1)/2$ values suffice for each, because $r_{ji}=1-r_{ij}$ and $n_{ji}=n_{ij}$. Also, the diagonal elements are undefined because i and j must always be different. To minimize storage space and redundant computation, only the upper-right diagonal of each matrix is stored, starting with the first row. This will become clear as the sample code is shown.

Before presenting a class for performing pairwise coupling, it is good to look at how the component models would be trained. The training set for each model should include only cases belonging to one of the two classes in which the model specializes. The file MULTCLAS.CPP includes a test program for comparing all of the algorithms in this chapter. A code fragment from this program that illustrates training the pairwise models is now shown. Note that this training set uses two predictor variables, and this fact is hard-coded into the demonstration test program.

```
imodel = 0;                                      // Identifies model in pair
for (i=0; i<nclasses-1; i++) {                   // First class in a pair
  for (j=i+1; j<nclasses; j++) {                 // Second class in pair

    ntrain_pair[imodel] = 0;                     // Will count training cases for this model
    for (k=0; k<nsamps; k++) {                   // For all training cases
      if ((x[(2+nclasses) * k + 2 + i] > 0.5)    // Case is first class?
      || (x[(2+nclasses) * k + 2 + j] > 0.5))    // Or second class?
        ++ntrain_pair[imodel];                   // Count nij
    }
    model_pairs[imodel] = new MLFN (ntrain_pair[imodel], 2, 1, nhid);
    model_pairs[imodel]->reset ();               // Initialize for cumulating training set

    for (k=0; k<nsamps; k++) {                    // For all training cases
      input[0] = x[(2+nclasses) * k];             // First predictor
      input[1] = x[(2+nclasses) * k + 1];         // Second predictor
      if (x[(2+nclasses) * k + 2 + i] > 0.5)      // Case is first class?
        input[2] = 1.0;                           // Train for high probability
      else if (x[(2+nclasses) * k + 2 + j] > 0.5) // Case is second class?
        input[2] = 0.0;                           // Train for low probability
      else                                        // This case is neither first or second class
        continue;                                 // So it doesn't belong in training set
      model_pairs[imodel]->add_case (input);      // Put case in training set
    }                                             // For all training cases
    model_pairs[imodel]->train ();                // Train model for (i,j) class pair
    ++imodel;                                     // Next model in collection of pairs
  } // For j, which is second class in pair
} // For i, which is first class in pair
```

Recall that we store and process only the upper-right diagonal of the matrices of class-pair information. The variable imodel is the index into the corresponding arrays. It starts at zero for $i=0$ and $j=1$, and it will be incremented for successive elements. The column index j changes fastest, starting at one past the current row index i each time.

In this particular implementation we need to find each n_{ij} before doing anything else because this information is required by the MLFN model constructor. Other implementations may be able to defer this counting until after the training set is built. We then construct a new MLFN model dedicated to this i,j class pair and call its reset() member function to prepare for building the training set. The MLFN code is in the file MLFN.CPP on my web site.

All nsamps cases in the grand training set are traversed. Each case consists of two predictor variables (which is imposed in this example but of course is not required in general) and nclasses output variables. As is standard in this chapter, exactly one of these output variables, the one corresponding to the case's class, is 1.0, and the others are 0.0. For each case in the grand training set, the two predictors are copied to the scratch vector inputs, and the class membership of the case is determined. If the case is in class i, we signify this by asking the MLFN model to learn the value 1.0, corresponding to high probability. If the case is in class j, we ask the MLFN model to learn the value 0.0, which signifies low probability. If the case is in neither class, it does not belong in the training set for this model, so it is skipped. Otherwise, the case is appended to the training set for this model. When the entire grand training set has been traversed to build the training set for the model specializing in distinguishing Class i from Class j, the model is trained.

The declaration and constructor for the class that performs pairwise coupling are as follows:

```
class Pairwise {

public:

   Pairwise (int nclasses, int *ntrain);
   ~Pairwise ();
   int classify (double *input, double *output);

private:

   int nout;      // Number of outputs (nclasses in constructor call)
   int npairs;    // Number of models (nclasses * (nclasses-1) / 2)
   int *nij;      // Number of training cases used for each model (ntrain); i<j
```

```
  double *rij;    // Models' predicted Prob (i given (i or j)); i<j
  double *uij;    // pi / (pi + pj); i<j
                  // For rij and uij, just subtract from 1 if i>j
};

Pairwise::Pairwise (
  int nclasses,  // Number of outputs (classes)
  int *ntrain    // Number of training cases used for each model
  )
{
  nout = nclasses;
  npairs = nclasses * (nclasses-1) / 2;
  nij = ntrain;

  rij = (double *) malloc (npairs * sizeof(double));
  uij = (double *) malloc (npairs * sizeof(double));
}
```

The classification member function, which does the real work, is shown next in sections. The first part of the classifier evaluates all models for the test case, saving the models' outputs in rij, which is r_{ij} in the text. The outputs are constrained to be greater than 0.0 and less than 1.0 in order to prevent numerical difficulties later. Then Equation (7.22) is used to compute starting estimates for the class probabilities.

```
int Pairwise::classify (double *input, double *output)
{
  int i, j, k, iclass, iter;
  double rr, best, numer, denom, sum, delta, oldval;

/*
  Evaluate all models for this case
*/

  for (i=0; i<npairs; i++) {
    model_pairs[i]->predict (input, &rr);
    if (rr > 0.999999)          // Prevent numerical difficulties later
      rr = 0.999999;
    if (rr < 0.000001)
      rr = 0.000001;
```

```
    rij[i] = rr;
    }
```

/*

Compute starting estimates as row means using Equation (7.22)

*/

```
  for (i=0; i<nout; i++)           // Class probabilities will go here
    output[i] = 0.0;

  k = 0;
  for (i=0; i<nout-1; i++) {        // First class in pair
    for (j=i+1; j<nout; j++) {      // Second class in pair
      rr = rij[k++];                // This model's Prob[i given i or j]; i<j
      output[i] += rr;              // Term for i<j
      output[j] += 1.0 - rr;        // And symmetric term
      }
    }

  for (i=0; i<nout; i++)           // Complete Equation (7.22)
    output[i] /= npairs;
```

After the starting estimates are in hand, the class probabilities (in output) are refined. Comments in the following code refer to the algorithm on page 377.

/*

We begin by computing uij, which approximates rij. This is Step 1.

*/

```
  k = 0;
  for (i=0; i<nout-1; i++) {        // First class in pair
    for (j=i+1; j<nout; j++)        // Second class in pair
      uij[k++] = output[i] / (output[i] + output[j]); // Equation (7.19)
    }
  for (iter=0; iter<10000; iter++) { // Cheap insurance against endless loop

    delta = 0.0; // Keeps track of max change in any output for convergence test
    for (i=0; i<nout; i++) {         // Correct one output at a time; Steps 2 and 3

      numer = denom = 0.0;    // Will cumulate for correction factor
```

```
    for (j=0; j<nout; j++) {        // Cumulate numer and denom of Equation (7.23)
      if (i < j) {
        k = (i * (2 * nout - i - 3) - 2) / 2 + j;  // Directly index above-diagonal element
        numer += nij[k] * rij[k];
        denom += nij[k] * uij[k];
        }
      else if (i > j) {                        // Storage assumes i<j, so reverse access
        k = (j * (2 * nout - j - 3) - 2) / 2 + i;     // Yeah, I know. Storing the whole matrix
        numer += nij[k] * (1.0 - rij[k]);             // would have been a lot easier.
        denom += nij[k] * (1.0 - uij[k]);             // Feel free. I like elegance.
        }
      } // For j, cumulating numer and denom for correction factor

    // Correct this one output (trial p) and renormalize to sum to one
    oldval = output[i];              // For convergence test
    output[i] *= numer / denom;      // Equation (7.23)
    sum = 0.0;
    for (j=0; j<nout; j++)
      sum += output[j];
    for (j=0; j<nout; j++)                   // Step 4, first part
      output[j] /= sum;

    if (fabs(output[i]-oldval) > delta) // How much did this output change?
      delta = fabs(output[i]-oldval);   // Keep track of max for Step 6

    // Recompute uij from the modified output; Step 4, second part.
    k = 0;
    for (i=0; i<nout-1; i++) {  // First class in pair
      for (j=i+1; j<nout; j++)  // Second class in pair
        uij[k++] = output[i] / (output[i] + output[j]);
      }
    } // For i

  if (delta < 1.e-6)      // If max change of any output is small
    break;                // No need to keep iterating; we have converged; Step 6

  } // For all iterations
```

The final step is trivial. Find the class having greatest probability and return its index.

```
/*
  We're done. Output the class having maximum probability.
*/

  for (i=0; i<nout; i++) {
    if ((i == 0) || (output[i] > best)) {
      best = output[i];
      iclass = i;
    }
  }
  return iclass;
}
```

Pairwise Threshold Optimization

We know from our description of ROC curves on page 48 that binary classifiers can often be improved by the simple expedient of optimizing the decision threshold. Sometimes moving the threshold increases the net classification accuracy. More often, an adjusted threshold shifts the costs of misclassification in such a way that the total cost of both types of error is reduced. When probabilities rather than arbitrary outputs are employed, as is the case for pairwise coupling, a method for adjusting a binary threshold is not obvious. Nonetheless, it can be done in an elegant manner by making use of the *logistic* or *logit* transform already discussed on page 341. In particular, define d_{ij} as shown in Equation (7.24).

$$d_{ij} = \ln \left[\frac{r_{ij}}{1-r_{ij}} \right] \tag{7.24}$$

Note that the domain of this function is [0, 1] and that $d_{ij}=0$ when $r_{ij}=0.5$. Small probabilities (r_{ij} near zero) correspond to large negative values of d_{ij}, while large probabilities correspond to large positive values for d_{ij}. Basing the binary classification decision on whether $r_{ij}>0.5$ is equivalent to basing it on whether $d_{ij}>0$.

It is the threshold of zero that we adjust in order to reduce the total error cost for the binary decision. Suppose we find that the binary classifier is improved by testing the decision rule $d_{ij} > t_{ij}$ where t_{ij} is some value other than zero. This is equivalent to testing $d_{ij} - t_{ij} > 0$. This shifted threshold can be embodied in a modified probability by inverting Equation (7.24) with the shifted output. This is shown in Equation (7.25).

$$r'_{ij} = \frac{e^{d_{ij}-t_{ij}}}{1 + e^{dij-t_{ij}}} \tag{7.25}$$

Thus, to perform pairwise coupling with binary models having implicitly shifted classification thresholds, the model's output must be modified. Apply Equation (7.24) to the original outputs (binary probabilities), offset the result by the optimal constant, and apply Equation (7.25) to produce the modified binary probabilities that will be used in the pairwise coupling algorithm.

A Cautionary Note

There is one situation in which pairwise coupling may suffer from a subtle difficulty. Suppose one of the classes is in some sense a catchall class, meaning that it has properties (as defined by the classification features) that are, on average, closer than the other classes to randomly chosen test cases. This class may be well distinguished from its competitors when a test case is clear. But especially if the test case is unusually noisy or not well defined in some other way, the case will on average be found to be closer to the catchall class than to the other classes. When such a case that is not a member of the catchall class is presented to a pairwise classifier, the tests in which the true class is represented may do fairly well but not excellently because of the noise. At the same time, the tests in which the catchall class is represented will probably assign relatively high probability to that class. This is because all of these tests except one, the one also involving the case's true class, were trained without knowledge of the test case's true class. These pairwise tests, instead of randomly favoring different classes, will all focus on the catchall class. The result is that pairwise combination will strongly favor catchall classes for noisy cases. The moral of the story is that when one class tends to be broadly representative, pairwise combination may not be advisable because it may exaggerate the problem.

Comparing the Combination Methods

The MULTCLAS program on my web site allows you to compare the performance of this chapter's model-combination algorithms under a variety of circumstances. You can specify the number of training cases, number of classes, and number of models to be used. The program will then run a specified number of Monte Carlo replications (1,000 in these examples) to compare mean performance (out-of-sample probability of misclassification) for each of the algorithms. As an added feature of the program, if four or more models are employed, the fourth model is deliberately useless to test how well each algorithm responds to such a situation. Also, if five or more models are employed, the fifth model, while generally good, occasionally produces wildly large or small numeric predictions. Finally, you can specify a spread between the classes, which determines the difficulty of the problem for the component models.

In the following examples, the tabulated names of the algorithms should be clear, with two exceptions. The first row in each table is labeled *Raw*. This is the average misclassification rate of the component models. This figure provides a baseline against which improvement (if any) due to combination may be compared. The last two rows, labeled *Intersection* and *Union*, are the rates for these two methods when the resultant set contains only one class. Cases whose resultant class contains more than one class are considered to be an error. This is, of course, a harsh penalty. However, given that only three classes are used here, it is reasonable for this demonstration.

The component models are simple feedforward networks having one hidden layer of four neurons. If you would like details of the models and training methods, you should consult the MULTCLAS.CPP file on my web site. Note that the simple training algorithm for these networks (in MLFN.CPP) is compromised when the separation between the classes is huge, far more than would be encountered in real life. Users experimenting with the program should keep this in mind.

Small Training Set, Three Models

We begin with the extreme situation of having only ten training cases with which to discriminate between three classes. Three models of equal power are employed, and three degrees of difficulty are tested. Misclassification rates are as follows:

	Hard	Med	Easy
Raw	.5107	.1940	.1353
Average	.5135	.2064	.1567
Median	.4960	.1642	.1044
MaxMax	.5290	.2273	.1895
MaxMin	.4996	.1466	.0724
Majority	.5011	.1737	.1177
Borda	.4961	.1607	.0958
Logistic	.5079	.1833	.1208
Logistic (Sep)	.5106	.1824	.1263
Local Acc	.4957	.1297	.0811
Fuzzy Int	.4907	.1287	.0700
Pairwise	.5105	.2397	.2562
Intersection	.6593	.3037	.2308
Union	.5494	.1944	.1325

First, note that many of the combination methods actually fare worse than a single model. A look at the *Average* and *Median* rows reveals the likely reason. With only ten training cases, at least one of the three component models will often be terribly ineffective. Combination schemes such as averaging, which are vulnerable to large errors, suffer. However, schemes like the median, which are relatively impervious to outliers, perform better. As is often the case, the fuzzy integral method wins in all situations.

Large Training Set, Three Models

Now we examine the more usual situation of having 200 training cases with which to discriminate between three classes. Three models of equal power are employed, and three degrees of difficulty are tested. Misclassification rates are as follows:

	Hard	Med	Easy
Raw	.3347	.1116	.0090
Average	.3327	.1108	.0012
Median	.3329	.1108	.0016
MaxMax	.3329	.1109	.0046
MaxMin	.3332	.1110	.0006
Majority	.3330	.1109	.0017
Borda	.3331	.1109	.0015
Logistic	.3344	.1115	.0012
Logistic (Sep)	.3345	.1117	.0076
Local Acc	.3357	.1116	.0001
Fuzzy Int	.3330	.1109	.0029
Pairwise	.3437	.1138	.0122
Intersection	1.0000	.8662	.0122
Union	1.0000	1.0000	.0176

Now that we have a decent size training set, nearly all of the combination methods (except the universally poor Intersection and Union) outperform a single model.

Interestingly, there is little variation among the methods. This should not be surprising, since we are in an ideal situation: models of equal power and a reasonably large training set. When everything is straightforward, the choice of combination method is not difficult.

Small Training Set, Three Good Models, One Worthless

The combination methods show their differences clearly when one of the component models is worthless, especially under the extreme situation of having only ten training cases with which to discriminate between three classes. Three degrees of difficulty are tested. Misclassification rates are as follows:

	Hard	Med	Easy
Raw	.5499	.3179	.2724
Average	.5267	.2499	.1978
Median	.5027	.1846	.1297
MaxMax	.5698	.3585	.2954
MaxMin	.5617	.3487	.2556
Majority	.5071	.1932	.1348
Borda	.5020	.1730	.1093
Logistic	.5082	.1930	.1239
Logistic (Sep)	.5093	.1982	.1333
Local Acc	.4961	.1219	.0781
Fuzzy Int	.4883	.1298	.0688
Pairwise	.5063	.2416	.2600
Intersection	.6615	.3161	.2483
Union	.5410	.2003	.1439

The first thing to observe is that even with one of the four models making random predictions, all of the serious combination methods outperform the mean performance of the individual models, often by large margins. This is powerful incentive to combine models. Also, note that the Fuzzy Integral method is a real winner, with Local Accuracy coming in a close second.

Large Training Set, Three Good Models, One Worthless

Now we examine what happens when the previous situation is modified by having a reasonably large training set (200 cases). Three degrees of difficulty are tested. Misclassification rates are as follows:

	Hard	Med	Easy
Raw	.4165	.2496	.1750
Average	.3336	.1108	.0017
Median	.3332	.1105	.0017
MaxMax	.3395	.1135	.0078
MaxMin	.3682	.1385	.0106
Majority	.3341	.1109	.0056
Borda	.3347	.1112	.0052
Logistic	.3350	.1112	.0012
Logistic (Sep)	.3387	.1336	.0082
Local Acc	.3447	.1138	.0001
Fuzzy Int	.3330	.1104	.0026
Pairwise	.3434	.1137	.0172
Intersection	1.0000	.8562	.0200
Union	1.0000	1.0000	.0194

When we have a decent number of training cases, model combination shines. In the medium difficulty scenario, most algorithms beat the raw models decently. And in the easy case, combination wins by over two orders of magnitude. Astounding. Once again, the Fuzzy Integral and Local Accuracy methods shine.

Small Training Set, Worthless and Noisy Models Included

The worst situation is when we have only ten training cases, and the power of our three good models is diluted by the presence of a worthless (random) model as well as a model that occasionally gives wild numeric predictions. Three degrees of difficulty are tested. Misclassification rates are as follows:

	Hard	Med	Easy
Raw	.5607	.3508	.3087
Average	.5917	.4697	.4375
Median	.5021	.1734	.1166
MaxMax	.5843	.4365	.3833
MaxMin	.5835	.4312	.3975
Majority	.5091	.1958	.1385
Borda	.5086	.1908	.1317
Logistic	.5105	.1928	.1192
Logistic (Sep)	.5167	.2045	.1454
Local Acc	.4958	.1148	.0763
Fuzzy Int	.4997	.1547	.0834
Pairwise	.5096	.2395	.2624
Intersection	.6677	.3278	.2564
Union	.5447	.2046	.1448

The inclusion of an often crazy model clearly separates the combination methods. As usual, the Fuzzy Integral performs universally well. However, the Local Accuracy method, with its inherent ability to differentiate between the models, is the winner by a small margin. Methods that make full use of all predictions perform badly, sometimes even worse than the mean performance of the component models.

Large Training Set, Worthless and Noisy Models Included

We now modify the previous example by employing 200 training cases, a reasonable number. Three degrees of difficulty are tested. Misclassification rates are as follows:

	Hard	Med	Easy
Raw	.4476	.2152	.2142
Average	.5579	.3956	.4026
Median	.3343	.0008	.0001
MaxMax	.3837	.0609	.0838
MaxMin	.5562	.3972	.4001
Majority	.3367	.0011	.0002
Borda	.3514	.0277	.0253
Logistic	.3358	.0011	.0001
Logistic (Sep)	.3435	.0022	.0006
Local Acc	.3508	.0005	.0000
Fuzzy Int	.3352	.0012	.0006
Pairwise	.3447	.0010	.0000
Intersection	1.0000	.0116	.0006
Union	1.0000	.0099	.0002

Surprisingly, the relatively primitive Median method was an outstanding performer. Still, the universally effective Fuzzy Integral did well, along with the Logistic method. The Local Accuracy method easily prevailed in the easy category. And of course, using nearly any combination algorithm improved performance relative to the raw models, often by phenomenal amounts.

Five Classes

The final example uses the prior situation (three good models, one worthless model, one wild model, 200 training cases) to separate five classes, instead of the three that have been used until now. Three degrees of difficulty are tested. Misclassification rates are as follows:

	Hard	Med	Easy
Raw	.3857	.3797	.3787
Average	.6342	.6372	.6390
Median	.1127	.0871	.0799
MaxMax	.1403	.1218	.1203
MaxMin	.6424	.6448	.6458
Majority	.1591	.1451	.1397
Borda	.2223	.2096	.2050
Logistic	.1063	.0827	.0769
Logistic (Sep)	.1877	.1870	.1871
Local Acc	.0690	.0439	.0374
Fuzzy Int	.1164	.0966	.0943
Pairwise	.0615	.0235	.0106
Intersection	.7009	.6435	.6143
Union	.9467	.7664	.6753

The first thing to jump out of this chart is that the columns show similar performance for most methods, despite the fact that the separation is quite varied. This seeming anomaly is because the underlying models are trained with a primitive algorithm that does not do well when there are this many classes and the separation is large. But the combination methods still greatly improve performance, and this model situation is where pairwise coupling shines, because the models can easily separate pairs of classes. Take note!

CHAPTER 8

Gating Methods

- Preordained Specialization

- After-the-Fact Specialization

- General Regression Gating

A *gating method*, as defined in this text, refers to an overseer that uses one or more *gate variables* to weight the opinions of two or more models in order to obtain an opinion superior to what could be obtained from a single model. These models may be numeric predictors or classifiers. What distinguishes the techniques of this chapter from the model combination methods described in prior chapters is that a gated combiner requires the use of a dedicated variable or variables to control the combination process, while the previously discussed techniques operate without the benefit of such outside help.

When may gating be beneficial? The prevailing trend of interest rates may impact the relative power of competing market-prediction models. The degree of histologic staining may affect the power of competing photomicrographic models. The possibilities are endless.

Preordained Specialization

The most primitive form of gating is when one variable is used to choose between two or more predefined specialist models. To see how this technique may arise, look at Figure 8-1.

© Timothy Masters 2018
T. Masters, *Assessing and Improving Prediction and Classification,*
https://doi.org/10.1007/978-1-4842-3336-8_8

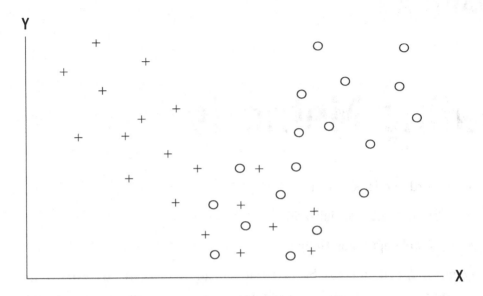

Figure 8-1. *A distribution requiring a specialist*

This figure depicts cases belonging to two classes. The horizontal axis represents a predictor variable X, and the vertical axis represents another predictor, Y. It is apparent that Y alone is worthless for distinguishing the classes. Also, X alone does only a fair job of discrimination. Some cases will be well separated, while others will be confused.

Of interest is the fact that Y does an excellent job of separating those cases for which X is a good classifier from those for which it is not. Any case having a large value for Y will be correctly classified based on X. In fact, if we were to set a threshold at about the midpoint of the range of Y, we could obtain perfect classification performance for those cases above the threshold. This is powerful incentive to split the dataset using Y, and devise two separate classification models, one for large values of Y and one for small values.

Of course, in this primitive example, little is gained by splitting the dataset if the remaining cases are hopeless. But it is often true that once the obviously easy cases are removed, the difficult cases can be handled by means of different variables and models. Also, it should be emphasized that we have illustrated the technique by employing a specific predictor variable X. Often, we will use exactly the same set of predictors to train two different models, with the model being selected based on the value of one or more variables that may or may not be members of the predictor set.

Learned Specialization

Careful examination of scatterplots will sometimes reveal not only a variable that is effective at splitting the dataset but a good splitting threshold as well. Unfortunately, such luck is the exception, not the rule. In most practical applications, at a minimum we will need to try various splitting points, retraining and testing each. Sometimes we even will need to try an assortment of splitting variables. This can be extremely time-consuming if the training algorithm is slow. However, the payoff may be substantial.

After-the-Fact Specialization

An interesting variation on the concept of specialization is when the choice of which model to use is made by examining the predictions of all of the contending models. Instead of choosing the model based on the value of one or more measured variables, we invoke all of the models and make our selection based on their outputs.

One primitive version of *after-the-fact specialization* is obtained when we are using two models to separate two classes. Each model chooses between the two classes. If the two models are in agreement, there is no problem. But what if the two models conflict on a test case? The sensible way to resolve the conflict is to pass through the training set and determine the most effective resolution rule. For example, we may find that when the first model chooses class *A* and the second model chooses class *B*, then the majority of the time the first model is correct. Simple tabulation enables us to define the best method of resolving conflicts.

Code for After-the-Fact Specialization

When model outputs are used as gate variables, complex situations can quickly arise. Multiple models, each having multiple outputs, separated by multiple thresholds, can lead to programming and logistical nightmares. To illustrate the principle of *after-the-fact specialization*, we will examine the most primitive numeric scenario: univariate numeric prediction with thresholds defined so as to divide the training set into equally sized "specialist" subsets for each model. It should be understood that more sophisticated methods, such as *general regression gating* described in the next section, are not only more broadly applicable but are also more powerful in most practical applications. Still, when a quick-and-dirty algorithm is required, the method shown here may suffice. Moreover, examination of this simple algorithm is instructive in that it provides a useful foundation for more advanced concepts. Note that because this technique is self-contained, it properly belongs in Chapter 6, *Combining Numeric Predictions*. But it is not as effective as most of the algorithms in that chapter, and it illustrates the ideas in this chapter. So, it is presented here.

The complete source code for this algorithm can be found in the file AFTERFAC.CPP on my web site. We begin with the class header:

```
class AfterFact {

public:

   AfterFact (int n, int ninputs, double *tset, int nmods, int ncat);
   ~AfterFact ();
   void predict (double *input, double *output);

private:
   int ncases;          // Number of cases
   int nin;             // Number of original case inputs
   int nmodels;         // Number of models (nmods in constructor call)
   int ncats;           // Number of 'specialist' categories
   double *outwork;     // Work vector nmodels long holds outputs for a case
   double *thresh;      // Nmodels by nfrac=ncats-1 array of fractile thresholds
   int *winners;        // Index of winning model for each bin kept here
};
```

This class keeps private copies of several allocated arrays. The outwork vector will be used every time the predict() member function is called. It is allocated in the constructor to avoid repeated allocation/deallocation cycles. The thresh vector contains the output thresholds that divide the training set into equal subsets. Finally, the winners vector identifies the best model for each subset.

We now look at the constructor. It allocates all required memory, computes the thresholds, and determines which model wins for each output category. The first section of the constructor is as follows:

```
AfterFact::AfterFact (
   int n,          // Number of training cases
   int ninputs,    // Number of inputs
   double *tset,   // Training cases, n by (ninputs+1)
```

```
    int nmods,    // Number of models in 'models' array
    int ncat      // Number of equal categories
    )
{

    int i, k, icase, imodel, ibin, nbins, index, klow, khigh, nthresh, ibest;
    int *bins;
    double *case_ptr, diff, x, best, *outptr, frac, *work;
    double *outputs; // Ncases by nmodels vector saves all model outputs

    ncases = n;
    nin = ninputs;
    nmodels = nmods;
    ncats = ncat;          // Number of categories
    nthresh = ncats - 1;   // Number of thresholds

/*
    The number of bins is ncats ** nmodels.
*/

    nbins = 1;
    for (i=0; i<nmodels; i++)
      nbins *= ncats;

/*
    Allocate memory that is needed throughout the life of the AfterFact object
*/

    outwork = (double *) malloc (nmodels * sizeof(double));
    thresh = (double *) malloc (nmodels * nthresh * sizeof(double));
    winners = (int *) malloc (nbins * sizeof(int));

/*
    Allocate scratch memory that is only needed during construction:
      outputs saves all model outputs
      work is scratch for fractile computation
      bins is a set of nbins vectors, each nmodels long
*/
```

```
outputs = (double *) malloc (ncases * nmodels * sizeof(double));
work = (double *) malloc (ncases * sizeof(double));
bins = (int *) malloc (nbins * nmodels * sizeof(int));

memset (bins, 0, nbins * nmodels * sizeof(int));
```

The first step in the previous code is to preserve private copies of required parameters. Naturally, the number of thresholds is one less than the number of categories that are defined by these thresholds. We also compute the number of bins as the number of categories raised to the power of the number of models.

Six vectors need to be allocated. The first three, already described, remain throughout the life of the class object. The other three are temporary. The first, outputs, preserves the output of every model for every training case. The second, work, will be used to sort the outputs of each model in order to find the thresholds that split the training set equally. The third, bins, will be used to count the number of times each model is the best within each bin. Immediately after allocation, we set the bin counts to zero.

The next section of the constructor passes through the entire training set. For each case, it executes all models and saves their outputs in outputs. Then, each model's output array (all training cases) is sorted by copying the data to work and sorting this vector with qsortd(). The nthresh thresholds are determined by finding the equally spaced entries in the sorted array. This code is as follows:

```
for (icase=0; icase<ncases; icase++) {
    case_ptr = tset + icase * (nin + 1); // Point to this case
    outptr = outputs + icase * nmodels;
    for (imodel=0; imodel<nmodels; imodel++)
        models[imodel]->predict (case_ptr, &outptr[imodel]);
}

for (imodel=0; imodel<nmodels; imodel++) {
    for (icase=0; icase<ncases; icase++)
        work[icase] = outputs[icase*nmodels+imodel];
    qsortd (0, ncases-1, work);
    for (i=0; i<nthresh; i++) { // Recall nthresh=ncats-1
        frac = (double) (i+1) / (double) ncats;
        thresh[imodel*nthresh+i] = work[(int)(frac*(ncases-1))];
    }
}
```

The next step looks complex but is mostly straightforward. For each case, we determine the bin corresponding to its model outputs. Then we determine which model's prediction is closest to the true value. Finally, we increment the counter for the model/bin combination. A lot of code is required to do this, but looks are deceptive. Most of the code is for the trivial binary search that locates an output relative to the threshold vector. This code is as follows:

```
for (icase=0; icase<ncases; icase++) {           // For each training case

  case_ptr = tset + icase * (nin + 1);           // Point to this case
  outptr = outputs + icase * nmodels;            // Corresponding model outputs

  ibin = 0;                                      // Cumulates bin
  index = 1;                                     // Indexes current layer

  for (imodel=0; imodel<nmodels; imodel++) {     // For each model
    x = outptr[imodel];                          // Its output for this case

    if (x <= thresh[imodel*nthresh])             // Bottom bin
      k = 0;

    else if (x > thresh[imodel*nthresh+nthresh-1])   // Topmost bin
      k = nthresh;

    else { // Keep thresh[klow] < x <= thresh[khigh]  // Somewhere in interior
      klow = 0;                                  // So find it with binary search
      khigh = nthresh - 1;

      for (;;) {
        k = (klow + khigh) / 2;
        if (k == klow) {
          k = khigh;
          break;
        }

        if (x <= thresh[imodel*nthresh+k])
          khigh = k;
        else
          klow = k;
      }
    }
```

```
      ibin += k * index;          // Locate within this model's layer
      index *= ncats;             // Advance to next layer

    } // For all outputs (models)
```

```
// We have the bin. Now find the best model for this case.
```

```
  for (imodel=0; imodel<nmodels; imodel++) {      // Determine best model
    diff = fabs (outptr[imodel] - case_ptr[nin]);    // Predicted minus true
    if ((imodel == 0) || (diff < best)) {
      best = diff;
      k = imodel;
      }
    }
```

```
  ++bins[ibin*nmodels+k];       // Count this model's win within this bin
  } // For all cases
```

The final step is to pass through each bin and locate the model that won most often in that bin. The winners are saved in winners. We can then free the three scratch vectors that are no longer needed.

```
for (ibin=0; ibin<nbins; ibin++) {  // For each bin
  k = 0;                              // Will point to winning model
  ibest = 0;                          // Best count
  for (imodel=0; imodel<nmodels; imodel++) {      // Check all models
    if (bins[ibin*nmodels+imodel] > ibest) {
      ibest = bins[ibin*nmodels+imodel];
      k = imodel;
      }
    }
  winners[ibin] = k;
  }
```

```
free (outputs);
free (work);
free (bins);
```

The structure of the bins array may be unclear, so here is an example of how it holds its information. Each of its elements is actually an array nmodels long. This array holds the count of how many times each model was the winner in this bin. Indexing of the bin elements is handled by two variables. As a case is processed, the initializations ibin=0 and index=1 are done. Now suppose that we have three models and four categories. Draw four dots in a horizontal row on a piece of paper. Each dot represents a category for the first model. Draw three more rows of four dots under the first row. You are looking at 16 dots arranged in a four-by-four square. Each row represents a category for the second model, so each of the 16 dots corresponds to a pair of categories for the first two models. Finally, imagine stacking four of these squares on top of one another. The result is a four-by-four-by-four cube of 64 dots, each of which represents the complete categorization of a case with respect to the three models.

Suppose the case is found to be in the second category (k=1) for the first model. At the end of the model-processing loop, ibin+=k*index will leave ibin at one, which is the appropriate column, and index will become four, which is the number of columns per row. If the case is then found to be in, say, the third category (k=2) for the second model, ibin will be incremented by 2*4=8. This pushes the bin down to the third row while leaving its column unchanged. At the same time, index will be changed to 16, which is the number of dots per layer. Lastly, suppose the case is in the third category (k=2) for the third model. This increments ibin by 2*16=32, which raises the bin to the third layer of dots while leaving its row and column unchanged. Thus, index keeps track of how many bin elements are in the prior dimensions so that multiplying by this quantity provides an increment that leaves prior dimensions unchanged.

The predict() member function makes a prediction. Again, it looks long and complicated. However, most of the code is dedicated to the binary search that locates the bin for the model outputs. This code, shown next, executes all models and finds the bin to which the outputs belong. Then, it checks the winner vector to see which model is most likely to be the best.

```
void AfterFact::predict (double *input, double *output)
{
  int k, klow, khigh, imodel, ibin, index, nthresh;
  double out;

  nthresh = ncats - 1;     // Number of thresholds in one less than categories
```

```
/*
    Invoke all models and find output of each.
    Then determine which bin the output set is in.
*/

    ibin = 0;
    index = 1;

    for (imodel=0; imodel<nmodels; imodel++) {          // For each model
        models[imodel]->predict (input, &out);          // Get its prediction
        outwork[imodel] = out;                          // And save for choosing best

        if (out <= thresh[imodel*nthresh])              // Bottom bin?
            k = 0;

        else if (out > thresh[imodel*nthresh+nthresh-1])   // Top bin?
            k = nthresh - 1;

        else { // Keep thresh[klow] < out <= thresh[khigh]  // Somewhere in interior
            klow = 0;                                   // So use binary search
            khigh = nthresh - 1;
            for (;;) {
                k = (klow + khigh) / 2;
                if (k == klow) {
                    k = khigh;
                    break;
                }
                if (out <= thresh[imodel*nthresh+k])
                    khigh = k;
                else
                    klow = k;
            }
        }
        ibin += k * index;
        index *= ncats;
    } // For all outputs (models)

    *output = outwork[winners[ibin]];
}
```

Some Experimental Results

The file AFTERFAC.CPP on my web site contains the complete source code for the class just described, as well as a program to test the method. This program allows the user to specify the number of cases in the training set, the number of bins, the number of models, and a noise level that determines the level of difficulty.

The following table shows the results obtained for a variety of circumstances in which three models of equal power are employed. The second-to-last column (*model mean*) is the average error variance for the three component models. The last column (*combined*) is the error using after-the-fact specialization.

Cases	Bins	Difficulty	Model Mean	Combined
10	2	easy	0.0216	0.0175
10	2	medium	85.2026	11.5342
10	2	hard	1501.7399	560.8960
500	2	easy	0.0012	0.0003
500	2	medium	0.2548	0.2559
500	2	hard	4.0829	4.0862
500	4	easy	0.0012	0.0003
500	4	medium	0.2548	0.2556
500	4	hard	4.0829	4.0872

These experiments were repeated with a fourth, useless (random prediction) model added. The results, shown in the following table, reveal a considerable change in behavior:

Cases	Bins	Difficulty	Model Mean	Combined
10	2	easy	905.9670	0.0112
10	2	medium	329.2057	6.8525
10	2	hard	2575.3047	4323.3319
500	2	easy	0.5118	0.0004
500	2	medium	0.7647	0.3981
500	2	hard	4.5755	5.5295
500	4	easy	0.5118	0.0017
500	4	medium	0.7647	0.4991
500	4	hard	4.5755	4.9930

Several aspects of after-the-fact specialization, as implemented here, are apparent:

- When the problem is easy (low noise), this technique is extremely useful. In one situation, it reduced the error variance by many orders of magnitude.

- When the problem is hard (high noise), not only does the technique almost never help, but it usually provides results that are inferior to using a single model. This is even true when all of the models are equally powerful, though the differences are admittedly small.

- At least in these tests, the difference in performance obtained by splitting into four bins instead of two is inconsistent and negligible. This is not surprising here, because either all of the models are similar in power or one of them is universally worthless. The main effect of this algorithm is to identify and reject the worthless model when it exists. There are undoubtedly some scenarios in which more than two categories would be useful.

- Finally, understand that it may not be entirely fair to call this algorithm gating. This is because the model outputs are used as the gate variables. Gating as it is most often understood employs an outside variable to act as the gate controller. Still, this algorithm illustrates the idea so well that it is an excellent example of the idea behind gating. It should be easy to modify the code to use an extraneous variable to choose the bin for each case.

General Regression Gating

An extremely powerful form of gating (sometimes called an *oracle*) is obtained by borrowing ideas from the *General Regression Neural Network* (GRNN). This method allows one or more variables to act as gates controlling variable admittance to the models being combined. In other words, both gating methods described so far require that one and only one model be selected for producing the output response. In contrast, general regression gating combines the outputs of *all* of the component models, weighting each in an optimal fashion. Moreover, the gate variables may be either extraneous measured values or model outputs. Thus, the technique described here subsumes the after-the-fact specialization presented in the previous section.

Before continuing in this section, please turn to page 296 and review the general regression neural network. For convenience, the fundamental GRNN equation is reproduced here as Equation (8.1). The distance function is given by Equation (8.2).

$$\hat{y}(\mathbf{x}) = \frac{\sum_{i=1}^{n} y_i \exp(-D(\mathbf{x}, \mathbf{x}_i))}{\sum_{i=1}^{n} \exp(-D(\mathbf{x}, \mathbf{x}_i))} \tag{8.1}$$

$$D_{\mathbf{x}}(\mathbf{x}, \mathbf{x}_i) = \sum_{j=1}^{p} \left(\frac{x_j - x_{ij}}{\sigma_j} \right)^2 \tag{8.2}$$

For maximum generality, this development is based on numeric prediction. It is assumed that we already possess two or more trained models. We also possess a dataset that will be used to train the GRNN combiner. Ideally, this training set should be different from that used to train the component models. However, in practice this may be impractical, and it is not strictly necessary.

Each training case includes the measured values of one or more gate variables, the outputs of the component models, and the known true value of the variable being predicted. Note that the values of the predictor variables for the component models are irrelevant; the models are black boxes, and their inputs play no role here. Note also that the outputs of one or more of the component models may themselves serve as gate variables. The training set contains n cases. Each case i (i=1, ..., n) consists of the following:

$x_{i,j}$ j=1, ..., p These *gate variables* ideally indicate the relative efficacy of the prediction models.

$q_{i,k}$ k=1, ..., m These are the outputs of the m contending prediction models.

y_i This is the desired output (the target value).

For the gate variables and model outputs, one subscript will be used when referring to a trial case:

x_j j=1, ..., p The values of the observed gate *variables*

q_k k=1, ..., m The computed outputs of the m contending prediction models

In the current context, the weighted Euclidean distance between a test case and a training case (Equation [8.2]) is determined by the gate variables. In other words, when a test case is evaluated, we want the GRNN gate to focus most of its attention on those training cases whose gate variables are most similar to the gate variables of the test case.

The error of model k in predicting the true value for case i is, by definition, $yi - qi, k$. If we use the GRNN defined in Equation (8.1) to predict the squared error of a model, the predicted squared error for model k is given by Equation (8.3).

$$\hat{e}_k(\mathbf{x}) = \frac{\sum_{i=1}^{n}(y_i - q_{i,k})^2 \exp(-D(\mathbf{x}, \mathbf{x}_i))}{\sum_{i=1}^{n}\exp(-D(\mathbf{x}, \mathbf{x}_i))} \tag{8.3}$$

Although there are an infinite number of ways the component models could be combined to produce a joint prediction, the simplest method is to let the final prediction be a linear combination of the outputs of the models.

$$\hat{y} = \sum_{k=1}^{m} w_k q_k \tag{8.4}$$

If the models have the (desirable) property that their predictions are unbiased, this property is preserved if and only if we impose the condition that the weights sum to unity, as shown in Equation (8.5). Even if the predictions are not unbiased, this condition is still desirable in nearly all situations.

$$\sum_{k=1}^{m} w_k = 1 \tag{8.5}$$

The linear combination of unbiased estimators having minimum mean squared error uses weights proportional to the reciprocal of each estimator's variance. If we use the predicted squared error in place of the variance, we arrive at the following formula for the weights:

$$w_k = \frac{1/\hat{e}_k}{\sum_{l=1}^{m} 1/\hat{e}_k} \tag{8.6}$$

If we know appropriate values for the sigma weights of Equation (8.2), we can make a GRNN-gated prediction by following these steps:

1) Use Equation (8.3) to estimate the prediction error that each model will suffer for this test case.

2) Use Equation (8.6) to compute the weights.

3) Evaluate all component models for this test case.

4) Use Equation (8.4) to combine these predictions into the final estimate.

The problem is that we do not automatically know good values for the sigma weights. We must use the training data to estimate effective sigma weights.

The best way to evaluate the quality of a trial sigma vector is by means of cross validation: remove a case from the training set to serve as a test case, apply the four steps just listed to make a prediction for this case, and compare the predicted value to the true value. Then return the case to the training set and extract another. Repeat this process for each case. Find the mean squared error across all such repetitions. This error indicates the quality of the trial sigma vector.

Any good derivative-free optimization algorithm may be used to find the set of sigma weights that provides minimum cross validation error. One technique that has been shown to be effective is differential evolution, described in [Price and Storn, 1997]. A faster method that provides good results in all but the most pathological situations is Powell's method. We will use this method here, even though differential evolution is universally applicable and superior in some rare situations involving multiple local extrema.

Code for GRNN Gating

The file GRNNGATE.CPP on my web site contains the complete source code for a GRNN gating class, along with a program to test the technique using synthetic data. We now examine the important parts of this code. The class declaration is as shown here:

```
class GRNNgate {

public:

   GRNNgate (int n, int n_gates, int nmods, double *gates,
               double *contenders, double *trueval);
   ~GRNNgate ();

   double trial (double *gates, double *contenders, int i_exclude, int n_exclude);
   void predict (double *gates, double *contenders, double *output);

private:
   int ncases;          // Number of cases
   int ngates;          // Number of gate variables
   int nmodels;         // Number of models (nmods in constructor call)
   double *tset;        // A copy of the training data is perpetually needed
   double *sigma;       // Learned sigma weights
   double *errvals;     // Used in trial

   friend double criter (double *params);
};
```

The outputs of the trained models will be referred to as *contenders*. Each of these models has already been trained to predict the dependent variable. The gate variables (often just one) will be used to differentially weight the contenders in making the final prediction.

The constructor is responsible for saving the training data (which will be needed for subsequent predictions) as well as finding optimal sigma weights. It is as shown here, and an explanation of its operation follows:

```
GRNNgate::GRNNgate (
    int n,                  // Number of training cases
    int n_gates,            // Number of gate variables
    int nmods,              // Number of models in external 'models' array
    double *gates,          // Training cases: gate variables, n by n_gates
    double *contenders,     // Training cases: contender variables, n by nmods
    double *trueval         // Training cases: n true values
    )
{

    int i;
    double *case_ptr, *params, *base, *p0, *direc;
    double x1, x2, x3, y1, y2, y3, err;

    ncases = n;
    ngates = n_gates;
    nmodels = nmods;

/*
    Allocate memory that is needed throughout the life of the GRNNgate object.
    Then copy the training data to a local area.
*/

    tset = (double *) malloc (ncases * (ngates+nmodels+1) * sizeof(double));
    sigma = (double *) malloc (ngates * sizeof(double));
    errvals = (double *) malloc (nmodels * sizeof(double));

    case_ptr = tset;
    for (i=0; i<ncases; i++) {
        memcpy (case_ptr, gates + i * ngates, ngates * sizeof(double));
        case_ptr += ngates;
```

```
      memcpy (case_ptr, contenders + i * nmodels, nmodels * sizeof(double));
      case_ptr += nmodels;
      *case_ptr++ = trueval[i];
      }

/*
   Allocate scratch memory that is only needed during construction:
*/

   params = (double *) malloc (ngates * sizeof(double));

/*
   Keep a local copy of this class pointer (needed by the criterion function).
   Call the optimizer, then call criter() one last time to set sigmas.
*/

   local_model = this;

   if (ngates == 1) {
     glob_min (-3.0, 3.0, 15, 0, 0.0, univar_crit,
             &x1, &y1, &x2, &y2, &x3, &y3);
     brentmin (10, 0.0, 1.e-5, 1.e-5,
             univar_crit, &x1, &x2, &x3, y2);
     params[0] = x2;
     }
   else {
     base = (double *) malloc ((2 * ngates + ngates * ngates) * sizeof(double));
     p0 = base + ngates;
     direc = p0 + ngates;

     for (i=0; i<ngates; i++)
       params[i] = 0.0;
     err = criter (params);
     if (err > 0.0) // Perfect model makes powell unnecessary
       err = powell (10, 0.0, 1.e-4, criter, ngates, params,
               err, base, p0, direc);
     free (base);
     }
```

```
  criter (params); // One last call to set sigma vector
  free (params);
}
```

Three blocks of memory will be required throughout the life of the object. All training data (gates, contenders, and true values of the predicted variable) must be preserved because the process of making subsequent predictions requires that general regression be used for intermediate prediction of the error of each model. The trained sigma vector will be referenced in that process. Finally, the errval vector will be needed as a scratch area each time a prediction is made. By reserving this vector during the life of the object, we avoid the need to repeatedly allocate and deallocate the memory.

One unusual aspect of the constructor is that it keeps a local copy of its *this* pointer in local_model. This lets the ordinary classless function criter() be passed as a parameter to a generic optimizer like powell(). There are several other ways of approaching this problem, but the method used here is simple and effective.

In the common situation of having a single gate variable, univariate optimization is employed. We first call glob_min() to find a trio of points encompassing the minimum and then refine with brentmin(). In the multiple-gate situation, we use powell(). Regardless, criter() must be called one last time to set the sigma vector according to the optimal parameters. The code for all three of these routines can be found in the file MINIMIZE.CPP on my web site.

The criterion function performs cross validation to test the quality of a trial sigma vector. Its code is as follows:

```
static double criter (double *params)
{
  int i, ngates, nmodels;
  double *inputs, out, diff, error, penalty;

  ngates = local_model->ngates;
  nmodels = local_model->nmodels;

/*
  Get the sigmas from the parameter vector
*/
```

```
penalty = 0.0;
for (i=0; i<ngates; i++) {
   if (params[i] > 8.0) {
      local_model->sigma[i] = exp (8.0);
      penalty += 10.0 * (params[i] - 8.0);
      }
   else if (params[i] < -8.0) {
      local_model->sigma[i] = exp (-8.0);
      penalty += 10.0 * (-params[i] - 8.0);
      }
   else
      local_model->sigma[i] = exp (params[i]);
   }

/*
   This is the main loop that does the cross validation
*/

   error = 0.0;

   for (i=0; i<local_model->ncases; i++) { // For each case in training set
      inputs = local_model->tset + i * (ngates+nmodels+1); // This case
      out = local_model->trial (inputs, inputs+ngates, i, 0);
      diff = inputs[ngates+nmodels] - out; // True minus predicted
      error += diff * diff;
      } // For each case

   return error / local_model->ncases + penalty;
}
```

Rather than directly optimizing each sigma, it is better to use the logarithm of sigma as the parameter being optimized. This linearizes the effect of variation, resulting in improved stability. However, the error surface of a GRNN gate can be very flat for extremely large or small values of sigma. Therefore, the first thing done by the criterion function is to exponentiate the parameter, while limiting its range. A penalty is introduced to encourage values of sigma that are not extreme.

We pass through the entire training set. For each case, a prediction is made, and this prediction is compared to the true value. The squared error is cumulated to serve as the error criterion.

412

A fold of cross validation is performed in the trial() function itself. This code, which makes a prediction from a test case, is as follows:

```
double GRNNgate::trial (double *gates, double *contenders,
                        int i_exclude, int n_exclude)
{
  int icase, imodel, ivar, idist, size;
  double psum, *dptr, diff, dist, *outptr, err, out;

  for (imodel=0; imodel<nmodels; imodel++)      // For each model
    errvals[imodel] = 0.0;                       // Will sum kernels here

  size = ngates + nmodels + 1;                   // Size of each training case

  for (icase=0; icase<ncases; icase++) {         // Do all training cases (Equation (8.3))

    idist = abs (i_exclude - icase);             // How close to excluded case?
    if (ncases - idist < idist)                  // Also check distance going
      idist = ncases - idist;                    // around end

    if (idist <= n_exclude)                      // If we are too close to excluded
      continue;                                  // Skip this tset case

    dptr = tset + size * icase;                  // Point to this case
    dist = 0.0;                                  // Will sum distance here

    for (ivar=0; ivar<ngates; ivar++) {          // All gates in this case
      diff = gates[ivar] - dptr[ivar];           // Input minus case
      diff /= sigma[ivar];                       // Scale per sigma
      dist += diff * diff;                       // Cumulate Euclidean distance
    }

    dist = exp (-dist);                          // Apply the Gaussian kernel
    dptr += ngates;                              // Contenders stored after gates
    outptr = dptr + nmodels;                     // True output last of all
    for (ivar=0; ivar<nmodels; ivar++) {         // For every contender
      err = dptr[ivar] - *outptr;                // Error of this contender
      errvals[ivar] += dist * err * err;         // Cumulate numerators of Equation (8.3)
    }
  } // For all training cases
```

```
/*
   Convert unnormalized squared errors to weights.
*/

   psum = 0.0;
   for (ivar=0; ivar<nmodels; ivar++) {    // Equation (8.6)
      if (errvals[ivar] > 1.e-30)           // Should nearly always be true
         errvals[ivar] = 1.0 / errvals[ivar];   // But fails for wild gate value
      else
         errvals[ivar] = 1.e30;             // Cheap insurance against dividing by zero
      psum += errvals[ivar];
      }

   for (ivar=0; ivar<nmodels; ivar++)
      errvals[ivar] /= psum;

/*
   Compute weighted output
*/

   out = 0.0;
   for (ivar=0; ivar<nmodels; ivar++)      // Equation (8.4)
      out += errvals[ivar] * contenders[ivar];

   return out;
}
```

We begin by initializing the errvals array to zero. The numerator of Equation (8.3) will be cumulated here for each model. Note that we do not need to compute the denominator of this equation because when the normalization factor in the denominator of Equation (8.6) is considered, we see that the denominator of Equation (8.3) disappears.

The next loop evaluates the sum in the numerator of Equation (8.3). Before including each term in the sum, though, sequential proximity of the test case to the training case is checked. Remember, this function can be used for predicting based on a member of the training set, as well as a totally unknown test case. By passing the sequence number i_exclude of each training case to the trial() routine, we can implement cross validation. A distance limit, n_exclude, is also passed. Normally, this would be set to zero, which excludes only the single case. However, some applications involve training sets that are serially correlated. Time series prediction is a common offender. This can be handled by excluding cases that are near the training case being tested.

If a training case passes the cross validation exclusion test, the equation is used to compute the weighted Euclidean distance separating the two cases. This distance is exponentiated for use in Equation (8.3). Then, for each model (contender) we find the prediction error and complete the computation of the numerator of Equation (8.3).

Equation (8.6) is used to compute the normalized weights. Note that although this equation explicitly assumes that we are using true error predictions (from Equation [8.3]), the fact that the denominator of that equation is not computed is of no import. The denominator would cancel across Equation (8.6).

The final step is to use Equation (8.4) to combine the outputs of each contender model into a single prediction. This is the function's return value.

Experiments with GRNN Gating

The GRNNGATE.CPP file on my web site includes a program that generates synthetic data for testing performance of the GRNN gate. This program allows you to specify the number of cases in the training set, number of models, and degree of difficulty. Four different types of gating are tested:

> *After-the-fact* uses the outputs of the component models as the gate variables. This facilitates performance comparisons with the primitive method described on page 395.

> *Original* uses the values of the original predictor variables (the variables used by the contender models) as the gates. Given the way the program is designed, there is no reason to believe that these variables will be effective gates. Therefore, this test demonstrates the effect of poor gate variables.

> *Random* takes this one step further by using random numbers as gates. This test demonstrates the effect of totally worthless gates.

> *Ratio* uses the log of the ratio of the prediction error of the first model relative to that of the second. Of course, this is the ultimate cheating method, because this quantity will never be known in a real application. If you knew the true value of the dependent

variable, there would be no need for prediction! If there are two models, this is a nearly perfect gate. Hence, performance with this gate provides an approximate lower limit on the obtainable error. If there are more than two models, this gate provides useful but incomplete information.

Four tests were performed. These include the use of three models having equal power, and augmentation of this set with a fourth model that makes random predictions. Also, easy and difficult situations were tested. All tests employed 500 training cases and 1,000 trials. Mean squared errors were as follows:

	3, Easy	3, Hard	4, Easy	4, Hard
Model mean	0.00123	4.07061	0.50875	4.56859
After-the-fact	0.00016	4.05860	0.00103	4.11057
Original	0.00013	4.05851	0.00060	4.11507
Random	0.00020	4.05816	0.00021	4.11812
Ratio	0.00013	4.03471	0.00022	4.07235

You should immediately notice that in every case GRNN gating improves over the use of a single model. Recall that such improvement was not universal with the primitive after-the-fact specialization described on page 395. You are encouraged to experiment with this program to verify that GRNN gating is a reliable performer, even with tiny training sets.

It is instructive to note what at first glance appears to be an anomaly: even worthless gating, obtained by using a random number as a gate, improves performance. In fact, with four models (three good and one worthless) and an easy problem, the mean model error of 0.51362 was reduced to 0.00021 with a random gate! How can this be? The answer is that the GRNN gating algorithm is able to discover the high mean error of the defective model and make use of this information, despite that the gate variable has no predictive power in regard to the error. In particular, for every test case, Equation (8.3) will produce a relatively large error prediction for the poor model, meaning that Equation (8.6) will produce a small weight for the model. The obvious moral of the story is that if there is reason to believe that the models may be unequal in power, a GRNN gate can be useful even if the gate variables are of little value.

CHAPTER 9

Information and Entropy

- Entropy
- Joint and Conditional Entropy
- Mutual Information
- Continuous Mutual Information
- Predictor Selection

An effective model takes one or more *predictor* variables, processes the information that they contain, and estimates a *predicted* variable that is ideally useful in some way. But even the most sophisticated model is helpless if it is not given the information it needs to make a good decision. In this chapter, we explore the concept of information content of a variable, and we present a variety of algorithms for assessing the amount and nature of this information.

Entropy

Suppose you have to send a message to someone, giving this person the answer to a multiple-choice question. The catch is, you are only allowed to send the message by means of a string of ones and zeros, called *bits*. What is the minimum number of bits that you need to communicate the answer? Well, if it is a true/false question, one bit will obviously do. If four answers are possible, you will need two bits, which provide four possible patterns: 00, 01, 10, and 11. Eight answers will require three bits, and so forth. In general, to identify one of K possibilities, you will need $\log_2(K)$ bits, where $\log_2(.)$ is the logarithm base two.

© Timothy Masters 2018
T. Masters, *Assessing and Improving Prediction and Classification*,
https://doi.org/10.1007/978-1-4842-3336-8_9

Working with base-two logarithms is unconventional. Mathematicians and computer programs almost always use *natural logarithms*, in which the base is $e \approx 2.718$. The material in this chapter does not require base two; any base will do. By tradition, when natural logarithms are used in information theory, the unit of information is called the *nat* as opposed to the *bit*. This need not concern us. For much of the remainder of this chapter, no base will be written or assumed. Any base can be used, as long as it is used consistently. Since whenever units are mentioned they will be bits, the implication is that logarithms are in base two. On the other hand, all computer programs will use natural logarithms. The difference is only one of naming conventions for the unit, at least as far as most of this chapter is concerned.

Different messages can have different degrees of utility. If you live in the midst of the Sahara Desert, a message from the weather service that today will be dry and sunny is of little value. On the other hand, a message that a foot of snow is on the way will be enormously interesting and hence valuable. A good way to quantify the value or *information* of a message is to measure the amount by which receipt of the message reduces uncertainty. If the message simply tells you something that you expected already, the message gives you little information. But if you receive a message saying that you have just won a million-dollar lottery, the message is valuable indeed, and not only in the monetary sense. The fact that its information is highly unlikely gives it special value.

Suppose you are a military commander. Your troops are poised to launch an invasion as soon as the order to invade arrives. All you know is that it will be one of the next 64 days. You have been told that tomorrow morning you will receive a single binary message: *yes* the invasion is today, or *no* the invasion is not today. Early the next morning, as you sit in your office awaiting the message, you are totally uncertain as to the day of invasion. It could be any of the upcoming 64 days, so you have six bits of uncertainty ($\log_2(64)=6$). If the message turns out to be *yes*, all uncertainty is removed. You know the day of invasion. Therefore, the information content of a *yes* message is six bits. Looked at another way, the probability of a *yes* message is 1/64, so its information is $-\log_2(1/64)=6$. It should be apparent that the value of a message is inversely related to its probability.

What about a *no* message? It is certainly less valuable than *yes*, because your uncertainty about the day of invasion is only slightly reduced. You know that the invasion will not be today, which is somewhat useful, but it still could be any of the remaining 63 days. The value of *no* is $-\log_2(63/64)$, which is about 0.023 bits. Information in bits or nats or any other unit can be fractional.

The *expected value* of a discrete random variable on a finite set (that is, a random variable that can take on one of a finite number of different values) is equal to the sum of the product of each possible value times its probability. For example, if you have a trading system that has a 0.4 probability of winning $1,000 and a 0.6 probability of losing $500, the expected value of a trade is 0.4 * 1000–0.6 * 500=$100. In the same way, we can talk about the expected value of the information content of a message. In the invasion example, the value of a *yes* message is 6 bits, and it has probability 1/64. The value of a *no* message is 0.023 bits, and its probability is 63/64. Thus, the expected value of the information in the message is (1/64) * 6+(63/64) * 0.023=0.12 bits.

The invasion example had just two possible messages, *yes* and *no*. In practical applications, we will need to deal with messages that have more than two values. Consistent, rigorous notation will make it easier to describe methods for doing so. Let χ be a set that enumerates every possible message. Thus, χ may be {*yes, no*}, or it may be {1, 2, 3, 4}, or it may be {*benign, abnormal, malignant*}, or it may be {*big loss, small loss, neutral, small win, big win*}. We will use X to generically represent a random variable that can take on values from this set, and when we observe an actual value of this random variable, we will call it x. Naturally, x will always be a member of χ. This is written as $x \varepsilon \chi$. Let $p(x)$ be the probability that x is observed. Sometimes it will be clearer to write this probability as $P(X=x)$. These two notations for the probability of observing x will be used interchangeably, depending on which is more appropriate in the context. Naturally, the sum of $p(x)$ for all $x \varepsilon \chi$ is one, since χ includes every possible value of X.

Recall that the information content of a particular message x is $-\log(p(x))$, and the expected value of a random variable is the sum, across all possibilities, of its probability times its value. The information content of a message is itself a random variable. So we can write the expected value of the information contained in X as shown in Equation (9.1). This quantity is called the *entropy* of X, and it is universally expressed as $H(X)$. In this equation, $0*\log(0)$ is understood to be zero, so messages with zero probability do not contribute to entropy.

$$H(X) = -\sum_{x \in \chi} p(x)\log(p(x)) \tag{9.1}$$

Returning once more to the military example, suppose that a second message also arrives every morning: mail call. On average, mail arrives for distribution to the troops about once every three days. The actual day of arrival is random; sometimes mail will arrive several days in a row, and other times a week or more may pass with no mail.

You never know when it will arrive, other than that you will be told in the morning whether mail will be delivered that day. The entropy of the *mail today* random variable is $-(1/3) \log_2 (1/3) - (2/3) \log_2 (2/3) \approx 0.92$ bits.

In view of the fact that the entropy of the *invasion today* random variable was about 0.12 bits, this seems to be an unexpected result. How can a message that resolves an event that happens about every third day convey so much more information than one about an event that has only a 1/64 chance of happening? The answer lies in the fact that entropy is an *average*. Entropy does not measure the value of a single message. It measures the expectation of the value of the message. Even though a *yes* answer to the invasion question conveys considerable information, the fact that the nearly useless *no* message will arrive with probability 63/64 drags the average information content down to a small value.

Let K be the number of messages that are possible. In other words, the set χ contains K members. Then it can be shown (though we will not do so here) that X has maximum entropy when $p(x)=1/K$ for all $x \in \chi$. In other words, a random variable X conveys the most information possible when all of its possible values are equally likely. It is easy to see that this maximum value is $\log(K)$. Simply look at Equation (9.1) and note that all terms are equal to $(1/K) \log(1/K)$, and there are K of them. For this reason, it is often useful to observe a random variable and use Equation (9.1) to estimate its entropy and then divide this quantity by $\log(K)$ in order to compute its *proportional entropy*. This is a measure of how close X comes to achieving its theoretical maximum information content.

It must be noted that although the entropy of a variable is a good theoretical indicator of how much information the variable conveys, whether this information is useful is another matter entirely. Knowing whether the local post office will deliver mail today probably has little bearing on whether the home command has decided to launch an invasion today. There are ways to assess the degree to which the information content of a message is useful for making a specified decision, and these techniques will be covered later in this chapter. For now, understand that significant information content of a random variable is a necessary but not sufficient condition for making effective use of that variable.

Entropy of a Continuous Random Variable

Entropy was originally defined for finite discrete random variables, and this remains its primary application. However, it can be generalized to continuous random variables. In this case, the summation of Equation (9.1) must be replaced by an integral, and the probability $p(x)$ must be replaced by the probability density function $f(x)$. The definition of entropy in the continuous case is given by Equation (9.2).

$$H(X) = -\int_{-\infty}^{\infty} f(x)\log(f(x))dx \qquad (9.2)$$

There are several problems with continuous entropy, most of which arise from the fact that Equation (9.2) is not the limiting case of Equation (9.1) when the bin size shrinks to zero and the number of bins blows up to infinity. In practical terms, the most serious problem is that continuous entropy is not immune to rescaling. One would hope that performing the seemingly innocuous act of multiplying a random variable by a constant would leave its entropy unchanged. Intuition clearly says that this should be so, for certainly the information content of a variable should be the same as the information content of ten times that variable. Alas, it is not so.

Moreover, estimating a probability density function $f(x)$ from an observed sample is far more difficult than simply counting the number of observations in each of several bins for a sample. Thus, Equation (9.2) can be very difficult to evaluate in applications. For these reasons, continuous entropy is avoided whenever possible. We will deal with the problem by discretizing a continuous variable in as intelligent a fashion as possible and treating the resulting random variable as discrete. The disadvantages of this approach are few, and the advantages are many.

Partitioning a Continuous Variable for Entropy

Entropy is a simple concept for discrete variables and a vile beast for continuous variables. Give me a sample of a continuous variable, and chances are I can give you a reasonable algorithm that will compute its entropy as nearly zero and an equally reasonable algorithm that will find the entropy to be huge and any number of intermediate estimators. The bottom line is that we first need to understand our intended use for the entropy estimate and then choose an estimation algorithm accordingly.

Our primary use for entropy is as a screening tool for predictor variables. Entropy has theoretical value as a measure of how much information is conveyed by a variable. But it has a practical value that goes beyond this theoretical measure. There tends to be a correlation between how well many models are able to learn predictive patterns and the entropy of the predictor variables. This is not universally true, but it is true often enough that a prudent researcher will pay attention to entropy.

The mechanism by which this happens is straightforward. Many models focus their attention roughly equally across the entire range of variables, both predictor and predicted. Even models that have the theoretical capability of zooming in on important areas will have this tendency because their traditional training algorithms can require an inordinate amount of time to refocus attention onto interesting areas. The implication is that it is usually best if observed values of the variables are spread at least fairly uniformly across their range.

For example, suppose a variable has a strong right skew. Perhaps in a sample of 1,000 cases, about 900 lie in the interval zero to one, another 90 cases lie in one to ten, and the remaining ten cases are up around a thousand. Many learning algorithms will see these few extremely large cases as providing one type of information and lump the mass of cases around zero to one into a single entity providing another type of information. The algorithm will find it difficult to identify and act on cases whose values on this variable differ by 0.1. It will be overwhelmed by the fact that some cases differ by a thousand. Some other models may do a great job of handling the mass of low-valued cases but find that the cases out in the tail are so bizarre that they essentially give up on them.

The susceptibility of models to this situation varies widely. Trees have little or no problem with skewness and heavy tails for predictors, although they have other problems that are beyond the scope of this text. Feedforward neural nets, especially those that initialize weights based on scale factors, are extremely sensitive to this condition unless trained by sophisticated algorithms. General Regression Neural Nets and other kernel methods that use kernel widths that are relative to scale can be rendered helpless by such data. It would be a pity to come close to producing an outstanding application and be stymied by careless data preparation.

The relationship between entropy and learning is not limited to skewness and tail weight. Any unnatural clumping of data, which would usually be caught by a good entropy test, can inhibit learning by limiting the ability of the model to access information in the variable. Consider a variable whose range is zero to one. One-third of its cases lie in {0, 0.1}, one-third lie in {0.4, 0.5}, and one-third lie in {0.9, 1.0}, with output values (classes or predictions) uniformly scattered among these three clumps. This variable has no real skewness and extremely light tails. A basic test of skewness and kurtosis would show it to be ideal. Its range-to- interquartile-range ratio would be wonderful. But an entropy test would reveal that this variable is problematic. The crucial information that is crowded inside each of three tight clusters will be lost, unable to compete with the obvious difference among the three clusters. The intra-cluster variation, which may be crucial to solving the problem, is so much less than the possibly worthless inter-cluster variation that most models would be hobbled.

When detecting this sort of problem is our goal, the best way to partition a continuous variable is also the simplest: split the range into bins that span equal distances. Note that a technique we will explore later, splitting the range into bins containing equal numbers of cases, is worthless here. All this will do is give us an entropy of $\log(K)$, where K is the number of bins. To see why, look back at Equation (9.1) on page 419. Rather, we need to confirm that the variable in question is distributed as uniformly as possible across its range. To do this, we must split the range equally and count how many cases fall into each bin.

The program ENTROPY on my web site illustrates this technique. Here are a few snippets from this program. The first step is to find the range of the variable (in work here) and the factor for distributing cases into bins. Then the cases are categorized into bins. Note that two tricks are used in computing the factor. We subtract a tiny constant from the number of bins to ensure that the largest case does not overflow into a bin beyond what we have. We also add a tiny constant to the denominator to prevent division by zero in the pathological condition of all cases being identical.

```
low = high = work[0];      // Will be the variable's range
for (i=1; i<ncases; i++) {   // Check all cases to find the range
  if (work[i] > high)
    high = work[i];
  if (work[i] < low)
    low = work[i];
  }

for (i=0; i<nb; i++)        // Initialize all bin counts to zero
  counts[i] = 0;

factor = (nb - 0.00000000001) / (high - low + 1.e-60);

for (i=0; i<ncases; i++) {   // Place the cases into bins
  k = (int) (factor * (work[i] - low));
  ++counts[k];
  }
```

Once the bin counts have been found, computing the entropy is a trivial application of Equation (9.1).

```
entropy = 0.0;
for (i=0; i<nb; i++) {                              // For all bins
   if (counts[i] > 0) {                             // Bin might be empty
      p = (double) counts[i] / (double) ncases;    // p(x)
      entropy -= p * log(p);                        // Equation (9.1)
      }
   }

entropy /= log(nb);                                 // Divide by max for proportional
```

Having a heavy tail is the most common cause of low entropy. However, clumping in the interior also appears in applications. We do need to distinguish between clumping of continuous variables because of poor design versus unavoidable grouping into discrete categories. It is the former that concerns us here. Truly discrete groups cannot be separated, while unfortunate clustering of a continuous variable can and should be dealt with. Since a heavy tail (or tails) is such a common and easily treatable occurrence and interior clumping is rarer but nearly as dangerous, it can be handy to have an algorithm that can detect undesirable interior clumping in the presence of heavy tails. Naturally, one could simply apply a transformation to lighten the tail and then perform the test shown earlier. But for quick prescreening of predictor candidates, a single test is nice to have around.

The easiest way to separate tail problems from interior problems is to dedicate one bin at each extreme to the corresponding tail. Specifically, assume that you want K bins. Find the shortest interval in the distribution that contains $(K-2)/K$ of the cases. Divide this interval into $K-2$ bins of equal width and count the number of cases in each of these interior bins. All cases below the interval go into the lowest bin. All cases above this interval go into the upper bin. If the distribution has a very long tail on one end and a very short tail on the other end, the bin on the short end may be empty. This is good, for it slightly punishes the skewness. If the distribution is exactly symmetric, each of the two end bins will contain $1/K$ of the cases, which implies no penalty. This test focuses mainly on the interior of the distribution, computing the entropy primarily from the $K-2$ interior bins, with an additional small penalty for extreme skewness and no penalty for symmetric heavy tails.

Keep in mind that passing this test does not mean that we are home free. This test deliberately ignores heavy tails, so a full test must follow an interior test. Conversely, failing this interior test is bad news. Serious investigation is required.

The ENTROPY program performs this test as well as the full test. Here is the code snippet that does the interior partitioning. Note that this algorithm assumes sufficient continuity so that ties are inconsequential. Because the quantity computed by this algorithm is just used for making subjective judgments in regard to good or poor entropy, exact figures are not needed. If the data contains such large blocks of ties that this method is invalidated, there is probably no point in coming up with a new algorithm to handle ties. There are issues other than entropy involved with the predictor!

```
ilow = (ncases + 1) / nb - 1;        // Unbiased lower quantile
if (ilow < 0)
  ilow = 0;

ihigh = ncases - 1 - ilow;           // Symmetric upper quantile

// Find the shortest interval containing 1-2/nbins of the distribution
qsortd (0, ncases-1, work);          // Sort cases ascending

istart = 0;                          // Beginning of interior interval
istop = istart + ihigh - ilow - 2;   // And end, inclusive
best_dist = 1.e60;                   // Will be shortest distance

while (istop < ncases) {             // Try bounds containing the same n of cases
  dist = work[istop] - work[istart]; // Width of this interval

  if (dist < best_dist) {            // We're looking for the shortest
    best_dist = dist;                // Keep track of shortest
    ibest = istart;                  // And its starting index
  }

  ++istart;                          // Advance to the next interval
  ++istop;                           // Keep n of cases in interval constant
}

istart = ibest;                      // This is the shortest interval
istop = istart + ihigh - ilow - 2;

counts[0] = istart;                  // The count of the leftmost bin
```

425

```
counts[nb-1] = ncases - istop - 1;   // and rightmost are implicit

for (i=1; i<nb-1; i++)               // Inner bins
   counts[i] = 0;

low = work[istart];                  // Lower bound of inner interval
high = work[istop];                  // And upper bound
factor = (nb - 2.00000000001) / (high - low + 1.e-60);

for (i=istart; i<=istop; i++) {      // Place cases in bins
   k = (int) (factor * (work[i] - low));
   ++counts[k+1];
   }
```

An Example of Improving Entropy

John decides that he wants to do intraday trading of the U.S. bond futures market. One variable that he believes will be useful is an indication of how much the market is moving away from its very recent range. As a start, he subtracts from the current price a moving average of the close of the most recent 20 bars. Realizing that the importance of this deviation is relative to recent volatility, he decides to divide the price difference by the price range over those prior 20 bars. Being a prudent fellow, he does not want to divide by zero in those rare instances in which the price is flat for 20 contiguous bars, so he adds one tick (1/32 point) to the denominator. His final indicator is given by Equation (9.3).

$$X = \frac{CLOSE - MA(20)}{HIGH(20) - LOW(20) + 0.03125} \tag{9.3}$$

Being not only prudent but informed as well, he computes this indicator from a historical sample covering many years, divides the range into 20 bins, and calculates its proportional entropy as discussed on page 422. Imagine John's shock when he finds this quantity to be merely 0.0027, about one-quarter of 1 percent of what should be possible! Clearly, more work is needed before this variable is presented to any prediction model.

Basic detective work reveals some fascinating numbers. The interquartile range covers −0.2 to 0.22, but the complete range is −48 to 92. There's no point in plotting a histogram; virtually the entire dataset would show up as one tall spike in the midst of a barren desert.

He now has two choices for eliminating those troublesome outliers: truncate or squash. The common squashing functions, *arctangent, hyperbolic tangent,* and *logistic,* are all comfortable with the native domain of John's indicator, which happens to be about –1 to 1 if extreme outliers are ignored. Figure 9-1 shows the result of truncating this variable at +/–1. This truncated variable has a proportional entropy of 0.83, which is decent by any standard. Figure 9-2 is a histogram of the raw variable after applying the hyperbolic tangent squashing function. Its proportional entropy is 0.81. Neither approach is obviously superior, but one thing is perfectly clear: one of them, or something substantially equivalent, must be used instead of the raw variable of Equation (9.3)!

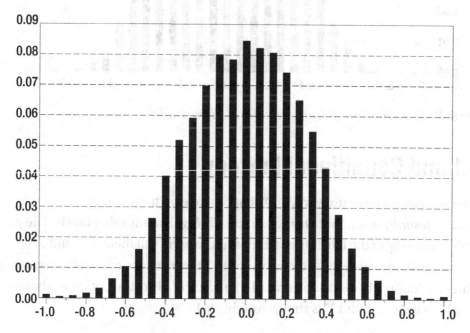

Figure 9-1. *Distribution of truncated variable*

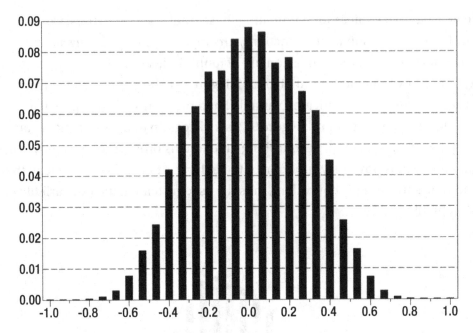

Figure 9-2. *Distribution of htan transformed variable*

Joint and Conditional Entropy

Suppose we have an indicator variable X that can take on three values. These values might be {*unusually low, about average, unusually high*} or any other labels. The nature or implied ordering of the labels is not important; we will call them 1, 2, and 3 for convenience. We also have an outcome variable Y that can take on two values: *win* and *lose*. After evaluating these variables on a large batch of historical data, we tabulate the relationship between X and Y as shown in Table 9-1.

Table 9-1. *Observed Counts and Probabilities, Theoretical Probabilities*

		Y win	Y lose	Marginal
		80	20	100
	1	0.16	0.04	
		0.12	0.08	
		100	100	200
X	2	0.20	0.20	
		0.24	0.16	
		120	80	200
	3	0.24	0.16	
		0.24	0.16	
Marginal		300	200	500

This table shows that 80 cases fell into Category 1 of X and also the *win* category of Y, 20 cases fell into Category 1 of X and also the *lose* category of Y, and so forth. The second number in each table cell is the fraction of all cases that fell into that cell. Thus, the (1, *win*) cell contained 0.16 of the 500 cases in the historical sample.

The third number in each cell is the fraction of cases that would, on average, fall into that cell if there were no relationship between X and Y. If two events are independent, meaning that the occurrence of one of them has no impact on the probability of occurrence of the other, the probability that they will both occur is the product of the probabilities that each will occur. In symbols, let $P(A)$ be the probability that some event A will occur, let $P(B)$ be the probability that some other event B will occur, and let $P(A,B)$ be the probability that they both will occur. Then $P(A,B)=P(A)*P(B)$ if and only if A and B are independent.

We can compute the theoretical probability of each X and Y event by summing the counts across rows and columns to get the *marginal* counts and then dividing each by the total number of cases. For example, in the $Y=win$ category, the total is 80+100+120=300 cases. Dividing this by 500 gives $P(Y=win)=0.6$. For X we find that $P(X=1)=(80+20)/500=0.2$. Hence, the probability of $(X=1, Y=win)$, if X and Y were independent, is $0.6*0.2=0.12$.

The observed probabilities for four of the six cells differ from the probabilities expected under independence, so we conclude that there might be a relationship between X and Y, though the difference is so small that random chance might just as well be responsible. An ordinary chi-square test would quantify the probability that the observed differences could have arisen from chance. But we are interested in a different approach right now.

Equation (9.1) on page 419 defined the entropy for a single random variable. We can just as well define the entropy for two random variables simultaneously. This *joint entropy* indicates how much information we obtain on average when the two variables are both known. Joint entropy is a straightforward extension of univariate entropy. Let χ, X, and x be as defined for Equation (9.1). In addition, let ¥, Y, and y be the corresponding items for the other variable. The joint entropy $H(X, Y)$ is based on the individual cell probabilities, as shown in Equation (9.4). In this example, summing the six terms gives $H(X, Y)\approx1.70$.

$$H(X,Y)=-\sum_{x\in\chi}\sum_{y\in¥}p(x,y)\log\big(p(x,y)\big) \tag{9.4}$$

It often happens that the entropy of a variable is different for different values of another variable. Look back at Table 9-1. There are 100 cases for which $X=1$. Of these, 80 have $Y=win$ and 20 have $Y=lose$. The probability that $Y=win$, given that $X=1$, which is written $P(Y=win|X=1)$, is 80/100=0.8. Similarly, $P(Y=lose|X=1)=0.2$. By Equation (9.1), the entropy of Y, given that $X=1$, which is written $H(Y|X=1)$, is $-0.8*\log(0.8) - 0.2*\log(0.2) \approx 0.50$ nats. (The switch from base 2 to base e is convenient now.) In the same way, we can compute $H(Y|X=2) \approx 0.69$, and $H(Y|X=3) \approx0.67$.

Hold that thought. Before continuing, we need to reinforce the idea that entropy, which is a measure of disorganization, is also a measure of average information content. On the surface, this seems counterintuitive. How can it be that the more disorganized a variable is, the more information it carries? The issue is resolved if you think about what is gained by going from not knowing the value of the variable to knowing it. If the variable is highly disorganized, you gain a lot by knowing it. If you live in an area where the weather changes every hour, an accurate weather forecast (if there is such a thing) is very valuable. Conversely, if you live in the middle of a desert, an accurate weather forecast is nearly always boring.

We just saw that we can compute the entropy of Y when X equals any specified value. This leads us to consider the entropy of Y under the general condition that we know X. In other words, we do not specify any particular X. We simply want to know, on average, what the entropy of Y will be if we happen to know X. This quantity, called the *conditional entropy of Y given X*, is an expectation once more. To compute it, we sum the product of every possibility times the probability of the possibility. In the example several paragraphs ago, we saw that $H(Y|X=1) \approx 0.50$. Looking at the marginal probabilities, we know that $P(X=1) = 100/500 = 0.20$. Following the same procedure for $X=2$ and 3, we find that the entropy of Y given that we know X, written $P(Y|X)$, is $0.2*0.50 + 0.4*0.69 + 0.4*0.67 = 0.64$.

Compare this to the entropy of Y taken alone. This is $-0.6*\log(0.6) - 0.4* \log(0.4)$ ≈ 0.67. Notice that the conditional entropy of Y given X is slightly less than that of Y without knowledge of X. In fact, it can be shown that $H(Y|X) \leq H(Y)$ universally. This makes sense. Knowing X certainly cannot make Y any more disorganized! If X and Y are related in any way, knowing X will reduce the disorganization of Y. Looked at another way, X may supply some of the information that would have otherwise been provided by Y. Once we know X, we have less to gain from knowing Y. A weather forecast as you roll out of bed in the morning gives you more information than the same forecast does after you have looked out the window and seen that the sky is black and rain is pouring down.

There are several standard ways of computing conditional entropy. The most straightforward way is direct application of the definition, as we did earlier. Equation (9.5) is the conditional probability of Y given X. The entropy of Y for any specified X is shown in Equation (9.6). Finally, Equation (9.7) is the entropy of Y given that we know X.

$$P\left(Y = y \middle| X = x\right) = \frac{P\left(Y = y, \ X = x\right)}{P\left(X = x\right)} \tag{9.5}$$

$$H\left(Y \middle| X = x\right) = \sum_{y \in Y} P\left(\left(Y = y \middle| X = x\right)\right) \log\left(P\left(Y = y \middle| X = x\right)\right) \tag{9.6}$$

$$H\left(Y \middle| X\right) = \sum_{x \in \chi} P\left(X = x\right) H\left(Y \middle| X = x\right) \tag{9.7}$$

A simpler method for computing the conditional entropy of Y given X is to use the identity shown in Equation (9.8). Although the proof of this identity is simple, we will not show it here. The intuition is clear, though. The entropy of (information contained in) Y given that we already know X is the total entropy (information) minus that due strictly to X. Rearranging the terms and treating entropy as uncertainty may make the intuition even clearer. The total uncertainty that we have about X and Y together is equal to the uncertainty we have about X plus whatever uncertainty we have about Y, given that we know X.

$$H(Y|X) = H(X, Y) - H(X) \qquad (9.8)$$

We close this section with a small exercise for you. Refer back to Table 9-1 on page 429 and look at the third line in each cell. Recall that we computed this line by multiplying the marginal probabilities. For example, $P(X=1)=100/500=0.2$, and $P(Y=win)=300/500=0.6$, which gives 0.2*0.6=0.12 for the $(1, win)$ cell. These are the theoretical cell probabilities if X and Y were independent. Using the Y marginals, compute to decent accuracy $H(Y)$. You should get 0.673012. Using whichever formula you prefer, Equation (9.7) or (9.8), compute $H(Y|X)$ accurately. You should get the same number, 0.673012. When *theoretical* (not observed) cell probabilities are used, the entropy of Y alone is the same as the entropy of Y when X is known. Ponder why this is so.

No solid motivation for computing or examining conditional entropy is yet apparent. This will change soon. For now, let's study its computation in more detail.

Code for Conditional Entropy

The source file MUTINF_D.CPP on my web site contains a function for computing conditional entropy using the definition formula, Equation (9.7). Here are two code snippets extracted from this file. The first snippet zeros out the array where the marginal of X will be computed, and it also zeros the grid of bins that will count every combination of X and Y. It then passes through the entire dataset, filling the bins.

```
for (ix=0; ix<nbins_x; ix++) {
   marginal_x[ix] = 0;
   for (iy=0; iy<nbins_y; iy++)
   grid[ix*nbins_y+iy] = 0;
   }
```

```
for (i=0; i<ncases; i++) {
  ix = bins_x[i];
  ++marginal_x[ix];
  ++grid[ix*nbins_y+bins_y[i]];
}
```

After the bins have been filled, the following code implements Equations (9.5) through (9.7) to compute the conditional entropy:

```
CI = 0.0;
for (ix=0; ix<nbins_x; ix++) {      // Sum Equation (9.7) for all x in χ

  if (marginal_x[ix] > 0) {         // Term only makes sense if positive marginal
    cix = 0.0;                      // Will cumulate H(Y|X=x) of Equation (9.6)
    for (iy=0; iy<nbins_y; iy++) {  // Sum Equation (9.6)
      pyx = (double) grid[ix*nbins_y+iy] / (double) marginal_x[ix];   // Equation (9.5)
      if (pyx > 0.0)                // 0 log(0) = 0
        cix += pyx * log (pyx);     // Equation (9.6)
    }
  }

  CI += cix * marginal_x[ix] / ncases; // Equation (9.7)
}
```

Mutual Information

John has four areas of expertise: football, beer, bourbon, and poker. Mary has three areas of expertise: gourmet cooking, martial arts, and poker. One night they meet at a hot game, decide that they make the perfect couple, and get married. Here are some statements about their expertise as a couple:

- John and Mary jointly have six areas of expertise: four from John, plus two from Mary (cooking, martial arts) that are beyond any supplied by John. Equivalently, they have three from Mary, plus three from John (football, beer, bourbon) that are beyond any supplied by Mary. See Equation (9.9).

- John and Mary jointly have six areas of expertise: four from John, plus three from Mary, minus one (poker) that they have in common and thus was counted twice. See Equation (9.10).

- John has three areas of expertise to offer (football, beer, and bourbon) if we already have access to whatever expertise Mary offers. These three are his four, minus the one that they share. See Equation (9.11).

- Similarly, Mary has two areas of expertise above and beyond whatever is supplied by John. See Equation (9.12).

Information (poker expertise here) that is shared by two random variables X and Y is called their *mutual information* and is written $I(X; Y)$. The following equations summarize the relationships among joint, single, and conditional entropy, and mutual information. Examination of Figure 9-3 may make the intuition behind these equations clearer.

$$H(X,Y) = H(X) + H(Y|X) = H(Y) + H(X|Y) \tag{9.9}$$

$$H(X, Y) = H(X) + H(Y) - I(X;Y) \tag{9.10}$$

$$H(X|Y) = H(X) - I(X;Y) \tag{9.11}$$

$$H(Y|X) = H(Y) - I(X;Y) \tag{9.12}$$

$$I(X;Y) = H(X) - H(X|Y) = H(Y) - H(Y|X) \tag{9.13}$$

$$I(X;Y) = H(X) + H(Y) - H(X,Y) \tag{9.14}$$

$$I(X;X) = H(X) \tag{9.15}$$

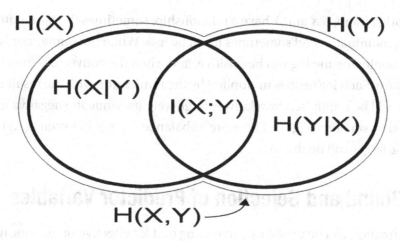

Figure 9-3. *Relationships between X and Y*

In Figure 9-3, the thin line that surrounds the entire diagram is the total joint entropy, $H(X,Y)$. The full ellipse on the left is the entropy of X, and the full ellipse on the right is the entropy of Y. The section where they overlap is their mutual information, and the two crescents are the conditional entropies. You should study this figure and understand how it relates to the equations before it.

Equation (9.13) or (9.14) may be used to compute the mutual information of a pair of variables. But it is often more convenient to use the official definition of mutual information. We will not prove that the definition given by Equation (9.16) concurs with the preceding equations, as it is tedious.

$$I(X;Y) = \sum_{x \in \chi} \sum_{x \in Y} p(x,y) \log \frac{p(x,y)}{p(x)p(y)} \tag{9.16}$$

There is simple intuition behind Equation (9.16). Recall that events X and Y are independent if and only if the probability of them both happening equals the product of each of them happening: $P(X, Y) = P(X)*P(Y)$. Thus, if X and Y in Equation (9.16) are independent, the numerator will equal the denominator in the log expression. The log of one is zero, so every term in the sum will be zero. The mutual information of a pair of independent variables will evaluate to zero, as expected.

On the other hand, if X and Y have a relationship, sometimes the numerator will exceed the denominator, and sometimes it will be less. When the numerator is larger than the denominator, the log will be positive, and when the converse is true, the log will be negative. Each log term is multiplied by the numerator, with the result that positive logs will be multiplied by relatively large weights, while the negative logs will be multiplied by smaller weights. The more imbalance there is between $p(x,y)$ and $p(x)*p(y)$, the larger will be the sum.

Fano's Bound and Selection of Predictor Variables

Mutual information can be useful as a screening tool for effective predictors. It is not perfect. For one thing, mutual information picks up any sort of relationship, even unusual nonlinear dependencies. This is fine as long as the variable will be fed to a model that can take advantage of such a relationship. But naive models may be helpless, missing the information entirely. Predictive information is a necessary but not sufficient condition.

Also, mutual information is bidirectional, meaning that it is just as sensitive to the ability of Y to predict X as it is to the ability of X to predict Y. In many cases, this is somewhat of a problem, as we are interested only in the latter ability, not the former. We will address this issue later in this chapter.

Finally, it can sometimes be the case that a single predictor alone is largely useless, while pairing it with a second predictor can work miracles. For example, neither weight nor height alone is a good indicator of physical fitness, but the two together provide valuable information. Therefore, any criterion that is based on a single predictor variable is potentially flawed. Algorithms given later will address this issue to some degree, though not perfectly.

Nonetheless, mutual information is widely applicable as a screening tool. In general, predictor variables that have high mutual information with the predicted variable will be good candidates for use with a model, while those with little or no mutual information will make poor candidates. Mutual information must not be used to create a final set of predictors. Rather, it is best used to narrow a large field of candidates into a smaller manageable set.

In addition to the obvious intuitive value of mutual information, it has a fascinating theoretical property that can quantify its utility. [Fano, 1961] shows that in a classification problem, the mutual information between a predictor variable and a predicted decision variable sets a lower bound on the classification error that can be obtained. Note that there is no guarantee that this accuracy can actually be realized in practice. Performance is dependent on the quality of the model being employed. Still, knowing the best that can possibly be obtained with an ideal model is useful.

Let Y be a random variable that defines a decision class from $¥=\{1, 2, ..., K\}$. In other words, there are K classes. Let X be a finite discrete random variable whose value ideally provides information that is useful for predicting Y. Note that we are not in general asking that the value of X be the predicted value of Y. (We will ask for this in a later modification.) X need not even have K values. In the example of Table 9-1 on page 429, $K=2$ (*win, loss*), and X has three values.

We have a model that examines the value of X and predicts Y. Either this prediction is correct or it is incorrect. Let P_e be the probability that the model's prediction is in error. The *binary entropy function* is defined by Equation (9.17), and Equation (9.18) is *Fano's bound* on the attainable error of the classification model.

$$h(p)=-p\log(p)-(1-p)\log(1-p) \tag{9.17}$$

$$P_e \geq \frac{H(Y)-I(X;Y)-h(P_e)}{\log(\max(K-1,2))} \tag{9.18}$$

Officially, the denominator of Fano's bound is just $\log(K-1)$, and the bound applies only to situations in which $K>2$. To accommodate two classes, the denominator has been modified as shown earlier. Details can be found in [Erdogmus and Principe, 2002].

One obvious problem with Equation (9.18) is that the probability of error appears on both sides of the equation. There are two approaches to dealing with this. Sometimes we will be able to come up with a reasonable estimate of the error rate, perhaps by means of an out-of-sample validation set and a good model. Then we can just blithely plug it into $h()$ in the numerator, rationalizing that the entropy and mutual information are also sample-based estimates. I've done it. In fact, I do it in one of the programs that will be presented later in this chapter. A more conservative approach is to realize that the maximum value of this term is $h(0.5)=\log(2)$. This substitution will ensure that the inequality holds, even though it will be looser than it would be if the exact value of P_e were known. Of course, if we already knew P_e, we wouldn't need the bound!

This, naturally, is a valid reason for not putting much store in computed values of Fano's bound. If we already have a model in mind, any dataset that we use to compute Fano's bound gives us everything we need to compute other, probably superior, estimates of the prediction error and assorted bounds. And if we don't have a model and hence resort to using $\log(2)$ in the numerator, the bound can be overly conservative.

437

The real purpose of Equation (9.18) is that it alerts us to the value of the mutual information between X and Y. Mutual information is not just an obscure theoretical quantity. *It plays a major role in imposing a ceiling over the prediction accuracy that can be obtained.* If we are comparing a number of candidate predictors, the denominator of Equation (9.18) will be the same for all competitors, and $H(Y)$, the entropy of the class variable, will also be constant. The error term, $h(P_e)$, may change a little, but $I(X, Y)$ is the dominant force. The minimum attainable error rate is inversely related to the mutual information. Therefore, candidates that have high mutual information with the class variable will probably be more useful than candidates with low mutual information.

Confusion Matrices and Mutual Information

Suppose we already have a set of predictor variables and a model that we use to predict a class. As before, Y is the true class of a case, and there are K classes. This time, we let X be the output of our model for a case. That is, X is the predicted value of Y.

Let's explore how mutual information relates to some three-by-three confusion matrices. Table 9-2 shows four examples. In each case, the row is the true class, and the column is the model's decided class. Thus, row i and column j contain the number of cases that truly belong to class i and were placed by the model in class j. Obviously we want the diagonal to contain as many cases as possible, because the diagonal represents correct classifications.

Mutual information quantifies a different aspect of performance than error rate. The top three confusion matrices in Table 9-2 all have an error rate of 13 percent. The first, *naive*, has very unbalanced prior probabilities. Class 3 makes up 80 percent of the cases. The model takes advantage of this fact by strongly favoring this class. The result is that the other two classes are mostly misclassified. But these errors do not contribute much to the total error rate because these other two classes make up only 20 percent of cases. Mutual information easily picks up the fact that the model has not truly solved the problem. The value of 0.173 is the lowest of the set, by far.

Table 9-2. *Assorted Confusion Matrices*

	4	0	6
naive	0	3	7
MI=0.173	0	0	80
	28	0	6
sure	0	26	7
MI=0.735	0	0	33
	29	2	3
spread	2	29	2
MI=0.624	2	2	29
	29	2	3
swap	2	2	29
MI=0.624	2	29	2

The *sure* and *spread* confusions have identical priors (34 percent, 33 percent, 33 percent) and equal error rates, 13 percent. Yet *sure* has considerably greater mutual information than *spread*. The reason for this difference is the pattern of errors. The *spread* confusion has its errors evenly distributed among the classes, while the *sure* confusion has a consistent pattern of misclassification. Even though both models make errors at the same total rate, with the *sure* model you know in advance what sorts of errors can be expected. In particular, if the model decides that a case is in class 1 or class 2, we can be sure that the decision is correct. This knowledge of error patterns is additional information above and beyond what the error rate alone provides, and the increased mutual information reflects this fact.

Finally, look at the *swap* confusion matrix. It is identical to the *spread* confusion matrix, except that for class 2 and class 3 the model has reversed its decisions. The error rate blows up to 67 percent, while the mutual information remains at 0.624, the same as *spread*. This highlights an important property of mutual information. It is not really measuring classification performance directly. Rather, it is measuring *transfer of useful information* through the model. In other words, we are measuring one or more predictor variables and then processing these variables by a model. The variables contain some information that will be useful for making a correct decision, as well as a great deal of irrelevant information. The model acts as a filter, screening out the noise while concentrating the predictive information. The output of the model is the information that has been distilled from the predictors. The effectiveness of the model at making correct decisions is measured by its error rate. But its ability to extract useful information from a cacophony of noise is measured by its mutual information. The fact that the *swap* model has high mutual information along with a high error rate reflects the fact that the model has done a good job of finding the needles in the haystack. Its decisions really do contain useful information. The requirement that a sentient observer may be needed to process this information in a way that helps us to achieve our ultimate goal of correct classification is something that is ignored by mutual information.

Extending Fano's Bound for Upper Limits

As in the prior section, assume that we have a confusion matrix. In other words, we have a model whose output X is a prediction of the true class Y. Fano's lower bound on the error rate, shown in Equation (9.18) on page 437, can be slightly tightened if we want. Also in this special case, we can compute an approximate upper bound on the classification error.

As was the case for the lower bound, there is little direct practical value in computing an upper bound using information theory. The data needed to compute the bound is sufficient to compute better error estimates and bounds using other methods. However, careful study of the upper bound not only confirms the importance of mutual information as an indicator of predictive power but also yields valuable insights into effective classifier design. We will see that if we can control the way in which the classifier makes errors, we may be able to improve the theoretical limits on its true error rate.

Both the tighter lower bound and the new upper bound depend on the entropy of the error given the decision. We saw in Equation (9.18) for the lower bound that the numerator contained the binary entropy function defined in Equation (9.17). If we are willing to assume even more detailed knowledge of the pattern of errors, we can compute the conditional error entropy using Equation (9.19). In this equation, $h(.)$ is the binary entropy function of Equation (9.17), and the quantity on which it operates is the probability of error given that the model has chosen class x. Because $H(e|X)$ is less than or equal to the binary entropy of the error, the lower bound given by Equation (9.20) is tighter than that of Equation (9.18).

440

$$H\left(e\middle|X\right)=\sum_{x\in\chi}P\left(X=x\right)h\left(P_e\middle|X=x\right) \tag{9.19}$$

$$P_e\geq\frac{H(Y)-I(X;Y)-H\left(e\middle|X\right)}{\log\left(\max(K-1,2)\right)} \tag{9.20}$$

The file MUTINF_D.CPP on my web site contains a function for computing the conditional error entropy of Equation (9.19). Here is a code snippet from this file to demonstrate the computation:

```
for (ix=0; ix<nbins_x; ix++) {      // For all decision classes
  marginal_x[ix] = 0;               // Will sum marginal distribution of X
  error_count[ix] = 0;              // Will count errors associated with each decision
}

for (i=0; i<ncases; i++) {          // Pass through all cases
  ix = bins_x[i];                   // The model's decision for this case
  ++marginal_x[ix];                 // Cumulate marginal distribution
  if (bins_y[i] != ix)              // If the true class is not the decision
    ++error_count[ix];              // Then this is an error, so count it
}

CI = 0.0;                           // Will cumulate conditional error entropy here
for (ix=0; ix<nbins_x; ix++) {      // For all decision classes
  if (error_count[ix] > 0 && error_count[ix] < marginal_x[ix]) { // Avoid degenerate math
    pyx = (double) error_count[ix] / (double) marginal_x[ix];  // P(e|X=x)
    CI += (pyx * log(pyx) + (1.0-pyx) * log(1.0-pyx)) * marginal_x[ix] / ncases;   // Eq 9.19
  }
}
```

To compute an upper bound for the error rate, we need to define the conditional entropy of Y given that the model chose class x, and this choice was an error. This unwieldy quantity is written as $H(Y|e, X=x)$, and it is defined by Equation (9.21). The upper bound on the error rate is then given by Equation (9.22).

$$H\left(Y\middle|e, X=x\right)=-\sum_{y\in Y, y\in x}\frac{P\left(Y=y\middle|X=x\right)}{P\left(e\middle|X=x\right)}\log\left[\frac{P\left(Y=y\middle|X=x\right)}{P\left(e\middle|X=x\right)}\right] \tag{9.21}$$

$$P_e \le \frac{H(Y) - I(X;Y) - H(e|X)}{\min_x \left[H(Y|e, X=x) \right]} \tag{9.22}$$

The key fact to observe from Equation (9.22) is that the denominator is the minimum of erroneous entropy over all values of x, the predicted class. If the errors are concentrated in one or a few predicted classes, this minimum will be small, leading to a large upper bound on the theoretical error rate. This tells us that we should strive to develop a model that maximizes the entropy over all erroneous decisions, as long as we can do so without compromising the mutual information that is crucial to the numerator of the equation. In fact, *the denominator is maximized (thus giving a minimum upper bound) when all errors are equiprobable.*

As was stated earlier, there is little or no practical need to compute this upper bound. It is of mainly theoretical interest. But if you want to do so, the code to compute the denominator of Equation (9.22), drawn from the file MUTINF_D.CPP, is as follows:

```
/*
   Compute the marginal of x and the counts in the nbins_x by nbins_y grid
*/
   for (ix=0; ix<nbins_x; ix++) {
      marginal_x[ix] = 0;
      for (iy=0; iy<nbins_y; iy++)
         grid[ix*nbins_y+iy] = 0;
      }

   for (i=0; i<ncases; i++) {
      ix = bins_x[i];
      ++marginal_x[ix];
      ++grid[ix*nbins_y+bins_y[i]];
      }

/*
   Compute the minimum entropy, conditional on error and each X
   Note that the computation in the inner loop is almost the same as in the
   conditional entropy. The only difference is that since we are also
   conditioning on the classification being in error, we must remove from
```

the X marginal the diagonal element, which is the correct decision.
The outer loop looks for the minimum, rather than summing.
*/

```
minCI = 1.e60;
for (ix=0; ix<nbins_x; ix++) {
   nerr = marginal_x[ix] - grid[ix*nbins_y+ix]; // Marginal that is in error
   if (nerr > 0) {
      cix = 0.0;
      for (iy=0; iy<nbins_y; iy++) {
         if (iy == ix) // This is the correct decision
            continue; // So we exclude it; we are summing over errors
         pyx = (double) grid[ix*nbins_y+iy] / (double) nerr;   // Term in Eq 9.21
         if (pyx > 0.0)
            cix -= pyx * log (pyx);                            // Sum Eq 9.21
      }
      if (cix < minCI)
         minCI = cix;
   }
}
```

Equation (9.22) will often give an upper bound that is ridiculously excessive, sometimes much greater than one. This is especially true if $H(e|X)$ is replaced by zero in the conservative analog to how we may replace this quantity by log(2) for the lower bound. As will be vividly demonstrated in Table 9-3 on page 451, this problem is particularly severe when the denominator of Equation (9.22) is tiny due to a grossly nonuniform error distribution. In this case, one can be somewhat (though only a little) aided by the fact that a naive classifier, one that always chooses the class whose prior probability is greatest, will achieve an error rate of $1-\max_x p(x)$, where $p(x)$ is the prior probability of class x. If there are K classes, and they are all equally likely, a naive classifier will have an expected error rate of $1-1/K$. If for some reason you do choose to use Equation (9.22) to compute an upper bound for the error rate, you should check it against the naive bound to be safe.

Simple Algorithms for Mutual Information

In this section we explore several of the fundamental algorithms used to compute mutual information. Later we will see how these can be modified and incorporated into sophisticated practical algorithms.

Equation (9.16) on page 435 is the standard definition of mutual information, although it is perfectly legitimate, and occasionally more efficient, to use any of the identities that preceded this equation. The file MUTINF_D.CPP contains a function that implements this definition. Here is a code snippet from this file, slightly modified for clarity:

```
/*
  Compute the marginals and the counts in the nbins_x by nbins_y grid
*/

  for (i=0; i<nbins_y; i++)
    marginal_y[i] = 0;

  for (i=0; i<nbins_x; i++) {
    marginal_x[i] = 0;
    for (j=0; j<nbins_y; j++)
      grid[i*nbins_y+j] = 0;
    }

  for (i=0; i<ncases; i++) {
    ix = bins_x[i];
    iy = bins_y[i];
    ++marginal_x[ix];
    ++marginal_y[iy];
    ++grid[ix*nbins_y+bins_y[i]];
    }

/*
  Compute the mutual information
*/

  MI = 0.0;

  for (i=0; i<nbins_x; i++) {
    px = (double) marginal_x[i] / (double) ncases;
```

```
for (j=0; j<nbins_y; j++) {
  py = (double) marginal_y[j] / (double) ncases;
  pxy = (double) grid[i*nbins_y+j] / (double) ncases;
  if (pxy > 0.0)
    MI += pxy * log (pxy / (px * py));     // Eq 9.16
  }
}
```

This algorithm assumes that the data is discrete. What if one or both of the variables are continuous? We saw on page 425 that the best way to partition a continuous variable to compute its entropy is to divide its range into bins based on equal spacing. This type of partitioning can produce unusually dense as well as unusually sparse bins, which is exactly what we want when we are estimating entropy. But for estimating mutual information, we would like the bin counts to reflect the relationship between the variables, rather than the marginal distributions. In the ideal situation, the marginal distribution of both variables would be uniform (all marginal bins would have equal counts) so that the counts in the grid represent the relationship between the variables to the maximum degree possible. This leads to a simple yet reasonably effective algorithm for computing the mutual information of a pair of continuous variables, or a continuous variable and a discrete variable. Later, on page 463, we will see a superior method for the case of two continuous variables. But for quick-and-dirty use, or for the case of one variable being continuous and one being discrete, equal-marginal partitioning is useful.

To this end, I have an automated partitioning algorithm that I use in my own work. I do not guarantee that it is optimal in any particular sense, largely because there are numerous competing definitions of optimality for partitions. On the other hand, it has always behaved well for me. In particular, if you specify a desired number of bins that is at least as large as the number of different values of the variable, it will return the actual number of bins and create a single bin for each different value. Also, if the variable has few or no ties and you specify a bin count that is small relative to the number of cases, it will compute bins whose counts are approximately or exactly equal. Finally, if the variable is continuous but has numerous ties, it will group cases into bins in a way that makes sense and seems to work well. The function is called as follows:

```
void partition (
  int n,            // Input: Number of cases in the data array
  double *data,     // Input: The data array
```

```
int *npart,          // Input/Output: Number of partitions to find; Returned as
                     // actual number of partitions, which happens if massive ties
double *bnds,        // Output: Upper bound (inclusive) of each partition
short int *bins      // Output: Bin id (0 through npart-1) for each case
)
```

The first step is to copy the data and sort it into ascending order. We need to preserve the indices of the original points, as we will need this information to assign cases to bins as the last step. Also, compute an integer array of ranks to identify ties. This is not strictly necessary, as we could simply use the floating-point data. But integer comparisons can be much faster than real comparisons on some hardware, which could make a difference for huge arrays.

```
for (i=0; i<n; i++) {
   x[i] = data[i];               // Copy the data for sorting
   indices[i] = i;               // Indices will be preserved here
   }

qsortdsi (0, n-1, x, indices);   // Sort ascending, also moving indices

ix[0] = k = 0;                   // Compute ranks, including ties

for (i=1; i<n; i++) {
   if (x[i] - x[i-1] >= 1.e-12 * (1.0 + fabs(x[i]) + fabs(x[i-1])))      // Check for effective tie
      ++k;      // If not a tie, advance the counter of unique values
   ix[i] = k;
   }
```

Compute an initial set of equal-size bins, ignoring ties for now. If there are no ties, this is all we need to do.

```
k = 0;                   // Will be start of next bin up
for (i=0; i<np; i++) {   // For all partitions
   j = (n - k) / (np - i);   // Number of cases in this partition
   k += j;               // Advance the index of next one up
   bin_end[i] = k-1;     // Store upper bound of this bin
   }
```

Iteratively refine the bin boundaries until no boundary splits a tied value into different bins. Note that the upper bound of the last partition is always the last case in the sorted array, so we don't need to worry about it splitting a tie, as there are no cases above it. All we care about are the np–1 internal boundaries. Each iteration does two things. First, it removes the first splitting bound that it finds. Then it attempts to replace this lost bound by inserting a new bound in a sensible way.

```
for (;;) {                      // Iterate until no ties are split across a boundary

  tie_found = 0;                // Flags if we found a split tie

  for (ibound=0; ibound<np-1; ibound++) {           // Check all boundaries
    if (ix[bin_end[ibound]] == ix[bin_end[ibound]+1]) {   // Splits a tie?

      // This bound splits a tie. Remove this bound.
      for (i=ibound+1; i<np; i++)
        bin_end[i-1] = bin_end[i];
      --np;                     // We just lost a bound
      tie_found = 1;            // Flag that we found a split tie and fixed it
      break;                    // Just remove one bad bound at a time
      }
    } // For all bounds, looking for a split across a tie

  if (! tie_found)              // If we got all the way through the loop
    break;                      // without finding a bad bound, we are done

  // The offending bound is now gone. Try splitting each remaining
  // bin. For each split, check the size of the smaller resulting bin.
  // Choose the split that gives the largest of the smaller.
  // Note that np has been decremented, so now np < *npart.

  istart = 0;
  nbest = -1;

  for (ibound=0; ibound<np; ibound++) {  // Check all bounds
    istop = bin_end[ibound];             // End of this bin

    // Now processing a bin from istart through istop, inclusive
    for (i=istart; i<istop; i++) {       // Try all possible splits of this bin
      if (ix[i] == ix[i+1])              // If this splits a tie
        continue;                        // Don't check it
```

```
        nleft = i - istart + 1;          // Number of cases in left half
        nright = istop - i;              // And right half

        if (nleft < nright) {            // If the left half is smaller
          if (nleft > nbest) {           // Keep track of the max
            nbest = nleft;               // This is the best so far
            ibound_best = ibound;        // And its base bound
            isplit_best = i;             // Its location in the base bin
            }
          }

        else {                           // Ditto when right half is smaller
          if (nright > nbest) {
            nbest = nright;
            ibound_best = ibound;
            isplit_best = i;
            }
          }
        }

      istart = istop + 1;                // Move on to the next bin
      } // For all bounds, looking for the best bin to split

    // The search is done. It may (rarely) be the case that no further
    // splits are possible. This will happen if the user requests more
    // partitions than there are unique values in the dataset.
    // We know that this has happened if nbest is still -1. In this case
    // we (obviously) cannot do a split to make up for the one lost above.

    if (nbest < 0)                       // If no further splits are possible
      continue;                          // Then don't do it!

    // We get here when the best split of an existing partition has been
    // found. Save it. The bin that we are splitting is ibound_best,
    // and the split for a new bound is at isplit_best.
```

```
for (ibound=np-1; ibound>=ibound_bes t; ibound--) // Move up old bounds
    bin_end[ibound+1] = bin_end[ibound];        // To make room for new one
bin_end[ibound_best] = isplit_best;             // The new split
++np;                                           // Count it

} // Endless search loop
```

At this point, the partitioning is complete. Return the bounds to the user. Also return the bin membership of each case.

```
*npart = np; // Return the final number of partitions
for (ibound=0; ibound<np; ibound++)
    bnds[ibound] = x[bin_end[ibound]];

istart = 0;                                     // The current bin starts here
for (ibound=0; ibound<np; ibound++) {           // Process all bins
    istop = bin_end[ibound];                    // Inclusive end of this bin
    for (i=istart; i<=istop; i++)
        bins[indices[i]] = (short int) ibound;
    istart = istop + 1;
}
```

The TEST_DIS Program

The file TEST_DIS.CPP is a program that illustrates the techniques discussed so far. It allows the user to specify properties for a pair of variables, and then it generates random datasets having the specified properties and computes mutual information and some related measures. This program is for demonstration and exploration only. Later in this chapter we will present a program that reads actual datasets and processes them. The TEST_DIS program is invoked by typing its name followed by five parameters:

TEST_DIS nsamples ntries type parameter ptie

- *nsamples*: Number of cases in the dataset

- *ntries*: Number of Monte Carlo replications

- *type*: Type of test

  - 0=bivariate normal with specified correlation

  - 1=discrete bins with uniform error distribution

  - 2=discrete bins with triangular error distribution

  - 3=discrete bins with cyclic error distribution

  - 4=discrete bins with attractive class error distribution

- *parameter*: Depends on type of test

  - 0: Correlation

  - >0: Error probability

  - *ptie*: If type=0, probability of a tied case, else ignored

The bivariate normal test generates two normally distributed random variables having the specified correlation. These continuous variables are partitioned into bins using the partition() subroutine presented in the prior section. Mutual information estimates will be strongly biased upward if more than a few bins are used because of randomly occurring false relationships. Therefore, this test is of little practical use other than clearly demonstrating this upward bias. In the next section, we will see a vastly superior method of computing mutual information for continuous variables.

All other tests generate a confusion matrix having the specified error probability. The uniform error test distributes the misclassifications to all erroneous bins with equal probability. The triangular test places most of the errors in the upper triangle. The cyclic test places the errors in a nearby class. The attractive test favors one or two unnaturally attractive classes. These all represent different types of model failure. Full details of the error distributions can be found in the source code.

A variety of numbers of bins are tested, depending on the number of cases that the user wants for each sample. The tests are repeated ntries times. For each test, it is possible to compute the theoretically correct mutual information. This enables the program to keep track of the bias and standard error of the mutual information estimates. It also computes loose and tight lower and upper bounds for misclassification error. The tight lower bounds are based on Equation (9.20) and the tight upper bounds on Equation (9.22). The loose lower bound is obtained by subtracting $h(0.5)=\log(2)$ in the numerator, as described on page 439, and the loose upper bound is obtained by not subtracting anything. The means of these bounds are computed across replications of the test. The program also counts how often the true value of the error rate falls outside the computed bounds. This demonstrates how the nature of the model's error distribution affects the width and quality of the bounds.

Table 9-3. *Some Results from the TEST_DIS Program*

	True	Est	Bias	StdE	|	Loose	|	|	Tight	|
1	2.85	2.80	0.05	0.06	0.00	0.24	0.08	0.11		
2	2.88	2.84	0.04	0.04	0.00	0.51	0.08	0.25		
3	3.07	3.07	0.00	0.01	0.00	0.66	0.02	0.11		
4	3.04	3.04	0.00	0.01	0.00	0.97	0.01	0.97		

Table 9-3 shows the results from four runs of the TEST_DIS program. In all tests, 10,000 cases were in each samples, the error rate was set at 0.1, and 32 bins were used. Observe that in all four scenarios, the estimated mutual information was very close to the true value, and the standard error of the estimate was only slightly greater than the bias, indicating that the estimates were very stable.

The loose error bounds, supposedly bounding the true value of 0.1, were universally worthless, being ridiculously wide. The tight bounds were very good (bounding 0.1 between 0.08 and 0.11) for the well-behaved model that had uniformly distributed errors. They deteriorated badly, though in different directions, for the triangular and cyclic error distributions. For a model with an attractive class, both the lower and upper bounds were totally worthless. Not shown in this table is the fact that the computed bounds never failed to enclose the true error rate, which at least is some consolation and great news for the well-behaved model.

The discussion of the TEST_DIS program is necessarily brief here. Careful study of the source code will show how the theoretical mutual information is computed, along with error bounds. Also, calling methods for the functions discussed earlier in the chapter are demonstrated.

Continuous Mutual Information

Near the beginning of this chapter we saw that entropy is fundamentally a property of finite discrete random variables, those that can take on only a finite number of fixed values. Entropy can be extended to continuous random variables by replacing summation with integration, but the continuous analog of entropy is of dubious worth in practical applications. Luckily, the situation is considerably better when it comes to mutual information. In prior sections we saw how the partition() function or something similar could be used to discretize a continuous variable into bins, and then the discrete mutual information could be computed from the bin counts. If both random variables are continuous, there are much better ways of estimating their mutual information, which is defined in Equation (9.23). (Note that if one variable is continuous and one is discrete, as would be the case when predicting a class based on a continuous predictor, the recommended method is to discretize the continuous variable into equal-sized bins and compute discrete mutual information.)

$$I(X;Y) = \iint f_{X,Y}(x,y) \log \frac{f_{X,Y}(x,y)}{f_X(x)f_Y(y)} dx dy \qquad (9.23)$$

One beautiful aspect of Equation (9.23) is that it is immune to transformations of the variables. Suppose $g(.)$ and $h(.)$ are one-to-one continuous differentiable functions over the domain of x and y, respectively. Let $x'=g(x)$ and $y'=h(y)$. Then $I(x;y)=I(x';y')$. This is in sharp contrast to continuous entropy, which is not even immune to linear rescaling, let alone nonlinear transformation. An immensely useful corollary of this property is that observed values of the variables can be transformed to ranks or to any predefined distribution prior to computing their mutual information. This simplifies and stabilizes numerical algorithms.

The Parzen Window Method

To use Equation (9.23), we need to know the joint and marginal density functions, $f_{X,Y}(.)$, $f_X(.)$, and $f_Y(.)$. Naturally, we almost never have any knowledge of these functions other than what our data sample provides. In most cases we aren't even willing to assume a functional form such as normality. The most common way of handling this difficult situation is to use a *Parzen window approximation*.

The intuition behind a Parzen window is that areas of the domain in which the probability density is large will manifest this in the data sample by the appearance of many cases in this area. Similarly, if the probability density is small in some area of the domain, few or no cases from this area will appear in the sample. This leads to a generalized binning of the samples. Instead of defining strict boundaries for bins and counting how many cases fall into each bin, we define a weighting function, a movable window that spans the sample. When we want to compute the probability density at some point in the domain, we center the window at that point and compute a weighted sum of the cases nearby. Cases that are close to the domain point receive a large weight, while further cases receive a small weight. Very distant cases receive no weight at all. This technique is called the method of *Parzen windows*, after its inventor.

The density approximation is simple for the one-dimensional case, which covers the marginal distributions. Let the sample values be $x_1, x_2, ..., x_n$. Assume that we have a weighting function $W(d)$, which should be large when d is near zero, and become smaller as d moves away from zero. Let σ be a scale factor. Then the Parzen density approximation is given by Equation (9.24).

$$f(x) = \frac{1}{n\sigma} \sum_{i=1}^{n} W\left(\frac{x - x_i}{\sigma}\right)$$

(9.24)

It should be clear that if the argument x has numerous cases nearby, the sum will be relatively large. Conversely, if there are no cases near x, the sum will be small. This is exactly what we want. The scale factor, sigma, determines the width of the window. If it is small, implying a narrow window, only cases in the immediate vicinity of x will impact the sum. If sigma is large, even distant cases will have an effect on the estimated density.

Parzen (1962) and Specht (1990a) provide a rigorous description of the properties that $W()$ must have in order for the Parzen method to be an effective density estimator. Here, we say only that these properties are reasonable: $W()$ must be bounded, go to zero rapidly as the argument goes away from zero, and integrate to unity (which is a fundamental property of a density function). The weight function favored by many is the ordinary Gaussian function of Equation (9.25).

$$W(d) = \frac{1}{\sqrt{2\pi}} e^{-d^2/2}$$

(9.25)

The Parzen density estimator is easily generalized to more than one dimension, as shown in Equations (9.26) and (9.27).

$$f(x_1,\ldots,x_p) = \frac{1}{n\sigma_1\ldots\sigma_p}\sum_{i=1}^{n}W\left(\frac{x_1 - x_{1,i}}{\sigma_1}, \ldots, \frac{x_p - x_{p,i}}{\sigma_p}\right) \tag{9.26}$$

$$W(d_1,\ldots,d_p) = \frac{1}{(2\pi)^{p/2}}e^{-\frac{1}{2}\sum_{1}^{p}d_i^2} \tag{9.27}$$

The file PARZDENS.CPP contains complete source code for computing Parzen density estimators in one, two, and three dimensions. Here we examine only a few snippets, modified for clarity when necessary, that illustrate the ideas just presented.

One aspect of the supplied code must be emphasized. Mutual information via the Parzen window method tends to be most stable when the variables have at least roughly normal distributions. For this reason, the Parzen window code applies a universal normalization transform before computing the density. The implication is that these routines *cannot* be used for general density computation. They are intended to be used only when integrating Equation (9.23), the definition of continuous mutual information. If you want to use them for other applications, you must remove the normalization code and compute the scale factor appropriately.

To estimate a normalized Parzen density in one dimension, create a ParsDens_1 object. The constructor header looks like this:

```
ParzDens_1::ParzDens_1 (
    int nd,            // Number of data points
    double *tset,      // The data array
    int div)           // Resolution divisor
```

The constructor first transforms the input data to a normal distribution. This is a standard statistical algorithm. To transform a dataset to a given distribution, first compute the cumulative distribution function (*CDF*) of the data and then map each point to the inverse CDF of the desired distribution. The sorting algorithm qsortdsi() swaps the indices along with the data.

```
for (i=0; i<nd; i++) {
    indices[i] = i;
    d[i] = tset[i];
    }
```

```
qsortdsi (0, nd-1, d, indices);

for (i=0; i<nd; i++)
   d[indices[i]] = inverse_normal_cdf ((i + 1.0) / (nd + 1));
```

The sigma scale factor in Equation (9.24) is represented by std in the code. It is equal to 2.0 divided by the user's specified resolution, div. The private variable var will be used in the density computation later. The integration routine will need to know the complete practical range of the variable. Since we know that the data now follows a standard normal distribution, it is trivial to compute these limits. Finally, we compute the normalizing factor of Equations (9.24) and (9.25) so that the function integrates to unity, an essential property of a density. The code to do all this is as follows:

```
std = 2.0 / div;
var = std * std;
high = 3.0 + 3.0 * std;
low = -high;
factor = 1.0 / (nd * sqrt (2.0 * PI * var));
```

If there are numerous data points, which is the rule in practice, the summation in Equation (9.24) is slow. For this reason, the code only uses Equation (9.24) when nd is small. For large values, the constructor evaluates the density using Equation (9.24) for a reasonable number of points, and then it constructs a cubic spline interpolating function. This spline is used in future calls to the density evaluation function. Since integration involves a huge number of function calls, the savings is enormous. The spline code is tedious and uninteresting, so it will not be discussed here. See PARZDENS.CPP and SPLINE.CPP for details.

After the constructor has been called, the density (in the normalized domain, not the original domain) is estimated by calling the density() member function. It uses the spline approximation, or it implements Equation (9.24) directly.

```
sum = 0.0;
for (i=0; i<nd; i++) {
   diff = x - d[i];
   sum += exp (-0.5 * diff * diff / var);
   }

return sum * factor;
```

The two-dimensional Parzen density code is a straightforward extension of the one-dimensional code, so it will not be shown here. It, too, uses interpolation to save time with large datasets. In this case, bilinear interpolation with quadratic extension is used. See PARZDENS.CPP and BILINEAR.CPP for details.

To compute the mutual information of a pair of variables using the Parzen window method, first create a MutualInformationParzen object. The constructor header and the most important line of code looks like this:

```
MutualInformationParzen::MutualInformationParzen (
   int n,                  // Number of cases
   double *depvals,    // They are here
   int div)                // Number of divisions, typically 5-10
{
dens_dep = new ParzDens_1 (n, depvals, div);
}
```

One of the two variables is supplied to the constructor. It is called depvals in the code, even though the inherent symmetry of mutual information means that there is no distinction between dependent and independent variables. The reason for this naming and for supplying one variable to the constructor is that this routine will often be used for evaluating the mutual information between a dependent variable and each of a set of candidates for independent variable. By doing as much processing as possible in the constructor, we avoid redundant computation later.

When we want to compute the mutual information between the dependent variable and a candidate predictor, the member function mutinf() is called. Its essential code, modified for clarity, is as follows:

```
this_dens_dep = dens_dep;

this_dens_trial = new ParzDens_1 (n, x, div);
this_dens_bivar = new ParzDens_2 (n, depvals, x, div);

criterion = integrate (this_dens_trial->low, this_dens_trial->high,..., outercrit);
```

The variables that start with this are statics local to the module, used to pass their data to local functions that the generic integration routine integrate() calls. This code does very little. It creates a univariate Parzen density for the candidate variable and a bivariate Parzen density for both variables. It then integrates outercrit() over the range of the candidate variable.

The real work of the algorithm is in the integration criterion routines outercrit() and innercrit(). These make up the integrand of Equation (9.23) and demonstrate a standard technique for double integration. The outer criterion, which is integrated over the range of the trial variable as shown in the prior code, itself integrates the inner criterion over the range of the dependent variable. The inner criterion needs both variables, as well as the density of the trial variable, so the two statics make it easy to pass this information from the outer criterion to the inner.

```
static double this_x, this_px; // Needed for two-dimensional integration

double outer_crit (double t)
{
  double val, high, low;

  high = this_dens_dep->high;
  low = this_dens_dep->low;
  this_x = t;
  this_px = this_dens_trial->density (this_x);
  val = integrate (low, high,..., inner_crit);
  return val;
}

double inner_crit (double t)        // Integrand of Equation (9.23)
{
  double py, pxy, term;
  py = this_dens_dep->density (t);
  pxy = this_dens_bivar->density (t, this_x);
  term = this_px * py;              // Denominator
  if (term < 1.e-30)               // Prevent dividing by zero
    term = 1.e-30;
  term = pxy / term;               // Will take log of this
  if (term < 1.e-30)               // Prevent taking log of zero
    term = 1.e-30;
  return pxy * log (term);
}
```

The code shown here is slightly different from the code on my web site. In addition to a few changes that clarify operation, there is a difference related to the fact that the Parzen code supplied with this text converts the data to a normal distribution. Since this is the case, it is both inefficient and slightly (though not seriously) inaccurate for the inner and outer criteria to use a one-dimensional Parzen window for the marginal distributions. We already know that they are normal, so the code on my web site replaces the Parzen window with direct evaluation of the standard normal density. Comments to this effect appear in the code. This is so that to experiment you can easily switch back and forth between the two methods.

Thus far we have conveniently pushed aside the issue of the scaling factor, sigma in Equations (9.24) and (9.26) and std in the code for the Parzen density. This is not a trivial issue. In fact, it is such a serious issue that many people avoid using Parzen windows to approximate mutual information. There are other algorithms, such as the excellent adaptive partitioning method shown in the next section. However, Parzen windows have a place in a complete toolbox. When the dataset contains just a few cases, perhaps several dozen, other methods are severely compromised. In this situation, a wide window will capture most of the important information in the distribution without running an inordinate risk of confusing random variation with true mutual information. Also, despite the fact that an excessively wide window will bias the computed mutual information downward, while an excessively narrow window will bias it upward, this bias will be reflected nearly equally in all candidate predictors. So if the purpose of computing mutual information is to evaluate the relative quality of predictor candidates, the ranking of the candidates will be only minimally impacted by the widow width, especially if the width is on the large side of optimal.

How does one choose a good window width? Ideally, one has software that plots a histogram with the Parzen density overlaid. By trying several different window widths, one can easily find the value that best captures the essence of the distribution. See, for example, Figures 9-4 through 9-7 on the next page. In the absence of such a tool, a decent rule of thumb for the Parzen window software supplied with this text is to use a division factor of about five for very small samples, ten if the sample contains several hundred cases, and 15 if there are more than a thousand cases.

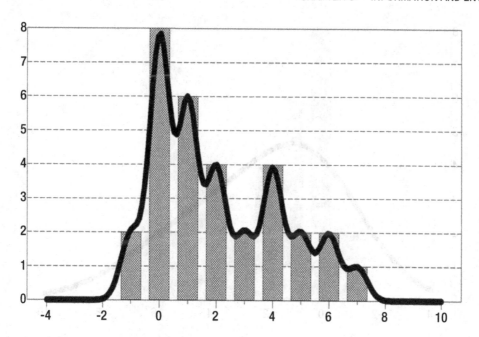

Figure 9-4. *Sigma is much too small*

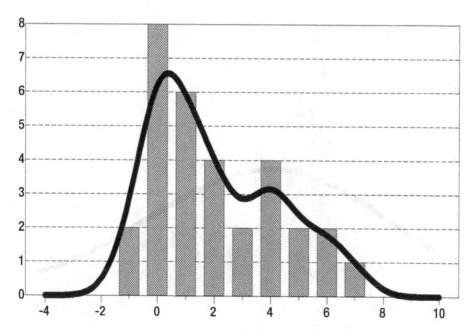

Figure 9-5. *Sigma is on the small side of optimal*

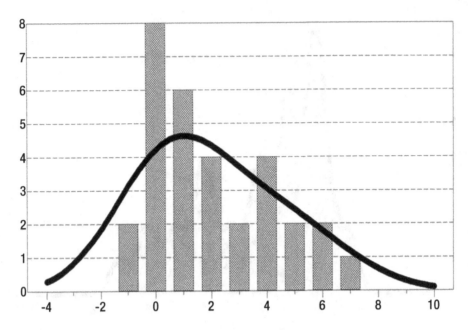

Figure 9-6. *Sigma is on the large side of optimal*

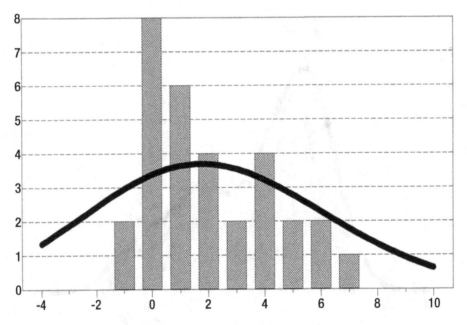

Figure 9-7. *Sigma is much too large*

Adaptive Partitioning

This section describes what is probably the best general-purpose algorithm for estimating the mutual information of two continuous variables. It is considerably more complex than the Parzen-window method just described, but the complexity is worthwhile. The algorithm is conceptually elegant and widely effective in practice. It also avoids the need to tweak a fussy parameter, which we must do for the Parzen window. It does involve two tunable parameters, but the algorithm is remarkably insensitive to their values, so in practice having to set these two parameters is almost never a problem.

Recall that the naive way to compute the mutual information of a pair of continuous variables is to partition the bivariate space into a checkerboard of bins by defining boundaries for each marginal distribution and then plugging the bin counts into the discrete formula for mutual information. This was discussed on page 447. The problem with the naive method is that it pays too much attention to areas of the bivariate domain that have few or no cases, while perhaps paying too little attention to dense areas where most of the information lies. The algorithm on page 447 partially solves this problem by at least ensuring that the marginals have equal sized bins. But it is nice to extend this property to two dimensions.

Figure 9-8 is a contour plot of the bivariate density of a pair of variables. Most cases lie in a J-shaped cluster, with fewer cases around the perimeter of the main pattern. No cases lie in the white areas. It should be obvious that if one were to divide this bivariate space into, say, 20 divisions for each variable, most of the 20*20=400 bins would be empty. This leads to serious problems with bias and error variance in the mutual information estimate.

[Darbellay and Vajda, 1999] present a beautiful algorithm that adaptively partitions the bivariate space in such a way that attention is focused on areas of high density. They also demonstrate that for a variety of distributions, their algorithm has much less error than naive algorithms.

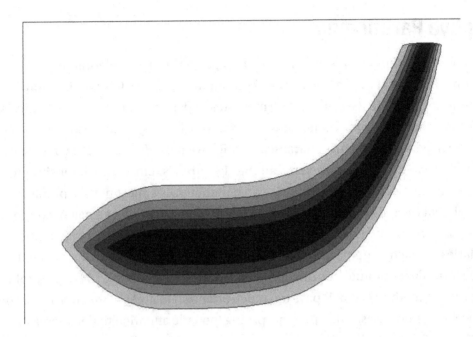

Figure 9-8. *A bivariate distribution*

Look at Figure 9-9. It shows the distribution of Figure 9-8 partitioned into a two-by-two grid. The upper-left block is empty, so it can be ignored. Each of the remaining three blocks is partitioned into a two-by-two grid, as shown in Figure 9-10. Two more blocks can be eliminated, one because it is empty and one because it is nearly empty. Partitioning again gives us Figure 9-11, in which several more blocks are eliminated. It should be apparent that eventually the entire focus will be on areas of support for the density.

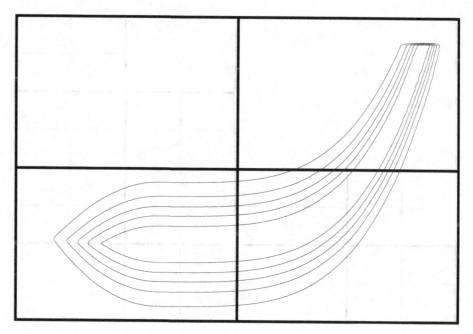

Figure 9-9. *First partitioning*

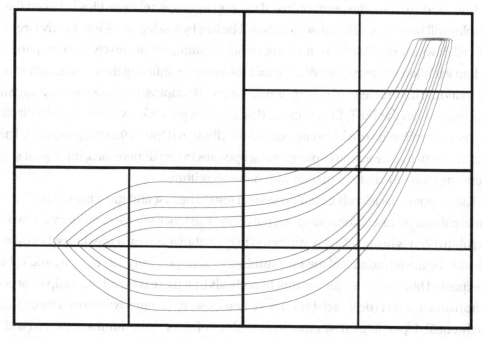

Figure 9-10. *Second partitioning*

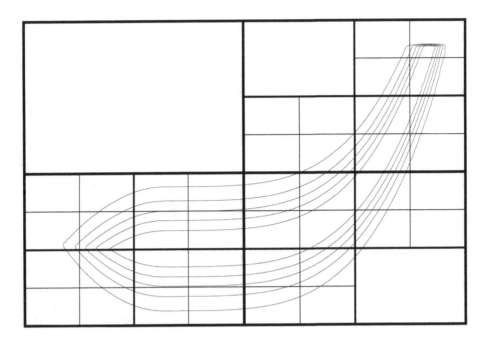

Figure 9-11. *Third partitioning*

How far do we take the partitioning? If we stop too soon, relationships between the two variables will be obscured because details will be lost by tossing cases into overly large bins. This will downwardly bias the mutual information estimate. Conversely, if we stop too late, random variation will masquerade as actual information, inflating the estimate of the mutual information. This problem, of course, is not unique to adaptive partitioning. Anyone who experiments with the TEST_DIS program, discussed on page 451, will see it vividly displayed with naive partitioning of a bivariate normal distribution (Option 0 in the program). The big difference is that since adaptive partitioning operates in two dimensions, intelligent stopping criteria are easier to implement than with naive algorithms.

The stopping decision is based on several tests. The first and most important is a simple chi-square test of the upcoming partition. The block whose candidacy for two-by-two subdivision is being tested is subjected to the subdivision on a trial basis. Let n_1, n_2, n_3, and n_4 be the bin counts of the four subdivisions, respectively. Let e_1, e_2, e_3, and e_4 be the expected bin counts under the null hypothesis that there is no relationship between the horizontal and vertical variables. These four expected counts will be exactly or almost exactly equal depending on whether the numbers of rows and columns are even (and hence exactly splittable in half) or odd (an exact split in half cannot be done). If the two variables are unrelated, the observed bin counts will equal the expected bin counts except for random variation. But if there is a relationship between the two variables, the counts will be skewed away from their expected values, with some bin being favored at the expense of another. The standard two-by-two chi-square test statistic is shown in Equation (9.28).

$$X^2 = \sum_{i=1}^{4} \frac{\left(\left|n_i - e_i\right| - 0.5\right)^2}{e_i}$$

(9.28)

If this test statistic fails to exceed the threshold corresponding to a small significance level, we conclude that the trial subdivision is probably pointless. However, it is possible that there really is a deterministic skewing of the data in the enclosing block, but a simple two-by-two subdivision fails to pick it up. This does not happen often, but it is still worth considering. For this reason, if the two-by-two chi-square test fails to detect a nonrandom distribution and if the enclosing block is relatively large, we subdivide into a four-by-four set of blocks and perform a chi-square test. If this test also fails to detect a nonrandom data distribution, we conclude that nothing is to be gained by subdividing the enclosing block, compute its contribution to the total mutual information, and henceforth ignore it.

But if either the original two-by-two chi-square test or the subsequent four-by-four test determines that the enclosing block is not uniform, we partition it into four smaller blocks. We check the size of each of these smaller blocks. If it is tiny, we compute its contribution to the total mutual information and declare that block finished. If it is still large enough for possible future splitting, we push it onto a stack of blocks to be explored, and continue processing.

When a block is determined to be finished, whether because it is small or because it is uniform, its contribution to the total mutual information is computed by using a discrete approximation to Equation (9.23) on page 454. This is shown in Equation (9.29), in which p_x is the fraction of the X marginal distribution encompassed by the X dimension of the block, p_y is the fraction of the Y marginal distribution encompassed by the Y dimension of the block, and p_{xy} is the fraction of the bivariate distribution encompassed by the area of the block.

$$MI\ Contribution = p_{xy} \log \frac{p_{xy}}{p_x p_y}$$

(9.29)

We will soon present a detailed discussion of the code that implements adaptive partitioning. But since it is quite complex, we begin with a simplified statement of the algorithm. Note that the code includes an optional provision to prevent splitting across tied data. It is senseless to define a subdivision in which some cases land on one side of the trial partition while other cases whose value on the variable are equal lie on the other side. It makes more sense to place all equal values on the same side of the boundary. However, truly continuous data will never have any ties, and this provision adds to the already severe complexity of the algorithm. For these reasons, the simplified statement here will ignore ties. The topic will be covered in the discussion of the code. The simplified algorithm is as follows:

Convert the data (*n* cases) to ranks.

Initialize *nstack*=1. This is the number of rectangles on the to-do stack. Also initialize this one stack entry to be the entire dataset. *Nstack* will be decremented when a rectangle is popped from the stack, and incremented when a rectangle is pushed onto the stack.

While *nstack* > 0 {

    Pop a rectangle from the stack

    Compute the X and Y boundaries for splitting the rectangle 2-by-2

    Compute the expected and actual bin counts in each of the four sub-rectangles

    Perform a 2-by-2 chi-square test. Set the flag *splitable* to true if the test found a significant disparity in bins counts, else false.

    If *splitable* = false and the rectangle is big {
        perform a 4-by-4 chi-square test.
        If the test finds a significant disparity, set *splitable* true.
        }

    If *splitable* = true {

        For each of the four sub-rectangles {

            If this rectangle is not tiny {
                Push it onto the stack
                Rearrange rectangle indices to reflect this partitioning
                }

            Else {
                Use Equation (9.29) to evaluate this sub-rectangle's contribution
            }
        }
    }

Else {

    Use Equation (9.29) to evaluate this current rectangle's contribution

    }

}

The complete code to implement the adaptive partitioning algorithm can be found in the file MUTINF_C.CPP on my web site. This code is quite complex, especially since keeping track of the nested rectangles in an efficient manner is tricky. Therefore, we will break it down into sections, slightly simplifying as needed, and discuss it one small part at a time.

One of the two core components of the program is an array called indices. It is initialized to the integers 1 through n. As the algorithm progresses and rectangles are subdivided, this array will be shuffled. At any time, we can define a rectangular block by pointing to its starting and ending elements in this array. This lets us efficiently handle nesting of rectangles. For example, we may have an enclosing block that starts at element 50 of indices and ends at element 89. It may consist of four smaller blocks, defined by elements 50-59, 60-69, 70-79, and 80-89, respectively.

The other core component is a stack of rectangles to be processed. Each stack entry has the following six members:

        Xstart, Xstop: Starting and ending (inclusive) ranks of X in the rectangle

        Ystart, Ystop: Starting and ending (inclusive) ranks of Y in the rectangle

        DataStart, DataStop: Rectangle's starting and ending elements of indices

The program begins by converting each of the two variables to integer ranks. It also keeps track of tied values so that later we can avoid splitting tied cases into different partitions. Note that rather than testing for exact equality, we test for values that are nearly equal in terms of double precision. This is a good habit in most programming environments, although you are free to be strict if desired. Here is the code for the x variable. The other variable, y, is treated similarly.

```
for (i=0; i<n; i++) {
   work[i] = xraw[i];                    // Copy the data, as we will sort it
   indices[i] = i;                       // Preserve the original locations
   }
qsortdsi (0, n-1, work, indices);        // Sort ascending, also moving indices

for (i=0; i<n; i++) {
   x[indices[i]] = i;                    // We now have ranks
   if (i < n-1 && work[i+1] - work[i] < 1.e-12 * (1.0 + fabs(work[i]) + fabs(work[i+1])))
      x_tied[i] = 1;                     // This case is tied with one above
   else
      x_tied[i] = 0;
   }
```

To initialize, the indices array is set equal to the entire dataset, and one rectangle, the entire dataset, is placed on the to-do stack. The stack entries are inclusive, so the last index is n–1.

```
for (i=0; i<n; i++)                      // For the entire dataset
   indices[i] = i;                       // These are the case indices

stack[0].Xstart = 0;                     // Lowest X rank in this rectangle
stack[0].Xstop = n-1;                    // And highest
stack[0].Ystart = 0;                     // Ditto for Y
stack[0].Ystop = n-1;
stack[0].DataStart = 0;                  // Index into indices of the first case in the rectangle
stack[0].DataStop = n-1;                 // And the last case
nstack = 1;                              // This is the top-of-stack pointer: One item in stack
```

The mutual information will be cumulated in MI. The program loops over the same code, processing one rectangle at a time, as long as there is at least one rectangle on the stack. The first step in the loop is to pop the rectangle off the stack.

```
MI = 0.0;                                // Will cumulate mutual information here
while (nstack > 0) {                     // As long as there is a rectangle to do

   // Get the rectangle pushed onto the stack most recently
   --nstack;                             // Pop the rectangle off the stack
```

```
fullXstart = stack[nstack].Xstart;          // Starting X rank
fullXstop = stack[nstack].Xstop;            // And ending
fullYstart = stack[nstack].Ystart;          // Ditto for Y
fullYstop = stack[nstack].Ystop;
currentDataStart = stack[nstack].DataStart;  // The cases start here
currentDataStop = stack[nstack].DataStop;    // And end here
```

Compute the center of this rectangle in preparation for the two-by-two trial split. This center will be the rightmost (largest) index in the left (smaller rank) subrectangle. If this case happens to be tied with the next one up, we don't want to split here, as such a split would put tied cases on opposite sides of the partition. So we set a flag to indicate whether we have this problem. If not, we are done. But if this exact center is tied, we attempt to move it off center as little as possible, stopping as soon as we find a split that is not tied. In the pathological case that we never succeed, the tie flag remains set. We will check it later. This code is repeated for the y variable. Here we show only the x code:

```
centerX = (fullXstart + fullXstop) / 2;      // Exact center, the ideal boundary
X_AllTied = (x_tied[centerX] != 0);          // Does it happen to be tied here?
if (X_AllTied) {                             // If so, try to move it
  for (ioff=1; centerX-ioff >= fullXstart; ioff++) { // Try to keep the offset small
    if (! x_tied[centerX-ioff]) {            // If this is not tied
      X_AllTied = 0;                         // We succeeded, so reset flag
      centerX -= ioff;                       // The new boundary is here
      break;                                 // Done searching
      }
    if (centerX + ioff == fullXstop)         // Quit if we hit the edge
      break;
    if (! x_tied[centerX+ioff]) {            // Try the other direction
      X_AllTied = 0;
      centerX += ioff;
      break;
      }
    }
  }
```

If either variable happens to be entirely tied, an ideally rare condition, the rectangle is declared to be nonsplittable. Otherwise, we trivially compute the starting and stopping indices of the four subrectangles defined by the split. The expected bin count in each partition is the total bin count times the fraction of the total x side and times the fraction of the total y side. The actual count in each partition is computed by tallying the number of cases that lie on each side of each center bound.

```
if (X_AllTied || Y_AllTied)            // If either variable is entirely tied
   splitable = 0;                      // No sense trying to split
else {
   trialXstart[0] = trialXstart[1] = fullXstart; // The four sub-rectangles
   trialXstop[0]  = trialXstop[1]  = centerX;
   trialXstart[2] = trialXstart[3] = centerX+1;
   trialXstop[2]  = trialXstop[3]  = fullXstop;
   trialYstart[0] = trialYstart[2] = fullYstart;
   trialYstop[0]  = trialYstop[2]  = centerY;
   trialYstart[1] = trialYstart[3] = centerY+1;
   trialYstop[1]  = trialYstop[3]  = fullYstop;

   // Compute the expected count in each of the four sub-rectangles
   for (i=0; i<4; i++)
      expected[i] = (currentDataStop - currentDataStart + 1) *            // Total count
         (trialXstop[i]-trialXstart[i]+1.0) / (fullXstop-fullXstart+1.0) *  // X fraction
         (trialYstop[i]-trialYstart[i]+1.0) / (fullYstop-fullYstart+1.0);   // Y fraction

   // Compute the actual count in each of the four sub-rectangles
   actual[0] = actual[1] = actual[2] = actual[3] = 0;
   for (i=currentDataStart; i<=currentDataStop; i++) { // All cases in this rectangle
      k = indices[i];          // Index of this case
      if (x[k] <= centerX) {   // Is it on the left side?
         if (y[k] <= centerY)  // Is it in the top half
            ++actual[0];
         else
            ++actual[1];
      }
```

```
else {
  if (y[k] <= centerY)
    ++actual[2];
  else
    ++actual[3];
  }
}
```

Compute the two-by-two chi-square test. If the actual counts are sufficiently different from the expected counts, declare the rectangle worth splitting.

```
testval = 0.0;                              // Will cumulate test statistic here
for (i=0; i<4; i++) {                       // The four sub-rectangles
  diff = fabs (actual[i] - expected[i]) - 0.5;  // Equation (9.28)
  testval += diff * diff / expected[i];
  }

splitable = (testval > chi_crit) ? 1 : 0; // Does it exceed the criterion?
```

It may sometimes be the case that the rectangle really does have a nonuniform data distribution, but the cases happen to be roughly equally distributed among the four subrectangles. We can usually avoid this trap by splitting it into a four-by-four set of 16 partitions. Of course, this makes sense only if the rectangle contains more than just a few cases. I don't bother checking for ties in this finer split because it would greatly complicate the code and this is a fairly rare occurrence anyway. The decision from the two-by-two split is the final decision the vast majority of the time. Moreover, ties will never occur in truly continuous data, so handling ties is a moot point in many or most situations.

```
if (! splitable && fullXstop-fullXstart > 30 && fullYstop-fullYstart > 30) {
  ipx = fullXstart - 1;   // Will be last index of prior sub-rectangle
  ipy = fullYstart - 1;   // Used for computing X and Y fractions
  for (i=0; i<4; i++) {   // Find the four x and y boundaries in this loop
    xcut[i] = (fullXstop - fullXstart + 1) * (i+1) / 4 + fullXstart - 1;       // Rightmost limit
    xfrac[i] = (xcut[i] - ipx) / (fullXstop - fullXstart + 1.0);               // Fraction in X direction
    ipx = xcut[i];                                                            // For next pass
    ycut[i] = (fullYstop - fullYstart + 1) * (i+1) / 4 + fullYstart - 1;      // Ditto for Y
    yfrac[i] = (ycut[i] - ipy) / (fullYstop - fullYstart + 1.0);
    ipy = ycut[i];
    }
```

```
// Compute expected counts
for (ix=0; ix<4; ix++) {
  for (iy=0; iy<4; iy++) {
    expected[ix*4+iy] = xfrac[ix] * yfrac[iy] * (currentDataStop-currentDataStart+1);
    actual44[ix*4+iy] = 0;
    }
  }

// Compute actual counts
for (i=currentDataStart; i<=currentDataStop; i++) { // All cases in rectangle
  k = indices[i];           // Index of this case
  for (ix=0; ix<3; ix++) {  // Compare x to all three inner boundaries
    if (x[k] <= xcut[ix])   // Stop before we cross incorrect boundary
      break;
    }
  for (iy=0; iy<3; iy++) {  // Ditto for Y
    if (y[k] <= ycut[iy])
      break;
    }
  ++actual44[ix*4+iy];      // Tally the count
  }

// Compute the chi-square test
testval = 0.0;
for (ix=0; ix<4; ix++) {
  for (iy=0; iy<4; iy++) {
    diff = fabs (actual44[ix*4+iy] - expected[ix*4+iy]) - 0.5;
    testval += diff * diff / expected[ix*4+iy];
    }
  }
splitable = (testval > 22.0) ? 1 : 0; // Discrepancy on four-by-four test?
  } // If trying 4x4 split
} // Else not all tied
```

If the rectangle is to be split, we now process the four subrectangles. If they are not tiny, push them onto the stack for processing later. Also, preserve the indices of the enclosing rectangle, because we will need them for rearranging the indices to reflect the partition.

```
if (splitable) {                          // If we are to split it

  for (i=currentDataStart; i<=currentDataStop; i++) // Preserve its indices
    current_indices[i] = indices[i];            // for rearrangement soon

  ipos = currentDataStart;                   // Will rearrange indices starting here
  for (iSubRec=0; iSubRec<4; iSubRec++) { // Check all 4 sub-rectangles

    if (actual[iSubRec] >= 3) { // Big enough to push onto stack for further splitting?
      stack[nstack].Xstart = trialXstart[iSubRec];
      stack[nstack].Xstop = trialXstop[iSubRec];
      stack[nstack].Ystart = trialYstart[iSubRec];
      stack[nstack].Ystop = trialYstop[iSubRec];
      stack[nstack].DataStart = ipos;
      stack[nstack].DataStop = ipos + actual[iSubRec] - 1;
      ++nstack;
```

The current, enclosing rectangle runs from currentDataStart through currentDataStop in indices. Rearrange these indices so that the subrectangle that we just pushed has all of its cases together in a contiguous string. If we don't push any of the four, we don't need to worry about them because we will not be processing them in the future.

```
      if (iSubRec == 0) {                    // Upper-left sub-rectangle
        for (i=currentDataStart; i<=currentDataStop; i++) { // All cases in rectangle
          k = current_indices[i];                 // Index of this case
          if (x[k] <= centerX && y[k] <= centerY)       // Is it in upper-left?
            indices[ipos++] = current_indices[i];       // If so, move it
          }
        }

      else if (iSubRec == 1) {
        for (i=currentDataStart; i<=currentDataStop; i++) {
          k = current_indices[i];
          if (x[k] <= centerX && y[k] > centerY)
            indices[ipos++] = current_indices[i];
          }
        }
```

```
      else if (iSubRec == 2) {
        for (i=currentDataStart; i<=currentDataStop; i++) {
          k = current_indices[i];
          if (x[k] > centerX && y[k] <= centerY)
            indices[ipos++] = current_indices[i];
          }
        }

      else { // iSubRec == 3
        for (i=currentDataStart; i<=currentDataStop; i++) {
          k = current_indices[i];
          if (x[k] > centerX && y[k] > centerY)
            indices[ipos++] = current_indices[i];
          }
        }
      } // If this sub-rectangle is large enough to be worth pushing
```

If this subrectangle is tiny, there is no reason to push it for an attempt at splitting further. Just compute its contribution to the mutual information using Equation (9.29).

```
      else { // This sub-rectangle is small, so get its contribution now
        if (actual[iSubRec] > 0) { // It only contributes if it has cases
          px = (trialXstop[iSubRec] - trialXstart[iSubRec] + 1.0) / n;
          py = (trialYstop[iSubRec] - trialYstart[iSubRec] + 1.0) / n;
          pxy = (double) actual[iSubRec] / n;
          MI += pxy * log (pxy / (px * py));   Equation (9.29)
          }
        } // Else this sub-rectangle is too small to push, so process it
      } // For all 4 sub-rectangles
    } // If splitting
```

The only other possibility is that the enclosing rectangle failed both the two-by-two and the four-by-four chi-square tests, meaning that it was so uniform that it was not worth splitting. In this case, process it using Equation (9.29).

```
    else { // Else the chi-square tests failed, so we do not split
      px = (fullXstop - fullXstart + 1.0) / n;
      py = (fullYstop - fullYstart + 1.0) / n;
```

```
    pxy = (currentDataStop - currentDataStart + 1.0) / n;
    MI += pxy * log (pxy / (px * py)); // Equation (9.29)
    }
  } // While rectangles in the stack
```

This algorithm requires the user to specify only two parameters: the threshold for the two-by-two chi-square test and that for the four-by-four. The latter is so uncritical that the threshold is hard-coded into the routine. The former is only slightly critical. Values between about four and eight suffice in a wide variety of circumstances. I use a value of six in all of my work, and I find this value to be universally applicable.

The TEST_CON Program

The file TEST_CON.CPP contains a complete program that demonstrates how to call the routines for using Parzen windows and adaptive partitioning to estimate mutual information for continuous variables. It also lets the user compare the performance of the two methods. It repeatedly generates a bivariate normal dataset with specified correlation and uses both methods to estimate their mutual information. The bias and standard error of the estimates is displayed. Later in this chapter we will present a practical program for reading datasets and analyzing mutual information. The TEST_CON program is for demonstration and experimentation only. The program is invoked as follows:

```
TEST_CON nsamps ntries correl ptie nosplit ndiv chi
```

        nsamps: Number of cases in the dataset

        ntries: Number of Monte Carlo replications

        correl: Correlation, 0-1

        ptie: Probability of a tie, 0-1 (0 is generally recommended)

        nosplit: If nonzero, adaptive partitioning prevents splits across ties

        ndiv: Number of divisions for the Parzen window width

        chi: Two-by-two chi-square threshold for adaptive partitioning

Predictor Selection Using Mutual Information

At last we come to the *raison d'etre* for this chapter. We have learned what mutual information is, why it is important, and how to compute it. Now we learn how to use it intelligently to select predictor variables that are likely to be effective and to reject those that will probably prove worthless. This can be enormously valuable when you have a massive number of candidates and need to whittle this universe down to a manageable number.

Maximizing Relevance While Minimizing Redundancy

Let $X_1, X_2, ..., X_M$ be a set of predictor candidates for predicting Y. Given some $m<M$, we want to find m members of this collection such that the subset, which we call S, has maximum *joint dependency* with Y. Joint dependency is an extension of mutual information in which one of the quantities is a collection of random variables rather than a single random variable. We can think of the joint dependency as the mutual information between S and Y, $I(S;Y)$. For convenience, let S be the first m candidates. Then this joint dependency is given by Equation (9.30), a straightforward extension of Equation (9.23).

$$I(S;Y) = \int \cdots \int f_{S,Y}(x_1, ..., x_m, y)$$
$$\log \frac{f_{S,Y}(x_1, ..., x_m, y)}{f_S(x_1, ..., x_m) f_y(y)} dx_1, ..., dx_m dy \tag{9.30}$$

Unfortunately, in practice this quantity is impossible to compute for $m>2$ and is often difficult even for $m=2$. The reason is that the multiple integration involves implicitly or explicitly partitioning the dataset in more than two dimensions, leading to excessive thinning of the density approximations. Consider the simplest case of $m=2$. Suppose there are 1,000 cases. We have a rectangular checkerboard for the two predictors, and a stack of these checkerboards to accommodate Y. Each case will have a position in this three-dimensional cube. If we were to partition each dimension into ten bins, we would have $10^3 = 1000$ bins, leading to an average of just one case per bin. If $m=3$, there would be an average of one-tenth of a case per bin! Clearly, there is no hope of implementing the direct approach to finding the optimal subset S if $m>2$, and probably no hope even for $m=2$ unless there are an enormous number of cases. The density approximations that are critical to the integrand are simply too inaccurate.

There is another problem, too. Combinatoric explosion is a standard nemesis of any predictor selection algorithm. If we are choosing m of M candidates, there are $M!/(m!(M-m)!)$ possible combinations. This is often so large that trying all of them is out of the question. A shortcut is needed.

There are several shortcuts in use. The simplest and most common is *first-order incremental search*, more commonly called *forward stepwise selection*. One first chooses the single best predictor. Then one finds the predictor that, when combined with the first, produces the maximum increment in whatever performance criterion is being evaluated. A third is added in the same way, and so forth.

It is theoretically possible for this method to fail, perhaps miserably. Suppose, for example, that variables 21 and 35 together do a superb job of predicting Y, although neither alone is any good. Maybe variable 17 is the best single predictor, while variable 19 provides the best incremental power. These two variables together may not come even close to being as good as 21 and 35. This is sad but often unavoidable. Other techniques do exist. Higher-order methods keep not just the best variable at each step but several of the best, which increases the likelihood of finding the optimal set. Backward selection starts by using all candidates and removing one at a time. However, first-order incremental search is by far the most efficient, making it the only practical choice in any application in which computational resources are limited. This is the approach used here, not only because of its efficiency but because of a fortuitous property of the algorithm when applied to joint dependency.

[Peng, Long, and Ding, 2005] exploit this property, providing a variable selection algorithm that is simple, elegant, and almost miraculously duplicates first-order incremental optimization of Equation (9.30), without ever having to evaluate the equation. We now present an intuitive development of this algorithm.

The *relevance* of a set of predictors S to a predicted variable Y is defined as the mean mutual information between Y and each predictor in S. This is shown in Equation (9.31), where $|S|$ is the number of predictors in the set.

$$Relevance(Y, S) = \frac{1}{|S|} \sum_{X_i \in S} I(Y; X_i) \tag{9.31}$$

It is tempting to simply maximize this quantity. We would begin by selecting the single predictor that has maximum mutual information with Y. Then we add the candidate that has second-max mutual information, and so forth, until we have m predictors in S. This would obviously maximize the relevance of S.

The problem with this simplistic approach is that it ignores the fact that S chosen this way will usually contain an enormous amount of redundancy. If two variables have high mutual information with Y, chances are they also have high mutual information with each other. It will probably be the case that if we simply choose a new variable that has high mutual information with Y, appending it to S will not improve the joint dependency between S and Y very much, because it won't add much information that is new.

The algorithm of [Peng, Long, and Ding, 2005] solves this problem by choosing the next variable as the one having maximum value of its mutual information with Y, minus its redundancy with the existing set of predictors. The definition of redundancy is shown in Equation (9.32). Note that the redundancy of a predictor candidate with S is the same as the relevance of this candidate with S. The only difference is the name of the quantity. The term *relevance* is used when referring to the predicted variable, while *redundancy* is used when referring to another predictor candidate.

$$Redundancy\left(X_j;S\right)=\frac{1}{|S|}\sum_{X_i \in S}I\left(X_j;X_i\right) \tag{9.32}$$

In summary, the algorithm begins by choosing the single predictor that has maximum mutual information with Y. Let S be this one variable. From then on, we add one new variable at a time by choosing the one that maximizes the criterion shown in Equation (9.33), stopping when we have the desired number m of predictors in S.

$$Criterion\left(X_j;S\right)=I\left(X_j;Y\right)-\frac{1}{|S|}\sum_{X_i \in S}I\left(X_j;X_i\right) \tag{9.33}$$

This algorithm makes obvious intuitive sense. At each step we want to simultaneously maximize the mutual information with Y while minimizing the average mutual information with the predictors already in S. What is not at all obvious is that *this algorithm will choose exactly the same variables as would be chosen if we were able to evaluate Equation (9.30)*, something that we have already seen to be practically impossible. The proof can be found in the original paper. All we do here is marvel that we can capitalize on this extraordinary result.

The MI_DISC and MI_CONT Programs

The files MI_DISC.CPP and MI_CONT.CPP contain complete programs that read a dataset containing predictor variables and a predicted variable and implement the Peng, Long, and Ding (*PLD*) algorithm for stepwise predictor selection. The MI_DISC program is primarily intended for discrete variables, or continuous variables that can be naively discretized. The MI_CONT program is intended strictly for continuous variables. They are invoked as follows:

MI_DISC datafile nin depname bins_dep bins_indep maxkept

>> *datafile*: Name of the text file containing the data

>> *nin*: Number of independent (predictor candidate) vars

>> *depname*: Name of the dependent (predicted) variable

>> *bins_dep*: Maximum number of dependent variable partitions
>>> If specified as zero, two bins are defined (>0 and <=0)

>> *bins_indep*: Ditto, but for independent variables

>> *maxkept*: Stepwise will allow at most this many predictors

MI_CONT datafile nin depname ndiv maxkept

>> *datafile*: Name of the text file containing the data

>> *nin*: Number of independent (predictor candidate) vars

>> *depname*: Name of the dependent (predicted) variable

>> *ndiv*: Normally zero, to employ adaptive partitioning
>>> Specify 5 to 15 or so to use Parzen windows. This is slow!

>> *maxkept*: Stepwise will allow at most this many predictors

The data file is an ordinary ASCII text file having one record (line) per case. The first line of the file contains the variable names. The names must not contain spaces, as spaces will be interpreted as delimiters. The predictor variable candidates must precede the predicted variable. The first *nin* variables will be used as the predictor candidates. The predicted variable can be anywhere after these candidates. Spaces, commas, and tabs will be treated as delimiters.

The MI_DISC program can be used to handle discrete or continuous variables. The number of bins can be set separately for the predictors and predicted variable. If a variable is discrete and the specified number of bins exceeds the number of actual values, there will be no problem. The number of bins will be automatically reduced to the correct value, separately for each variable. So it is always safe to specify a huge number of bins, confident that the program will reduce the number appropriately. Also note that if the number of bins is specified as zero, the program will do a binary split, placing all positive values in one bin and all zero or negative values in the other bin.

A Contrived Example of Information Minus Redundancy

An example with synthetic data can clarify the concept of predictor selection based on mutual information minus redundancy. Five random variables named RAND0 through RAND4 are generated. Then SUM12 is defined as the sum of RAND1 and RAND2. SUM34 is similarly defined. Finally, these two sums are themselves summed to define SUM1234. Note that RAND0 is not used. The MI_CONT program produces the following output for predicting SUM1234 from the other seven variables:

```
Initial candidates, in order of decreasing mutual information

          Variable    Information
            SUM12      0.35980
            SUM34      0.35791
            RAND1      0.14820
            RAND3      0.14383
            RAND4      0.14023
            RAND2      0.13998
            RAND0      0.00000
```

Variables so far	Relevance	Redundancy	Criterion
SUM12	0.35980	0.00000	0.35980

```
Searching for an additional candidate...
```

Variable	Relevance	Redundancy	Criterion
RAND0	0.00000	0.00000	0.00000
RAND1	0.14820	0.47706	-0.32887
RAND2	0.13998	0.48002	-0.34004

RAND3	0.14383	0.00000	0.14383
RAND4	0.14023	0.00000	0.14023
SUM34	0.35791	0.00000	0.35791

Variables so far	Relevance	Redundancy	Criterion
SUM12	0.35980	0.00000	0.35980
SUM34	0.35791	0.00000	0.35791

Searching for an additional candidate...

Variable	Relevance	Redundancy	Criterion
RAND0	0.00000	0.00000	0.00000
RAND1	0.14820	0.23853	-0.09033
RAND2	0.13998	0.24015	-0.10016
RAND3	0.14383	0.23635	-0.09252
RAND4	0.14023	0.23796	-0.09774

Final set	Relevance	Redundancy	Criterion
SUM12	0.35980	0.00000	0.35980
SUM34	0.35791	0.00000	0.35791

Observe that SUM12 and SUM34, each of which makes up half of the variance of the predicted variable (SUM1234), have the most mutual information. The four individual components, RAND1 through RAND4, have less mutual information, and RAND0 has none at all.

SUM12 is picked first. When the remaining candidates are tested, RAND1 and RAND2 have high redundancy with the predictor already selected (no surprise there!). The other candidates have no redundancy.

SUM34 is picked next. In the search for a third predictor, RAND0 has neither mutual information nor redundancy with the current set of SUM12 and SUM34. The remaining candidates (RAND1 through RAND4) all have such high redundancy that their criterion is negative. The search stops, with SUM12 and SUM34 being chosen as the optimal predictor set for predicting SUM1234.

A Superior Selection Algorithm for Binary Variables

If the predicted variable and all predictor candidates are binary, (Fleuret, 2004) presents a stepwise selection algorithm that seems to be superior to the *PLD* algorithm. Recall that the *PLD* algorithm has the fabulous property that its selections are identical to those that would be obtained by forward stepwise selection based on the optimal but impossible Equation (9.30). Nonetheless, also recall that forward stepwise selection is itself suboptimal. The optimal method is to examine every possible combination of predictors, a task that is usually impractical, even if we could evaluate the criterion of Equation (9.30), which of course we cannot. So, there is room for improvement.

Actually, the Fleuret algorithm described in this section can theoretically be used for any discrete variables, not just binary. It's just that unless the number of cases is huge, the algorithm fails because of sparse bins. For this reason, it is typically implemented only for binary data.

We need to introduce the notion of conditional mutual information. Recall from Equation (9.13) on page 436 that the mutual information shared by two variables is equal to the entropy of one of them minus its entropy conditional on the other. This is shown in Equation (9.34). Intuitively, this means that the information shared by X and Y is equal to the information in Y minus the information content of Y that is above and beyond that provided by X. Equivalently, the total information in Y is equal to that which is shared with X plus that which is above and beyond X.

$$I(X;Y) = I(Y;X) = H(Y) - H(Y|X) \qquad (9.34)$$

Now suppose that we already possess some information in the form of the value of variable Z. We can then talk about the mutual information of X and Y given that we know Z, written as $I(X;Y|Z)$. If Z happens to be totally unrelated to X and Y, its knowledge will have no impact on the mutual information of X and Y. At the other extreme, it may be that X and Y share a lot of information, but Z happens to completely duplicate this shared information. In this case, $I(X;Y)$ will be large, but $I(X;Y|Z)$ will be zero. Conditional mutual information can be computed with Equation (9.35). Observe that this is a simple extension of Equation (9.34), obtained by conditioning all terms on Z.

$$I(X;Y|Z) = I(Y;X|Z) = H(Y|Z) - H(Y|X,Z) \qquad (9.35)$$

Conditional mutual information allows us to approach the problem of redundancy from a different direction. Recall from the *PLD* algorithm that our goal is to find a variable from among the candidates that has maximum mutual information with *Y* and minimum mutual information with the predictors already selected. We now have an excellent tool. Suppose *X* is a candidate for inclusion and *Z* is a variable that is already in *S*, the set of predictors chosen so far. The conditional mutual information of *X* and *Y* given *Z* measures how much the candidate *X* contributes to predicting *Y* above and beyond what we already get from *Z*. A good candidate will have a large value of $I(X;Y|Z)$ for every *Z* in *S*. If there is even one variable *Z* in *S* for which $I(X;Y|Z)$ is small, there is little point in including this candidate *X*, because it contributes little beyond what is already contributed by that *Z*. This inspires us to choose the candidate *X* that has the maximum value of the criterion shown in Equation (9.36).

$$Criterion(X, Y, S) = \min_{Z \in S} I(X;Y|Z) \tag{9.36}$$

Equation (9.35) is a good intuitive definition of conditional mutual information, but it is not the easiest way to compute it. A better way is Equation (9.37).

$$I(X;Y|Z) = H(X,Z) + H(Y,Z) - H(Z) - H(X,Y,Z) \tag{9.37}$$

The file MUTINF_B.CPP contains complete source code to evaluate this equation for *X*, *Y*, and *Z* arrays. This code is simple but very tedious, so I will not reproduce it in its entirety here. The easiest approach, though not necessarily the most efficient, is to use nested logical expressions to tally the two-by-two-by-two bin counts. This is done as follows:

```
n000 = n001 = n010 = n011 = n100 = n101 = n110 = n111 = 0;
for (i=0; i<n; i++) {
  if (x[i]) {
    if (y[i]) {
      if (z[i])
        ++n111;
      else
        ++n110;
    }
```

```
      else {
        if (z[i])
          ++n101;
        else
          ++n100;
        }
      }
    else {
      if (y[i]) {
        if (z[i])
          ++n011;
        else
          ++n010;
        }

      else {
        if (z[i])
          ++n001;
        else
          ++n000;
        }
      }
    }
```

Once the eight bins counts are tallied, computing the four terms in Equation (9.37) is straightforward. For example, $H(Z)$ can be computed with the following code:

```
nz0 = n000 + n010 + n100 + n110;
nz1 = n - nz0;
if (nz0) {
  p = (double) nz0 / (double) n;
  HZ = p * log (p);
  }
else
  HZ = 0.0;
```

```
if (nz1) {
  p = (double) nz1 / (double) n;
  HZ += p * log (p);
  }
```

The other four terms are computed similarly. See the code for details. It should be noted that [Fleuret, 2004] discusses faster ways of summing the bin counts. Since the variables are all binary, values of X, Y, and Z can be encoded as bits in integers. By using logical conjunctions of these integers, along with table lookups, the bin counts can be found very quickly. I have not found speed to be a problem, so I have not implemented their algorithm.

The interesting part of the variable selection procedure is the stepwise algorithm. We begin by selecting the candidate that has maximum mutual information with Y. After that, for each step we evaluate the criterion of Equation (9.36) for each remaining candidate and choose the candidate having the greatest criterion.

Fleuret describes a cute trick for avoiding having to check every candidate against every Z, which can consume enormous amounts of time if there are a lot of variables in the kept set S. When a new Z is tested in computing the minimum across all Zs in S, the minimum obviously cannot increase. So if the minimum across Z so far is already less than the best candidate criterion so far, there is no point in continuing to test more Zs for the candidate. This candidate has already lost the competition for this round. Of course, we need to keep track of, for each candidate, the place where we have stopped testing it against Zs. This is because on a later round of adding a variable, the best so far may be small, and a candidate whose testing was stopped early on a prior round may need to be tested against more Zs to see if it might be the best now. A tentative winner cannot be confirmed until it has been checked for all Zs, but a loser can be eliminated early.

The file MI_BIN.CPP contains a complete program to read a dataset and perform stepwise selection of predictor variables using the Fleuret algorithm. This program is very similar to the MI_CONT and MI_DISC programs already discussed, so it will not be described here. See the code for usage details. However, examination of a simplified snippet helps to understand proper implementation of the algorithm. The following loop is invoked after one variable, that having maximum mutual information with Y, has been picked. At this time, scores[icand] has been initialized to the mutual information between that candidate and Y, and last_indices[icand] has been initialized to –1 for all candidates.

```
while (nkept < maxkept) {                          // While still adding predictors

  bestcrit = -1.e60;                               // Will be criterion of the best candidate
  for (icand=0; icand<n_indep_vars; icand++) { // Try all candidates
    for (i=0; i<nkept; i++) {                      // Is this candidate already in kept set?
      if (kept[i] == icand)                        // If it's there
        break;                                     // Quit searching for it
    }
    if (i < nkept)                                 // If this candidate 'icand' is already kept
      continue;                                    // Skip it

    // Compute I(Y;X|Z) for each Z in the kept set, and keep track of min
    // We've already done them through last_indices[icand], so start
    // with the next one up. Allow for early exit if icand already loses.

    for (iz=last_indices[icand]+1; iz<nkept; iz++) { // Continue checking all Zs
      if (scores[icand] <= bestcrit)     // Has this candidate already lost?
        break;                           // If so, no need to keep doing Zs
      j = kept[iz];                      // Index of variable in the kept set
      temp = mutinf_b (ncases, bins_dep, bins_indep + icand * ncases,
              bins_indep + j * ncases);  // I(Y;X|Z)
      if (temp < scores[icand])          // Keep track of min across all Zs
        scores[icand] = temp;
      last_indices[icand] = iz;          // Also remember how far we've checked
    } // For all kept variables, computing min conditional mutual information

    criterion = scores[icand];           // Equation (9.36), possibly abbreviated
    if (criterion > bestcrit) {          // Did we just set a new record?
      bestcrit = criterion;              // If so, update the record
      ibest = icand;                     // Keep track of the winning candidate
    }
  } // For all candidates

  // We now have the best candidate
  kept[nkept] = ibest;
  crits[nkept] = bestcrit;
  ++nkept;
} // While adding new variables
```

486

Screening Without Redundancy

Sometimes you would rather have your model training algorithm choose the exact set of predictors it prefers, rather than trusting an information-based process to choose the ideal set. It often is the case that the redundancy-based algorithms just presented whittle down the candidate set into a disturbingly small subset, leaving the model training algorithm little wiggle room. When this happens, you should prefer using mutual information to simply screen for predictive power, without regard to redundancy in the selected set. This will provide a larger (usually much larger) subset of candidates from which the model's training algorithm can choose its own nonredundant predictors.

This would normally be a trivial task, not worth presenting as a special case in this text. The program would just compute the mutual information for each candidate and present them to the user, sorted from most predictive to least. However, the MI_ONLY.CPP program available on my web site takes the selection process one giant step further than the naive approach. We'll now discuss the program and present some code snippets from it.

We already know that *selection bias* is a powerful force that threatens the experimenter with unjustifiably optimistic results. This was discussed in the context of model development on page 9. It applies just as well to selection of predictor candidates from a master list. Suppose all of the candidates are worthless. If there are a lot of them, one or more of them could well have decent mutual information with the predicted variable just by random good luck. Moreover, we already know that continuous mutual information suffers from the fact that it is relative. Its absolute magnitude generally has little or no meaning. Thus, even if all candidates are worthless, one of them is going to appear at the top of the list! It would be nice to know the answer to two questions:

1) If all of the candidates were truly worthless, what is the probability that the best performer could have done as well as it did by pure luck?

2) How many of the best performers are worth keeping? In other words, where do we draw the performance line that distinguishes the keepers from the rejects?

If either every predictor candidate is independent and identically distributed across the dataset or the predicted variable is similarly independent and identically distributed, then we can usually get decent approximations to the answers to these two questions, or at least the first. In fact, we can even occasionally (though not often) go one step further and find meaningful lower and upper bounds for the probability in question 1 for every candidate, not just the best.

487

It is worth emphasizing that the algorithm to be described falls apart if there is significant serial correlation (or any other dependency) among both the predicted variable as well as one or more of the predictor candidates. In most practical applications, the predictor candidates are hopelessly dependent, so the key is the predicted variable. If it has anything beyond tiny dependency (typically serial correlation), the test will become anti-conservative: the computed p-values will be smaller than the correct values. This is dangerous.

The technique for estimating the probability that the best candidate's performance could have arisen from pure luck (the *p-value*) is simple: destroy any relationship by randomly permuting the predicted variable, and then perform the candidate selection process. Do this again and again, hundreds or even thousands of times. Compute the fraction of the time that the mutual information of the best candidate after permutation equals or exceeds that of the original, unpermuted test. This is a good approximation of the probability that random good luck could have produced a best performance at least as great as that observed for the original, unpermuted data.

This is an enormously useful quantity. The actual magnitude of the best candidate's continuous mutual information has little meaning, as already discussed. But if we know the probability that a value as good as that observed could have arisen from a worthless but lucky candidate, we have something valuable. If this probability is large, we immediately have doubt about the quality of this candidate, even if it is the best candidate. And of course, this outcome instantly casts the same doubt on the other, inferior candidates. Conversely, if the p-value is small, our faith in this candidate is bolstered, though of course this faith does not automatically carry through to lesser candidates.

What about the other candidates, those with lesser mutual information? We would love to have p-values for them as well. Unfortunately, I am not aware of any algorithm for doing so. But there is one frequently useful thing that we can do, and another computation that is occasionally of some use.

We should always, for each candidate, count how many times the *best* permuted performer's mutual information exceeds that of the candidate. In other words, we compute the p-value of each candidate based on the permutation distribution of the *best* candidate in each permutation. (Note that the best candidate in each permutation will change randomly from permutation to permutation.) Since we will then be comparing the mutual information of the k'th best actual performer with the mutual information of the *best* permuted performer, the deck is stacked against all but the best original performer. It hardly seems fair to permute the data, find the best performer, and compare it to, say, the fifth-best original performer!

But this nevertheless gives us one useful piece of information. If this p-value is small, we can have confidence in the quality of the candidate. In particular, we state without proof that the p-value of each performer computed by comparing it to the distribution of the best permuted values is an upper bound on the p-value for the null hypothesis that this candidate is worthless. So, attaining a small p-value in this manner for any candidate is good news for that candidate.

The bad news is that we may unjustifiably reject truly good candidates. This p-value is an upper bound, so it may be that the true p-value is much smaller and we should have faith in this candidate instead of rejecting it based on a poor p-value. But half a glass is better than no glass at all.

You may be tempted to take the obviously (not!) better approach to computing p-values for candidates other than the best: base the computation on the same rank within the permutations. For example, suppose we want to compute the p-value for the third-best candidate. For each permutation, compare the mutual information of the third-best permuted value to that of the third-best original, unpermuted value. Actually, the intuition here is not terribly bad. This algorithm will work *if all candidates are worthless*. In other words, if none of the candidates has true predictive power, the p-value of every candidate computed by matching ranks this way will have a uniform distribution, just as we would expect for a valid statistical test.

This algorithm fails miserably, though, if even one of the candidates has some true quality. To see why, we'll take the simple situation of two candidates, one of which has predictive power. Suppose X1 has strong mutual information with the predicted variables, and X2 has none at all. Now consider what happens when we take a random sample from the population. It is virtually guaranteed that X2 will be in second place. This would not be the case if X1 were also worthless. In that situation, whenever X2 exceeded X1, the less worthy X1 would land in second place. The second-place score would always be the lower of the two. But when X1 is a great predictor, X2 will still be in second place, even when it is unusually good by random good luck.

This same effect does not happen in the permutations. The second-place finisher is always the lesser of the two. So in this "X1 is good" situation, we are comparing the value of the always-second-place X2, which is prejudiced to be unusually large, against a permutation distribution which has no such prejudice. As a result, the p-value of X2 will be unfairly small.

We state without proof that the p-value computed by comparing the same rank, as just described, is a lower bound on the true p-value. Given that we already know how to compute an upper bound (test the original unpermuted value against the permutation distribution of the best candidate), we can easily bound the true p-value between these two limits. Moreover, we know that when all candidates are truly worthless, the lower bound computed by matching ranks is the ultimate in tight bounds: it is the correct value (within limits of randomness, of course)!

But don't get excited yet. The bad news is that the presence of even modest amounts of true predictive quality in one or more candidates causes the matched-rank p-value to plunge. So you often end up with a lower bound so low that it is worthless. Nonetheless, this quantity is trivial to compute along with the best-rank upper bound, so you might as well find it.

It's time to look at the algorithm as implemented in MI_ONLY.CPP. We'll divide it into several sections, listing each and walking through the code. Here is how the algorithm begins:

```
for (irep=0; irep<nreps; irep++) {

  for (i=0; i<ncases; i++)        // Get the dependent (predicted) variable
    work[i] = data[i*nvars+idep];

// Shuffle dependent variable if in permutation run (irep>0)

  if (irep) {           // If doing perm uted runs, shuffle
    i = ncases;        // Number remaining to be shuffled
    while (i > 1) {    // While at leas t 2 left to shuffle
      j = (int) (unifrand () * i);
      if (j >= i)
        j = i - 1;
      dtemp = work[--i];
      work[i] = work[j];
      work[j] = dtemp;
    }
  }

  mi_adapt = new MutualInformationAdaptive (ncases, work, 1, 0.1);
```

The main outer loop does nreps replications, the first of which handles the original, unpermuted data, and the remainder of which handle permuted data. The dataset contains ncases cases, each of which consists of nvars variables, with the first n_indep_vars of them being predictor candidates. The index of the dependent (predicted) variable is idep.

If we are past the original, unpermuted replication, we shuffle the predicted variable, which is now in work. Finally, we create the object that will compute the mutual information measures. There is one thing to notice about this object. Earlier, we stated that using 6.0 as the splitting threshold is good. Here we use the tiny value 0.1 instead. The reason is that larger values such as 6.0 cause the algorithm to return a mutual information of exactly zero whenever the actual value is tiny. In most applications this is good, because tiny mutual information is usually just noise. But here we want tiny values to register, because otherwise most permutations would return a value of zero. This would produce valid but potentially confusing p-values. Picking up noise as if it were real mutual information is no problem in this particular application, because the original data and all permutations will be treated the same way and the computed p-values will properly account for this.

The next section of code computes and saves the mutual information of each candidate with the predicted variable. We will walk through this code fragment in conjunction with the one following.

```
for (icand=0; icand<n_indep_vars; icand++) { // Try all candidates
  for (i=0; i<ncases; i++)
    work[i] = data[i*nvars+icand];

  criterion = mi_adapt->mut_inf (work, 1);
  save_info[icand] = criterion; // We will sort this when all candidates are done

  if (irep == 0) {               // If doing original (unpermuted), save criterion
    index[icand] = icand;        // Will need original indices when criteria are sorted
    crits[icand] = criterion;
    mcpt_max_counts[icand] = mcpt_same_counts[icand ]= mcpt_solo_counts[icand] = 1;
  }
  else {
    if (criterion >= crits[icand])
      ++mcpt_solo_counts[icand];
  }
} // Initial list of all candidates

delete mi_adapt;
```

One thing to note in the previous code is the "else" clause in which the criterion for each candidate after permutation is compared to the criterion before permutation. This lets us compute a "solo" p-value for each candidate, its p-value in isolation. We'll discuss this more later.

Here is the final block of code, which does most of the bookkeeping for the permutation test:

```
if (irep == 0) // Find the indices that sort the candidates per criterion
   qsortdsi (0, n_indep_vars-1, save_info, index);

else {
   qsortd (0, n_indep_vars-1, save_info);
   for (icand=0; icand<n_indep_vars; icand++) {
      if (save_info[icand] >= crits[index[icand]])
         ++mcpt_same_counts[index[icand]];
      if (save_info[n_indep_vars-1] >= crits[index[icand]]) // Valid only for largest
         ++mcpt_max_counts[index[icand]];
      }
   }

} // For all reps
```

If this is the first (irep==0) replication, the unpermuted one, we sort the criteria of each candidate in ascending order of mutual information. We simultaneously move the indices of the candidates. If you look back at the prior code fragment, you will see that the criteria are always (all replications) saved in save_info, which is sorted here, and the indices are saved in index. Thus, after this sort, index[0] will be the index of the candidate having smallest mutual information, and this quantity will be in both save_info[0] and crits[index[0]]. The remaining values will be similarly positioned according to the sorting.

If this is beyond the first replication, into the permutations, we must count the number of times each permuted value equaled or exceeded the original, unpermuted value. Hence, we must sort the permutation results, which are also in save_info. We keep two separate counts. The lower limit for the p-value is counted using the same rank in the permutation results as the rank in the original data, while the upper limit is counted using the maximum mutual information for this permutation, which is in save_info[n_indep_vars-1] after sorting.

Looking back at the prior code fragment, you will see that these two counters are initialized to one, not zero. This is because we need to count the original result as well as the permutation results, and obviously any value is greater than or equal to itself, and the maximum is greater than or equal to those inferior. So it would be silly to initialize these counters to zero and then add one to them in the first (irep==0) replication.

Finally, although it is almost trivial, here is the code that prints the results of the screening.

```
fprintf (fp, "\nPredictors, in order of decreasing mutual information");
fprintf (fp, "\n");
fprintf (fp, "\n          Variable Inform ation Solo pval Min pval Max pval");

for (icand=0; icand<n_indep_vars; icand++) { // Do all candidates
  k = index[n_indep_vars-1-icand];    // Index of sorted candidate
  fprintf (fp, "\n%31s %11.5lf %12.4lf %10.4lf %10.4lf", names[k], crits[k],
           (double) mcpt_solo_counts[k] / nreps,
           (double) mcpt_same_counts[k] / nreps,
           (double) mcpt_max_counts[k] / nreps);
}
```

This loop prints results from largest mutual information to smallest. To do so, it gets the candidate indices in reverse sorted order, starting with the last (largest). Note that the smallest possible p-value is the reciprocal of the number of replications.

The primary motivation for this algorithm is the need to account for selection bias, the troubling effect of unwarranted good luck being confused with true power. The main statistic used in this regard is the p-value based on the distribution of the maximum criterion in each permutation. This serves as an exact (within limits of random variation) measure of the quality of the best actual candidate, and it is an upper bound on the p-value for the remaining candidates. A much less useful adjunct statistic is the p-value based on matching ranks. This is a lower bound on the true p-value, but in practice it drops off too rapidly to be of much use in most applications.

But there is a third statistic computed here, the *solo p-value*. For each individual candidate, this is the probability that mutual information at least as large as that obtained could have been produced by pure luck if the candidate in truth had no mutual information with the predicted variable. It is tempting to rely on this statistic to judge the value of each candidate. At the risk of being overly pedantic, we must yet again emphasize that the problem with the solo p-value is that it suffers from selection bias. If all candidates were truly worthless, the solo p-values would still have a uniform distribution. If there are 100 candidates, it is likely that at least one of them would have a solo p-value of 0.01, a very significant result. Thus it is easy to be misled by these values.

However, if a reasonably large number of replications are done, the solo p-value does have one use, even if this use is a bit suspect. If the p-value is not small, say at least 0.2 or so, then we may hesitantly conclude that the candidate is worthless or nearly so. "But wait!" you might scream. One cannot accept a null hypothesis, and judging a candidate to be worthless based on an insignificant p-value is just that. Uh, well, yes. Technically, this is true. But the fact of the matter is that if numerous replications are done, the test is usually powerful enough that a worthy candidate will be recognized by its small solo p-value. We could get into Bayesian arguments here, but we will not. The fact of the matter is that in most practical applications, if a large number of replications fail to produce an at least modestly significant p-value, the candidate is probably of no practical use. But the converse is not true for solo p-values, as already emphasized.

The MI_ONLY program was run on the same synthetic dataset used in the example on page 482. To review, five random variables named RAND0 through RAND4 are generated. Then SUM12 is defined as the sum of RAND1 and RAND2. SUM34 is similarly defined. Finally, these two sums are themselves summed to define SUM1234. RAND0 is not used. The MI_ONLY is invoked to predict SUM1234 from the other seven variables. It produces the output shown on the next page.

Variable	Information	Solo pval	Min pval	Max pval
SUM12	1.00322	0.0010	0.0010	0.0010
SUM34	0.99298	0.0010	0.0010	0.0010
RAND2	0.78251	0.0010	0.0010	0.0010
RAND1	0.78226	0.0010	0.0010	0.0010
RAND3	0.77040	0.0010	0.0010	0.0010
RAND4	0.76725	0.0010	0.0010	0.0010
RAND0	0.62558	0.9180	0.5730	1.0000

Observe that the mutual information attributed to the totally worthless RAND0, 0.62558, is not much less than that of the other predictor candidates. Examination of only the mutual information would let the developer rank the candidates according to quality, but it would do little to reveal the true extent of their relative quality. However, the p-values do a fabulous job of distinguishing the good from the bad.

We conclude this section with a brief preview of a topic covered in the next section. There, we will point out that mutual information is symmetric in the sense that it is as sensitive to the ability of the dependent variable to predict the independent as the converse. This fact raises questions about the wisdom of using a symmetric statistic in an application in which one-direction prediction is the goal. Lest you be troubled when this fact is discussed later and wonder why a symmetric statistic is used for predictor selection, understand that when a single dependent variable is involved, as is the case here, there is no problem. The order of predictor selection will be the same with both the mutual information and the asymmetric statistic discussed there. I leave it as an exercise for you to figure out why as you read the next section. Hint: Consider the denominator.

Asymmetric Information Measures

Mutual information is symmetric in the sense that $I(X;Y) = I(Y;X)$. In other words, mutual information shows how much information two variables carry in common. This may be troubling when one's goal is to use one variable, say X, to predict another, say Y. Their mutual information is based as much on the ability of Y to predict X as the ability of X to predict Y. This becomes an especially serious problem when one wants to speak of *causality,* the value of one variable causing a change in the probability distribution of another variable. This section will discuss two common approaches to investigating asymmetric information.

Uncertainty Reduction

Please turn back to page 435 and look at Figure 9.3, a depiction of the relationship between two variables. The two overlapping circles represent the uncertainty inherent in each variable before its value is known. Their region of overlap represents the information that is in common between them. Now suppose we have a predictor X that can take on three values, and a predicted variable Y that can take on two values. Table 9-4 shows an extreme example of asymmetric information.

Table 9-4. *Asymmetric Predictive Information*

	Y=1	Y=2
X=1	41	0
X=2	38	0
X=3	0	92

We see that there are 41 cases for which X=1 and Y=1, but no cases for which X=1 and Y=2. Examination of the other entries shows that X is a perfect predictor of Y; if we know X, then we know Y with absolute certainty. This is likely a useful thing to know about our data. But the converse is not true. When Y=1, our knowledge of whether X is one or two is essentially a coin toss. If our goal is to use X to predict Y, inclusion of this asymmetry in our test statistic may be counterproductive.

This can be visualized in Figure 9-3 on page 435. Call one of the entropy circles Y. Now consider how much of that circle is encompassed by the overlapping region. If the overlap encompasses most of the Y circle, then the mutual information between X and Y eliminates most of the uncertainty in Y. Conversely, if the overlap is only a small portion of the Y circle, the mutual information does little to reduce the uncertainty in Y. Note that the relationship between the overlap and the X circle (its entropy or uncertainty) plays no direct role in this computation.

This concept can be quantified by comparing the entropy of Y, which is written as $H(Y)$, with the conditional entropy of Y given that we know X, which is written as $H(Y|X)$. If these two quantities are equal, then X contributes nothing to our knowledge of Y; it has no predictive power. Conversely, if $H(Y|X)$ is zero, meaning that knowledge of X removes all uncertainty of Y, then X is a perfect predictor of Y.

The relative amount by which uncertainty in Y is reduced by knowledge of X can be expressed as shown in Equation (9.38). We have already seen the identity shown in Equation (9.39). Employing this identity in the definition gives the usual computation formula shown in Equation (9.40).

$$Uncertainty\ reduction = \frac{H(Y)-H(Y|X)}{H(Y)} \tag{9.38}$$

$$H(Y|X)=H(X,Y)-H(X) \tag{9.39}$$

$$Uncertainty\ reduction = \frac{H(X)+H(Y)-H(X,Y)}{H(Y)} \tag{9.40}$$

The file STATS.CPP provided on my web site contains a small subroutine for computing uncertainty reduction. It is listed here. Little explanation is needed because this subroutine is a direct implementation of the basic information formulas. A brief summary of its operation follows the code listing.

```
void uncert_reduc (
   int nrows,          // Number of rows in data
   int ncols,          // And columns
   int *data,          // Nrows by ncols (changes fastest) matrix of cell counts
   double *row_dep,    // Returns asymmetric UR when row is dependent
   double *col_dep,    // Returns asymmetric UR when column is dependent
   double *sym,        // Returns symmetric UR
   int *rmarg,         // Work vector nrows long
   int *cmarg          // Work vector ncols long
   )
{
   int irow, icol, total;
   double p, numer, Urow, Ucol, Ujoint;

   if (nrows < 2 || ncols < 2) { // Careless user!
     *row_dep = *col_dep = *sym = 0.0;
     return;
     }

   total = 0;

   for (irow=0; irow<nrows; irow++) {
     rmarg[irow] = 0;
     for (icol=0; icol<ncols; icol++)
       rmarg[irow] += data[irow*ncols+icol];
     total += rmarg[irow];
     }

   for (icol=0; icol<ncols; icol++) {
     cmarg[icol] = 0;
```

```
    for (irow=0; irow<nrows; irow++)
      cmarg[icol] += data[irow*ncols+icol];
    }

  Urow = 0.0;
  for (irow=0; irow<nrows; irow++) {
    if (rmarg[irow]) {
      p = (double) rmarg[irow] / (double) total;
      Urow -= p * log (p);
      }
    }

  Ucol = 0.0;
  for (icol=0; icol<ncols; icol++) {
    if (cmarg[icol]) {
      p = (double) cmarg[icol] / (double) total;
      Ucol -= p * log (p);
      }
    }

  Ujoint = 0.0;
  for (irow=0; irow<nrows; irow++) {
    for (icol=0; icol<ncols; icol++) {
      if (data[irow*ncols+icol]) {
        p = (double) data[irow*ncols+icol] / (double) total;
        Ujoint -= p * log (p);
        }
      }
    }

  numer = Urow + Ucol - Ujoint;
  if (Urow > 0)
    *row_dep = numer / Urow;
  else
    *row_dep = 0.0;

  if (Ucol > 0)
    *col_dep = numer / Ucol;
```

```
  else
    *col_dep = 0.0;

  if (Urow + Ucol > 0)
    *sym = 2.0 * numer / (Urow + Ucol);
  else
    *sym = 0.0;
}
```

The first block of code cumulates the row marginals as well as the total case count. The second block cumulates column marginals. The next three blocks compute the row, column, and joint entropies, respectively. Finally, Equation (9.40) is used to compute the uncertainty reduction in each direction. The pooled symmetric measure computed last is not often used.

Transfer Entropy: Schreiber's Information Transfer

In 2000, Thomas Schreiber published a seminal paper on modern information theory: "Measuring Information Transfer." His paper, [Schreiber, 2000], showed how one could measure a form of causality, the transfer of information from one time series to another. Later, [Vicente et al, 2011] provided some additional practical applications of Schreiber's information transfer. We now present the basic algorithm, along with code for computing information transfer (often also called *transfer entropy*).

Both of these papers discuss methods for dealing with the curse of dimensionality that plagues this computation when data is limited. These specialized algorithms come with problems of their own, and the ideal algorithm to choose is strongly application-dependent. For this reason, here we will stick with the original and most straightforward algorithm. If you are dealing with limited data and want to experiment with alternative algorithms, you should see these two papers for suggestions.

By the way, it is worth mentioning up front that the long-popular *Grainger Causality* is a special case of transfer entropy in which one assumes that the underlying model is linear autoregressive with Gaussian noise. If you are willing to accept these often restrictive assumptions, then Grainger Causality might be preferable to transfer entropy due to its more efficient use of data. However, in many applications these assumptions are too onerous to be applicable.

What is causality? Rather than digging into a deep theoretical discussion, we'll simply restate Granger's two rules:

1) The cause precedes the effect.

2) The cause contains unique information, not available in any other variable.

Note that the second rule is generally impossible to verify in practice, because we cannot know for sure whether there are other variables related to the causative which are not aware of. Still, it's nice to consider this rule in the context of an application.

To quote [Vicente et al, 2011], who in turn quotes an earlier source, "A signal X is said to cause a signal Y if the future of Y is better predicted by adding knowledge from the past and present of signal X than by using the present and past of Y alone." The code presented later shifts this back in time by one measurement period, developing the measure of causality in terms of the present value of Y being impacted by past values of X and Y. This alternate approach is more amenable to data analysis. But the traditional mathematical development which predicts future values of Y will be used in the explanations here in order to remain consistent with tradition. Obviously, the two approaches are equivalent and differ only in starting and ending subscripts.

Realize that what we are talking about here is not the mutual information between signal Y and prior values of X. On first thought one might believe that this mutual information, which involves only values of X prior to the current value of Y, is a good way to quantify information transfer from X to Y. However, [Schreiber, 2000] shows that this approach has limited value and numerous problems.

An algorithm for estimating information transfer would ideally have at least the following four properties. Transfer entropy satisfies them all to a reasonable degree.

1) It should not require the investigator to describe the nature of the expected interaction in advance of analysis. This property allows the algorithm to be useful for investigation.

2) It should respond to common nonlinear causality modes, including purely nonlinear effects. Methods that respond only to linear components of causality, such as Granger's, are seriously limited in applicability.

3) It should not be limited to just one delay for the causality. Different delays should be detectable.

4)   It should be reasonably robust against crosstalk, the phenomenon of a signal or noise component that appears simultaneously in X and Y. Many sources of data suffer this effect. For example, EEG measurements have common-mode noise, and equities share market- wide swings.

To rigorously present the algorithm, we need a compact notation for signifying the current and recent historical values of a time series. In particular, at time t we will represent the k most recent values of X (including the current value) as $X_t^{(K)} = (X_t, X_{t-1}, ..., X_{t-k+1})$, and similarly for Y.

We also need a brief detour to discuss the *Kullback-Liebler distance* between two discrete probability distributions. Suppose P and Q are discrete probability distributions over some domain indicated by i. Then the Kullback-Liebler distance between P and Q is given by Equation (9.41).

$$D(P\|Q) = \sum_i P(i)\log\left(\frac{P(i)}{q(i)}\right)$$ (9.41)

A little intuition about this definition is in order. Suppose, for example, that the two distributions are identical. In other words, the probability of every possible event is the same in both distributions. In this case, the ratio will be one for every i, and the log of one is zero. So the K-L distance will be zero. Now suppose that for some event the probability under P of that event is much larger than under Q. The ratio is greater than one, so the log will be positive, and the weight will be unusually large, resulting in a large contribution to the sum. Conversely, suppose for some event its probability under Q is much larger than its probability under P. Now the ratio will be less than one, the log will be negative, but the weight will be small, so only a small value will be subtracted from the sum. The more the two distributions diverge, the greater will be the sum.

We state without proof that this sum can never be negative, which is a nice property for a distance! But it is not symmetric: $D(P \| Q)$ does not necessarily equal $D(Q \| P)$. Rather, the *K-L* distance measures the amount of information lost when the distribution Q is used to approximate In most applications, P is the (assumed) true distribution of the data, while Q is some experimental approximation of P, perhaps based on a proposed model or other tentative explanation of P.

We are now ready to proceed. Recall that we know current and historical values of Y, and this knowledge gives us some ability to predict the next value of Y. Our goal in computing information transfer is to measure the degree to which the additional knowledge of current and historical values of X adds to our ability to predict the next Y. Equivalently, we will measure the amount of predictive information that is lost by denying ourselves knowledge of X.

Suppose we are at observation time t. If we have knowledge of the historical values of both X and Y, then we can write the probability of the next $(t+1)$ value of Y as $p(y_{t+1}|y_t^{(n)}, x_t^{(m)})$, where n and m may be different (we may know different lengths of X and Y history). But if we do not know X, then the probability of the next value of Y is $p(y_{t+1}|y_t^{(n)})$. If X has no causative effect on Y, then these two probabilities are equal for all possible outcomes. But if X does have causative effect, then they will differ.

We are now in a position to define transfer entropy. Recall that the Kullback-Liebler distance $D(P \| Q)$ measures the amount of information lost when the distribution Q is used to approximate P. The actually observed data provides $p(y_{t+1}|y_t^{(n)}, x_t^{(m)})$. What if we were to approximate this with the probability distribution that lacks access to X, namely, $p(y_{t+1}|y_t^{(n)})$? The former plays the role of P, and the latter plays the role of Q. Because of the conditional probabilities, we must sum across the conditions. The information lost by denying knowledge of X is the transfer entropy from X to Y, and it is defined as shown in Equation (9.42).

$$Transfer\ entropy = \sum p\left(y_{t+1}, y_t^{(n)}, x_t^{(m)}\right) \log\left(\frac{p\left(y_{t+1}|y_t^{(n)}, x_t^{(m)}\right)}{p\left(y_{t+1}|y_t^{(n)}\right)}\right) \tag{9.42}$$

We can define the required conditional probabilities in terms of primitive probabilities, shown here using our current notation:

$$p\left(y_{t+1}|y_t^{(n)}, x_t^{(m)}\right) = \frac{p\left(y_{t+1}, y_t^{(n)}, x_t^{(m)}\right)}{p\left(y_t^{(n)}, x_t^{(m)}\right)} \tag{9.43}$$

$$p\left(y_{t+1}|y_t^{(n)}\right) = \frac{p\left(y_{t+1}, y_t^{(n)}\right)}{p\left(y_t^{(n)}\right)} \tag{9.44}$$

The file TRANS_ENT.CPP on my web site computes transfer entropy. It differs from the presentation just shown in one small way. The mathematical presentation uses the current and prior values of X and Y to predict the next value of Y in order to conform to already published work. But in programming terms it is easier to use strictly historical values of X and Y to predict the current value of Y. These two approaches are obviously equivalent, differing only in subscripts.

There is one feature in the program that adds versatility but is not represented in the mathematical presentation given earlier. So, to make sure everything is clear, here is a rigorous statement of the problem addressed by the program:

y: The series being predicted.

x: The series whose causative nature is being evaluated.

n: The length of each series.

nbins_y: The number of values that y can take on.

nbins_x: The number of values that x can take on.

yhist: The number of historic y observations used for prediction.

xhist: The number of historic x observations used for prediction.

xlag: See the problem statement and the comment that follows.

We are given two series, x and y, each having n cases. It is assumed that $p(y[i])$ is a function of $y[i-1], y[i-2], ..., y[i-yhist]$. But does $x[i-xlag], x[i-xlag-1], ..., x[i-xlag-xhist+1]$ influence the conditional state probabilities of y? This function measures the extent to which this occurs.

The traditional version of transfer entropy computation has xlag=1, meaning that the value of x concurrent with y is not allowed to participate in influencing y. However, many applications employ a dataset in which the X series is already implicitly lagged with respect to Y. For example, most model-based trading datasets compute X based strictly on the current and prior values of the market, and they compute Y based strictly on future values of the market. Rather than requiring the user to shift the data series or adjust addressing, this routine lets the user set xlag=0 to account for X already being lagged.

Note that we have nbins_x ^ xhist * nbins_y ^ (yhist+1) cells in the probability matrix corresponding to $(y_{t+1}, y_t^{(yhist)}, x_t^{(xhist)})$. (The symbol ^ means "raised to the power.") This blows up very, very quickly. For this reason, the majority of applications will use xhist=yhist=1 and have both nbins_x and nibins_y at most three, and often just two.

To clarify the program code, we use three single letters to represent the otherwise complex terms in the algorithm.

a: The current value of y, which is being predicted

b: The yhist historic values of y

c: The xhist historic values of x

Using this compact notation, the transfer entropy of Equation (9.42) is expressed in the much less fierce Equation (9.45). Corresponding to Equations (9.43) and (9.44), we have p(a|b,c) = p(a,b,c) / p(b,c) and p(a|b) = p(a,b) / p(b).

$$Transfer\,entropy = \Sigma p(a,b,c) \log\left(\frac{p(a|b,c)}{p(a|b)}\right) \tag{9.45}$$

Now that this simpler notation is in place, we can present the routine in segments. It is called as shown here. Note that the values in x and y range from zero through nbins_x-1 and nbins_y-1, respectively.

```
double trans_ent (
    int n,          // Length of x and y
    int nbins_x,    // Number of x bins.
    int nbins_y,    // Ditto y
    short int *x,   // Independent variable, which impacts y transitions
    short int *y,   // Dependent variable
    int xlag,       // Lag of most recent predictive x: 1 for traditional, 0 for concurrent
    int xhist,      // Length of x history. At least 1
    int yhist,      // Ditto y
    int *counts,    // Work vector (see comments in code for length)
    double *ab,     // Ditto
    double *bc,     // Ditto
    double *b       // Ditto
    )
```

The first step is to compute several frequently used constants: nx=nbins_x^xhist and ny=nbins_y^yhist. This is done as follows:

```
nx = nbins_x;
for (i=1; i<xhist; i++) // Number of bins for X history
   nx *= nbins_x;

ny = nbins_y;
for (i=1; i<yhist; i++) // Number of bins for Y history
   ny *= nbins_y;

nxy = nx * ny;        // Total number of history bins
```

Count the number of cases that lie in each of the possible bins determined by the X history, the Y history, and the current value of Y. The counts are kept in an array with X history changing fastest, then Y history, and current Y changing last. We make sure not to start so early in the array that a negative subscript would be used.

```
memset (counts, 0, nxy * nbins_y * sizeof(int));
istart = xhist + xlag - 1;
if (yhist > istart)
   istart = yhist;

for (i=istart; i<n; i++) {

   // Which of the nbins_x ^ xhist X history bins does this case lie in?
   ix = x[i-xlag];
   for (j=1; j<xhist; j++)
      ix = nbins_x * ix + x[i-j-xlag];

   // Which of the nbins_y ^ yhist Y history bins does this case lie in?
   iy = y[i-1];
   for (j=2; j<=yhist; j++)
      iy = nbins_y * iy + y[i-j];

   ++counts [ y[i] * nxy + iy * nx + ix ]; // Increment the correct bin
   }

total = n - istart;
```

The next step is to compute the marginal probabilities, which will be used in later computation. This is just basic summation.

```
for (i=0; i<nbins_y*ny; i++)
   ab[i] = 0.0;
for (i=0; i<nx*ny; i++)
   bc[i] = 0.0;
for (i=0; i<ny; i++)
   b[i] = 0.0;

for (ia=0; ia<nbins_y; ia++) {
   for (iy=0; iy<ny; iy++) {
     for (ix=0; ix<nx; ix++) {
        p = (double) counts [ia * nxy + iy * nx + ix] / (double) total;
        ab[ia*ny+iy] += p;
        bc[iy*nx+ix] += p;
        b[iy] += p;
        }
     }
}
```

Finally, we compute the transfer entropy. This is just a straightforward implementation of the defining equations.

```
trans = 0.0;
for (ia=0; ia<nbins_y; ia++) {
   for (iy=0; iy<ny; iy++) {
     for (ix=0; ix<nx; ix++) {
        p = (double) counts [ia * nxy + iy * nx + ix] / (double) total;       // p(a,b,c)
        if (p <= 0.0)
           continue;
        numer = p / bc[iy*nx+ix];           // p(a | b,c)
        denom = ab[ia*ny+iy] / b[iy];       // p(a | b)
        trans += p * log (numer / denom);    // Equation (9.45)
        }
     }
}
```

We close this section by noting that my web site contains a program called TRANSFER.CPP that uses transfer entropy to sort a list of predictor candidates. This is practically identical to the MI_ONLY.CPP program, so we will not bother listing it here. However, we will note one crucial difference between the two programs. MI_ONLY shuffles the dependent variable to do the Monte Carlo permutations. This is the efficient way to do it, as there is only one dependent variable, while there are many independent candidates. But when data for transfer entropy is shuffled, one cannot take this approach. The reason is that shuffling the dependent variable would destroy any predictive power associated with its *own* historical values, when all we want to destroy is the relationship with the independent variable. Therefore, we must shuffle each candidate. Examination of the code will make clear how this is done.

References

Breiman, Leo (1994). "Bagging Predictors." *Technical Report No. 421*, Dept. of Statistics, University of California at Berkeley.

Cho, S. and Kim, J. (1995). "Multiple Network Fusion Using Fuzzy Logic." *IEEE Transactions on Neural Networks.* 6:2 pp. 497-501.

Darbellay, G. and Vajda, I. (1999) "Estimation of the Information by an Adaptive Partitioning of the Observation Space." *IEEE Transactions on Information Theory.* 45:4, pp. 1315-1321.

Efron, Bradley (1982). The Jackknife, the Bootstrap, and Other Resampling Plans. Society for Industrial and Applied Mathematics, Philadelphia, PA.

Efron, Bradley (1983). "Estimating the Error Rate of a Prediction Rule: Some Improvements on Cross Validation." *Journal of the American Statistical Association* 78: 316-331.

Efron, Bradley, and Tibshirani, Robert (1993). An Introduction to the Bootstrap. Chapman and Hall, New York.

Erdogmus, D. and Principe, J. (2002) "Insights on the Relationship Between Probability of Misclassification and Information Transfer Through Classifiers." *IJCSS* 3:1 pp. 40-55.

Fano, R. (1961) Transmission of Information: A Statistical Theory of Communications, MIT Press and John Wiley and Sons, NY, 1961.

Fleuret, F. (2004) "Fast Binary Feature Selection with Conditional Mutual Information" *Journal of Machine Learning Research* vol. 5 pp. 1531-1555.

© Timothy Masters 2018
T. Masters, *Assessing and Improving Prediction and Classification,*
https://doi.org/10.1007/978-1-4842-3336-8

REFERENCES

Freund, Yoav, and Schapire, Robert (1997). "A Decision-Theoretic Generalization of On-Line Learning and an Application to Boosting." *Journal of Computer and System Sciences* 55(1): 119-139.

Hashem, Sherif (1997). "Optimal Linear Combinations of Neural Networks." *Neural Networks* 10(4): 599-614.

Hastie, T. and Tibshirani, R. (1996) "Classification by Pairwise Coupling." Available from their Websites at Stanford University and the University of Toronto.

Kittler, J., Hatef, M., Duin, R., and Matas, J. (1998). "On combining Classifiers." *IEEE Transactions on Pattern Analysis and Machine Intelligence* 20(3): 226-239.

Lehmann, E. L. (1975). Nonparametrics. Holden-Day, San Francisco.

Masters, Timothy (1993). Practical Neural Network Recipes in C++. Academic Press, New York.

Masters, Timothy (1994). Signal and Image Processing with Neural Networks. John Wiley & Sons, New York.

Masters, Timothy (1995a). Advanced Algorithms for Neural Networks. John Wiley & Sons, New York.

Masters, Timothy (1995b). Neural, Novel, and Hybrid Algorithms for Time Series Prediction. John Wiley & Sons, New York.

Masters, T. and Land, W. (1997) "A New Training Method for the General Regression Neural Network", *SMS'97 conference proceedings*, Orlando, FL. Oct. 12-15, 1997, Vol. 3, pp.1990-1995.

Masters, T., Land, W. and Maniccam, S. (1998) "An Oracle Based on the General Regression Neural Network." *IEEE International Conference on Systems, Man, and Cybernetics*. Oct. 11-14. pp. 1615-1618.

Paparoditis, E. and Politis, D. (2002) "The Tapered Block Bootstrap for General Statistics from Stationary Sequences." *Econometrics Journal*, vol. 5, pp. 131-148.

Parzen, E. (1962). "On estimation of a Probability Density Function and Mode." *Annals of Mathematical Statistics*, 33:1065-1076.

Pedrycz, W. (1993). Fuzzy Control and Fuzzy Systems, Second, Extended Edition. Research Studies Press, Somerset, England.

Peng, H., Long, F., and Ding, C. (2005) "Feature Selection Based on Mutual Information: Criteria of Max-Dependency, Max-Relevance, and Min- Redundancy." *IEEE Transactions on Pattern Analysis and Machine Intelligence.* 27:8, pp.1226-1237.

Politis, D., and Romano, J. (1994) "The Stationary Bootstrap." *Journal of the American Statistical Association*, 89:428, 1303-1313.

Politis, D. and White, H. (2004) "Automatic Block Length Selection for the Dependent Bootstrap." *Econometric Reviews.* vol. 23, pp. 53-70.

Price, K. and Storn, R. "Differential Evolution." *Dr. Dobb's Journal*, April 1997, pp. 18-24.

Ryan, Thomas (1997). Modern Regression Methods. John Wiley & Sons, New York.

Schreiber, Thomas (2000). "Measuring Information Transfer." *Physical Review Letters*, 85:2, pp. 461-464.

Schapire, Robert (1997). "Using Output Codes to Boost Multiclass Learning Problems." *Machine Learning: Proceedings of the Fourteenth International Conference.*

Schapire, Robert, and Singer, Yoram (1998). "Improved Boosting Algorithms Using Confidence-Rated Predictions." *Proceedings of the Eleventh Annual Conference on Computational Learning Theory.*

Specht, Donald (1988). "Probabilistic Neural Networks for Classification, Mapping, or Associative Memory." *IEEE International Conference on Neural Networks*, San Diego, CA.

Specht, Donald (1990a). "Probabilistic Neural Networks." *Neural Networks*, 3:109 118.

Specht, Donald (1990b). "Probabilistic Neural Networks and the Polynomial Adeline as Complementary Techniques for Classification." *IEEE Transactions on Neural Networks*, 1:1 111-121.

Specht, Donald (1991). "A General Regression Neural Network." *IEEE Transactions on Neural Networks*, 2:6 568-576.

Specht, Donald (1992). "Enhancements to Probabilistic Neural Networks." *International Joint Conference on Neural Networks* (Baltimore, MD).

Specht, Donald F., and Shapiro, Philip D. (1991). "Generalization Accuracy of Probabilistic Neural Networks Compared with Back-Propagation Networks." Lockheed Missiles & Space Co., Inc. Independent Research Project RDD 360, I-887-I-892.

REFERENCES

Sugeno, M. (1977). Fuzzy Measures and Fuzzy Integrals, A survey. *Fuzzy Automata and Decision Processes*. North Holland, Amsterdam. Pp. 89-102.

Taniguchi, Michiaki, and Tresp, Volker (1997). "Averaging Regularized Estimators." *Neural Computation* vol. 9 pp. 1163-1178.

Vince, Ralph (1992). The Mathematics of Money Management. John Wiley & Sons, New York.

Vincent, Wibral, Lindner, and Pipa (2011). "Transfer Entropy - A Model- Free Measure of Effective Connectivity for the Neurosciences." *Journal of Computational Neuroscience*, 30:1, pp. 45-67.

White, H. (2000) "A Reality Check for Data Snooping." *Econometrica*, 68:5, pp. 1097-1126.

Woods, K, Kegelmeyer, W., and Bowyer, K. (1997). "Combination of Multiple Classifiers Using Local Accuracy Estimates." *IEEE Transactions on Pattern Analysis and Machine Intelligence*. 19:4, pp. 405-410.

Index

Get the eBook for only $5!

Why limit yourself?

With most of our titles available in both PDF and ePUB format, you can access your content wherever and however you wish—on your PC, phone, tablet, or reader.

Since you've purchased this print book, we are happy to offer you the eBook for just $5.

To learn more, go to http://www.apress.com/companion or contact support@apress.com.

Apress®

Printed in the United States
By Bookmasters